英汉·汉英
棉花专业实用手册

English-Chinese / Chinese-English
Cotton Professional Practical Manual

冯 勇　段永生　连素梅　主编

化学工业出版社

·北京·

本书前半部分为英汉、汉英词汇部分，精选了与棉花种植、生长、加工、包装、仓储、贸易、棉纺生产以及各环节的检验工作等方面相关的专业词汇5800多条，可为棉花行业相关人员深入学习国外先进技术及知识提供更为专业的辅助工具。后半部分实用文件收录了中纺棉花进出口公司购棉合同（简称中纺条款）、国际棉花协会棉花规章与规则、中国国际经济贸易仲裁委员会仲裁规则、中国棉花协会购棉条款、进口棉花检验监督管理办法、棉花质量监督管理条例、世界主要产棉国家和地区、棉花生长常用农药名称以及世界主要产棉国和地区分级标准9条重要内容，可以帮助国内外棉花管理及检验机构、国内外棉花贸易及生产企业更好地了解中国及国际棉花贸易及检验规则，进一步规避风险，减少贸易摩擦及纠纷，共同促进全球棉花产业健康有序发展。

图书在版编目（CIP）数据

英汉·汉英 棉花专业实用手册/冯勇，段永生，连素梅主编. —北京：化学工业出版社，2014.10
ISBN 978-7-122-21624-3

Ⅰ.①英… Ⅱ.①冯… ②段… ③连… Ⅲ.①棉花-词汇-手册-英、汉 Ⅳ.①S562-62

中国版本图书馆CIP数据核字（2014）第189579号

责任编辑：刘俊之　　　　　　　　装帧设计：刘丽华
责任校对：边　涛

出版发行：化学工业出版社
　　　　　（北京市东城区青年湖南街13号　邮政编码100011）
印　　刷：北京永鑫印刷有限责任公司
装　　订：三河市胜利装订厂
850mm×1168mm　1/32　印张11¾　字数396千字
2015年1月北京第1版第1次印刷

购书咨询：010-64518888（传真：010-64519686）
售后服务：010-64518899
网　　址：http://www.cip.com.cn
凡购买本书，如有缺损质量问题，本社销售中心负责调换。

定　　价：86.00元　　　　　　　　　　　　版权所有　　违者必究

《英汉·汉英 棉花专业实用手册》编辑委员会名单

主　任　　冯　勇　　段永生　　连素梅
副主任　　李　静　　张玉冰　　张香云

编　委　　郭春海　　李　朋　　郝冬生　　唐一凡　　张建宏
　　　　　赵淑忠　　李树荣　　袁志清　　刘宇辉　　郭　维
　　　　　商建霄　　周征宇　　陈宝喜　　裴新华　　董绍伟
　　　　　刘　帅　　李妍彩　　林　闪　　李俊美　　张念祖
　　　　　王　琨　　张汉民　　李宏涛　　郭会清　　傅丹华
　　　　　刘　莉　　马增梅

序 言

　　棉花是关系国计民生的重要物资，与人们的生活密不可分。中国是世界上最大的棉花生产国、消费国和进口国，中国棉花年产量占世界的四分之一，消费量占世界的三分之一，进口量占世界棉花贸易量的四分之一。随着我国纺织业的迅猛发展，棉花作为主要的纺织原料，生产及进口数量逐年递增，棉花产业进入快速发展期。

　　与棉花有关的产业链包括棉花育种、种植、籽棉采摘、脱籽加工、打包成型、仓储、贸易、棉纺生产以及各环节的检验监管等，属劳动密集型产业，从业人员占据比例较高。为了适应棉花产业国际化的迅速发展，便于从事棉花专业及相关产业的工作人员、技术人员、科研人员和院校师生阅读和翻译相关书刊，学习国外棉花产业先进科学技术，努力赶超世界先进水平，河北出入境检验检疫局组织有关专家编写了《英汉·汉英 棉花专业实用手册》一书。

　　该书可以为从事棉花专业及相关产业人员深入学习国外先进技术及知识提供更为专业的辅助工具，也可以帮助国内外棉花管理及检验机构、国内外棉花贸易及生产企业更好地了解中国及国际棉花贸易及检验规则，从而进一步规避风险，减少贸易摩擦及纠纷，共同促进我国棉花专业及相关产业和国际棉花贸易的繁荣。

河北出入境检验检疫局局长

2014 年 4 月

编者的话

从帮助进口棉花企业翻译合同等原始单据,到指导企业遵循贸易规则并解决一些进口棉花贸易摩擦,我们决定利用自身资源优势为从事棉花及相关产业的同志们提供一部棉花专业知识手册。历时11年,经过单词积累、资料搜集、词汇收集及分类整理、贸易规则查询整理核对、多次校对通稿、审稿,《英汉·汉英 棉花专业实用手册》终于和大家见面了。在编写过程中,编写组参考了进口棉花贸易中英文版合同、发票、提单、植物检疫证书、原产地证书、质量证书等原始单据以及大量相关专业书籍,词汇量较大,贸易规则实用性很强。

本书包括英汉、汉英词汇以及实用文件。其中英汉、汉英词汇部分,精选了与棉花种植、生长、加工、包装、仓储、贸易、生产以及各环节检验监管工作等方面相关的专业词汇5800多条。本书实用文件收录了与国内外棉花贸易、棉花检验监管及棉花生长有关的9项重要内容。

本书英汉、汉英词条对照清晰醒目,便于查找,书后9个实用文件更增强了本书的实用性。

本书适合于国内外棉花检验管理机构、进出口棉花贸易专业人员阅读,可用作大专院校的参考书,也是棉花企业、公司管理人员特别是在华从事棉花相关工作外国朋友必备的工具书。

本书在编写过程中得到了国家质检总局检验监管司、河北省纤维检验局以及河北省农科院棉花研究所有关领导和专家的大力支持,在此一并感谢!

由于编者水平有限,本书在编纂过程中难免会出现错漏之处,敬请广大读者批评指正。

2014 年 4 月

使用说明

一、内容及释义

1. 本书收集有关棉花专业方面的词汇 5800 多条。

2. 英汉部分按照英文字母顺序编排，汉英部分按照汉语拼音顺序编排。

3. 英汉词条中给出汉语解释若干个，每个汉语解释中间用","隔开；汉英词条中给出英语解释若干个，每个英语解释中间用";"隔开。

4. 英汉部分及汉英部分中"（）"表示另种解释或补充，"[ ]"表示缩略。

5. 本书有 9 个实用文件，实用文件 1~3 和实用文件 5~8 为汉英对照，以中文为准；实用文件 4 为英汉对照，以英文为准；实用文件 9 为汉英对照，以英文为准。

二、编排及检索

1. 正文英汉部分按照英文字母 A~Z 顺序排列，词目中第一字母相同的，按第二字母排序，以此类推。

2. 正文汉英部分按照拼音字母 a~z 顺序排列，同音字按四声（阴平、阳平、上声、去声）顺序排列，同声字按照笔画排列，笔画少者在前，笔画多者在后，笔画数相同的，按起笔笔形横（一）、竖（丨）、撇（丿）、点（、）、折（乛）的次序排列，起笔笔形相同的按第二笔笔形的次序排列，以此类推。

目录
Content

英汉部分 ·· 1

汉英部分 ·· 55

实用文件 ··· 113

  1 中纺棉花进出口公司购棉合同 ······················· 114
  1 Cotton Purchase Contract of China National Textile Import & Export Co.
  2 中国棉花协会棉花买卖合同 ·························· 124
  2 Cotton Purchase Contract of China Cotton Association
  3 中国国际经济贸易仲裁委员会仲裁规则 ············· 144
  3 China International Economic and Trade Arbitration Commission (CIETAC) Arbitration Rules
  4 国际棉花协会棉花规章与规则 ······················· 188
  4 Bylaws and Rules of the International Cotton Association Limited
  5 进口棉花检验监督管理办法 ·························· 328
  5 Administrative Measures on Inspection and Supervision of Import Cotton
  6 棉花质量监督管理条例 ································ 342
  6 Regulations on Supervision & Administration of Cotton Quality
  7 世界主要产棉国家和地区 ····························· 358
  7 Major Cotton Producing Country and Region in the World
  8 棉花生长常用农药名称 ································ 359
  8 Name of Cotton Pesticides
  9 世界主要产棉国和地区分级标准 ···················· 361
  9 Grading Standards of the World's Major Cotton-Producing Countries and Region

参考文献 ··· 364

英汉部分
English-Chinese

A

a magnifying glass 放大镜
a person having a quarantinable infectious disease 染疫人
a person suspected of having a quarantinable infectious disease 染疫嫌疑人
a power of attorney 委托书，代理权
abide 遵守，守约
abnormal fibers 异状纤维，疵点
abroad 国外
abrogate 取消，废除
absent 未检出
absolute dry condition 绝对干燥状态
absolute dry weight 绝对干重
absolute humidity 绝对湿度
absolute intensity 绝对强度
absolute temperature 绝对温度
absorb 吸收
absorbefacient 吸收的，吸收剂
absorbency 吸收性
absorbent cotton 脱脂棉，药棉
absorbent quality 吸收性，吸湿性
abstergent 洗净剂
Acala cotton 阿卡拉棉，爱字棉（美国陆地棉种）
acaricide 杀螨剂
accelerant 催速剂，促进剂
accept 接受，认可
acceptance 接受报价，货物验收，（票据的）承兑
acceptance of a case 案件受理
acceptance sampling 收货取样，进仓取样，验收取样
accident error 偶然误差
accord with 符合，与…一致
according to 按照，依照，根据
account 账户，账目
account payable 应付款
account receivable 应收款
accountable 负有责任的，有说明义务的
accounting unit 核算单位
accuracy 准确度，精确度
acerose 针状的
acknowledge 确认，承认，收悉
acknowledged rule of technology 公认的技术规则
acotyledon （植）无子叶植物
Acrain cotton 阿克雷恩棉（苏丹产）
Acri cotton 阿克里棉（叙利亚产）
active ingredient 活性成分
activity sampling 快速取样
actual arrival 实际到货
actual fineness 实际细度
actual moisture regain 实际回潮率
actual weight 实际重量
adaptability 适应性
adaxial 近轴的（植）

addenda 附录，附加物
additional clause 附加条款
additional tax 附加税
address 地址
adjustment 调整
administration 管理，行政机关，局
administrator 管理人，行政主管
admissible error 允差，容许误差
admixture 混合物，掺和剂
adulteration 掺假，伪劣品
advance 进展，预付（款）
advance payment 预付货款
advance shipment 提前装船
advanced cultivar 选育品种
advanced freight [A/F] 预付运费
adverse 恶劣
advice 通知，意见
advice of arrival 到货通知，抵港通知
advice of drawing 汇票通知单
advice of shipment 装船通知
adviser in agriculture 农业顾问
adviser on legal 法律顾问
advising bank 通知银行
advisory committee 咨询委员会
aerial port 空运港，空运场
affect 影响
affirm 断言，证实，批准
affix 黏附，贴附，签署
Afifi cotton 阿菲菲棉（埃及产）
afloat 飘浮的，（票据等）可流通
African cotton 非洲棉
afterflame 续燃
afterglow 阴燃
agent 代理，代理人

agio 贴水
agiotage 汇兑业务
agreement 协议，一致，认可
agribusiness 农业综合企业
agricultural economy 农业经济
agricultural pharmacology 农药学
agricultural product 农产品
agriculture 农业
agrimotor 农用拖拉机
agrobiology 农业生物学
agrology 农业土壤学
agronomy 农学，作物学
agrostology 草本学
agrotechnique 农业技术
agrotechny 农产品加工学
air compressor 空气压缩机，气泵
air conditioning system 空气调节系统，空调
air current 气流
air express service 航空快运
air flow tester 气流式细度试验仪
air mail 航空邮件
air pressure 气压
air spinning 气流纺
air stream cleaner 气流杂质机，气流净棉机
air transport 空运
airflow 气流
air-flow instrument 气流仪
airfreight charge 航空运费
airway bill 航空运单
Alabama cotton 阿拉巴马棉（美国产）
Alagoas cotton 阿拉戈斯棉（粗的巴西棉）

Alaska 阿拉斯加（美国州名）
alert information 预警信息
Alexandrette cotton 亚历山大勒特棉（叙利亚产）
Alexandria cotton 亚历山大棉（埃及产）
algongon 棉花（西班牙名称）
aliquot sample 等分试样
all risks [AR] 全险，所有险
Allanseed cotton 阿伦种棉（美国产）
all-cotton 全棉的
Allen cotton 阿伦棉（美国产）
alliance 联盟，同盟，联合
allogamy 异花授粉
allowance 允许，津贴
allowance error 允许误差
allowances and rebates 折让和回扣
almanac 历书，年鉴
alteration 变化，修改
Ambassador cotton 大使棉（美国产）
ambient air 周围空气，环境空气
ambient conditions 环境条件
ambient temperature 室温，环境温度
ambiguity 不明确
amend 改正，修正
amendment advice 更改通知单
amendment charge 更改费
American Upland Cotton 美国陆地棉
amino acids 氨基酸
ammonium hydroxide 氨水
amount 货值，总金额，总计
analysis 分析
anatomy 解剖
Andes cotton 秘鲁棉（别名）
Andrews cotton 安德鲁斯棉（一种海岛棉）
angiosperm 被子植物
Anguilla cotton 安圭拉棉（美洲产）
animal carcass 动物尸体
annex 附件，附录
annotation 注释，注解
annual 年度
annual policymaking 年度政策制定
anomaly 异常，不规则
anther color 花药色
Anthonomus Grandis Boheman 墨西哥棉铃象
anthrax 炭疽
anthrax bacillus 炭疽杆菌
anticorrosive 防腐剂
anti-dumping 反倾销
antigas mask 防毒面具
antiseptic 抗菌剂，防腐剂
aphicide 杀蚜虫剂
aphid 蚜虫
appeal 上诉，要求
appendix 附录
applicant 申请人
application 申请，应用
apply 使用，应用，申请
appraise 鉴定，估价，评价
appraiser 鉴定人
appropriate 适当的
approve 批准，认可，赞同
approximation 近似值，近似法
April [Apr] 四月
arable 可耕地
Aracaju cotton 阿拉卡儒棉（巴西产）
araneid 蜘蛛
arbitrage 套利，套汇

arbitral tribunal 仲裁庭
arbitration 仲裁，裁决
arbitration commission 仲裁委员会
arbitration documents 仲裁文件
arbitration proceedings 仲裁程序
arbitration rules 仲裁规则
arbitrator 仲裁员，仲裁人
are [A.] 公亩（等于100平方米或0.15市亩）
area 面积
arise 产生，发现
aritificial 人工
Arizona [ARIZ] 亚利桑那（美国主要产棉区）
Arizona-Egyption cotton 亚利桑那埃及种棉（美国产）
Arkensas rowden cotton 阿肯色改良棉（美国产）
arrearages 欠款，债，落后，拖延
arrival notice [A.N.] 抵港通知，到货通知单
article 项目，物品，条款
as far as known 就所知而言
as follows 如下
as per 依据，按照
asbestos 石棉
ash content 灰分含量
ashes 灰分
Asian Development Bank [ADB] 亚洲开发银行
Asiatic cotton 粗绒棉
Aspero cotton 阿斯派罗棉（秘鲁产）
Assam cotton 阿萨姆棉（印度产）
assess 对…征收（税款、罚款等）

asset 财产
assign 分配，安排
Assili cotton 阿西利棉（埃及产）
assistance 协助
association 协会
at sight 一见就，即期，见票即付
at the request of 应…请求
Atlantic states cotton 大西洋棉（美国东部产）
attach 贴，附上
attached photo 附带照片
attachment 附件
attorney 代理人
attributable 归责于
attribute 属性，把…归因于
auction 拍卖
auctioneer 拍卖商
audit 审核，审计，查账
Audrey Peterkin cotton 彼得色棉（美国产）
August [Aug] 八月
authentic 权威性的，真实的
authority 官方，权威，当局
authorized officer 授权签字人，权威签字人
authorized representatives 授权代表
authorized signatory 授权签字人，权威签字人
autumn 秋季
Ava cotton 阿瓦棉（印度产）
available 有效地，可用的
average 海损，平均
average fineness 平均细度

average length　平均长度
avoidance　废止，避免，使无效
avoirdupois　常衡

award　判定，判决，裁决书，授权
axil　腋（植物）
axile　轴上的（植物）

B

backlight　背光
back-up　备份
bactericide　杀菌剂
badness condition　不良状态
Baer diagram　拜氏纤维长度分布图
Baer stapler　拜氏纤维长度分析仪
Bahama cotton　巴哈马棉（西印度群岛及美国产短绒棉）
Bahia cotton　巴伊亚棉（巴西产）
Bahmia cotton　巴米亚棉（埃及产海岛棉）
Bailey cotton　贝利棉（美国产）
Baindix cotton　班迪克斯棉（土耳其产）
Bairaiti cotton　拜拉迪棉（印度产）
balance　秤，差额
Balao cotton　巴洛棉（巴西产）
bale　包，件
bale by bale　逐个棉包
bale cotton　成包原棉
bale hoop cotter　拆包钳
bale management　棉包管理
bale packing　成包
bale tare　（棉包）包皮重量
bale ties　打包铁皮，打包结

baling　打包
ballast　压舱物
Bamia cotton　巴米亚棉（埃及产）
Bancroft cotton　班克罗夫特棉（美国产）
band　带，带状物，箍
Bani cotton　巴尼棉（印度产）
bank credit　银行信贷
bank discount　银行贴现
bank draft [B/D]　银行汇票
bank rate　（银行规定的）银行利率
bankroll　资产
Bankukri cotton　班库克里棉（美国产）
banned azo colorants　禁用偶氮燃料
Bar Sakel cotton　塞克尔棉（简称"S"棉，苏丹产）
Barakat cotton　巴拉卡特棉（简称"B"棉，苏丹产）
Barbadoes cotton　巴巴多斯棉（美国产）
Barbe　巴布（长度）
Barcelona cotton　巴塞罗那棉（哥伦比亚产）
Barnes cotton　巴恩斯棉（美国产）
Barnett cotton　巴涅特棉（美国产）

Barsi-Natar cotton 巴尔西纳塔棉（印度产）
barter 易货贸易
basis cotton 基础级棉（据以计价）
basis for claim 索赔依据
Bates'bigboll cotton 贝氏大铃棉（美国产）
beaker 烧杯
bear 承担
bearer 持票人
Below Good Ordinary [B.G.O] 级外（美棉分级标准等级）
Below Good Ordinary Light Spotted [B.G.O Lt.Sp] 级外淡点污棉（美棉分级标准等级）
Below Good Ordinary Spotted [B.G.O Sp] 级外点污棉（美棉分级标准等级）
Below Good Ordinary Tinged [B.G.O Tg] 级外淡黄染棉（美棉分级标准等级）
Below Good Ordinary Yellow Stained [B.G.O Y.S] 级外黄染棉（美棉分级标准等级）
belt 带，带状物；地带，区
Ben Smith cotton 史密斯棉（美国产）
beneficial 有利的，有益的
beneficiary 受益人，收款人
benefit 利益，好处，受益
Berar cotton 贝拉尔棉（印度产）
Berbische cotton 伯比舍棉（巴西棉）
berth 停泊，泊位
Bhoga cotton 波加棉（印度产）
Bhownuggar cotton 波努加棉（印度产）

Big Boffe cotton 大波菲棉（巴西产）
Big Boll cotton 大铃棉（美国产）
Bihar cotton 比萨尔棉（印度产）
bill 票据，账单
bill holder 持票人
bill of entry [B/E] 报关单
bill of exchange [B/E] 汇票
bill of lading [B/L] （海运）提单
bill of the customs clearance for entry cargo 入境货物通关单
bill of the customs clearance for exit cargo 出境货物通关单
bills payable [B/P] 应付票据
bills receivable [B/R] 应收票据
bind 捆，绑
biocatalyst 生物催化剂
biochemistry 生物化学
biocide 杀虫剂
black seed cotton 黑籽棉
blank test 空白试验
blast 鼓风
blight 枯萎病，凋枯病
blind fire 暗火
blot 污点，疵点
bobbin 线筒，缠线管（纺织用）
bobbinet 六角网眼纱（纺织用）
Bolivar County cotton 波利瓦棉（美国产）
boll （棉、亚麻的）铃，圆荚，成铃
boll color 铃色
boll opening date 吐絮期
boll opening degree 吐絮率
boll setting type 铃着生方式

boll shape 铃形
boll trip 铃尖突起程度
boll weight 铃重
bolls anthracnose 棉铃炭疽病
bolls per plant 单株成铃数
Bombasin cotton 巴西棉（旧名）
bonded warehouse 保税库
bonded zone 保税区
border trade 边境贸易
borne 结果，承担
bottom 底
Boweds cotton 波维丝棉（美国产）
Boyd prolific cotton 博伊德棉（美国产）
bract ally 苞叶联合
bract tooth 苞齿
bract withered 苞叶自落
Brady cotton 布雷迪棉（美国产）
Bragg long staple 布雷格长绒棉（美国产）
branch 分公司，分支机构，分部
brand 商标，品牌
Brannon cotton 布兰农棉（美国产）
Brazilian cotton 巴西棉
breach 违约
break 启封
break ends 断头（纺纱）
break off an engagement 解约
break the law 犯法
breakage 破损，破裂，断头率（纺纱）
breakbulk 零担
breaking elongation 断裂伸长
breaking length 断裂长度

breaking load 断裂负荷
breaking strength 断裂强力
breaking test 断裂试验
breed 繁育，繁殖
breeding 育种
breeding institute 选育单位
breeding line 品系
breeding methods 选育方法
brightness 亮度，光泽，白度
Broach cotton 布罗奇棉（印度产）
broken and damaged cargo list 残损货物单
broken seed 破籽
bronchitis 支气管炎
Brooks improved cotton 布鲁克斯改良棉（美国产）
Brown cotton 布朗棉（美国产）
Brown Egyptian cotton 金黄色埃及棉（埃及产）
brush 毛刷
bulk 体积，大量，大批
bulk sale 整批销售
bulletin 公告，公报
Bumble cotton 大蜂棉（美国产）
bump cotton 硬块棉
bumper harvest 丰收
bundle 束，捆
bundle fiber strength 束纤维强度
bunker adjustment factor [BAF] 燃油调整费
Burma cotton 缅甸棉（缅甸产）
burned cotton 火烧棉
burning behavior 燃烧性能
Bush cotton 布什棉（美国产）

bushel 蒲式耳（容量单位）
business accounting 核算
buttery cotton 轻油色棉
button 按钮
buyer 买方

buyer order 订购单
buyer's agent 买方代理
by type 凭小样（成交）
by-effect 副作用
by-product 副产品

C

cabin seat 舱位
cable transfer 电汇
calcium magnesium phosphate 钙镁磷肥
calculate 计算
calendar 日历，日程表
calendar days 日历天，自然日
calibration 校准，校正
calibration cotton standard sample 标定棉花样品，校准棉花标准样品
California [CAL] 加利福尼亚（美国主要产棉区）
call to pay respects 拜谒
Cambodia cotton 坎博迪亚棉（印度产）
camlet 羽纱
cancel 取消
cancel after verification 核销
Canebrake cotton 坎布拉克棉（美国产）
capacity 容积，能力，容量
captain 船长
Caragach cotton 卡拉加棉（土耳其产）

card strips 盖板花，斩刀花（梳棉机用）
cardinal number 基数
carding machine 梳棉机
cargo 货物（船装载）
cargo damage survey 货损检验，货损鉴定
cargo in case 集装箱装货物
cargo manifest 载货清单
cargo ship 货船
cargo survey for general average 海损鉴定
cargo vessel 货船，货轮
Carioba cotton 卡里奥巴棉（巴西产）
Carolina Pride cotton 卡罗来纳棉（美国产）
Carolina Queen cotton 卡罗来纳皇后棉（叙利亚产）
carriage by land 陆运
carriage by road 公路运输
carrier 承运人，运载工具，载体
carry 搬运

carrying 载重
case 案件，事实，情况，容器，箱
cash 现金，现款
cash before delivery [CBD] 交货前付款
cash debit 现金收入
cash deposit 保证金
cash on delivery [C.O.D] 货到付款
cash payment 付现
cash sale 现货
Catacaos cotton 卡塔卡欧斯棉（秘鲁产）
category 种类，类别，范畴
cause 原因
cause of damage 致损原因
Cauto cotton 考托棉（古巴产）
Cawnpore-American 坎普尔美种棉（印度产）
Ceara cotton 西阿拉棉(巴西和墨西哥产)
ceiling outlet 天花板出风口
cell wall 细胞壁（纤维）
cell wall thinness 胞壁厚度（纤维）
celloidin 火棉
cellucotton 棉絮，棉团
cellulose 纤维素
cent 美分
centigrade 摄氏温度
centimeter 厘米
Ceratitis capitata 地中海实蝇
certificate 证书，执照
certificate of analysis 分析证书
certificate of commodity inspection 商检证书
certificate of disinfection 消毒证书
certificate of fumigation 熏蒸证书
certificate of origin 原产地证书
certificate of Packing 包装证书
certificate of quality 质量证书
certificate of quantity 数（重）量证书
certificate of quarantine 检疫证书
certificate of revision 更正证书
certificate of sampling 取样证书
certificate of sealing 加封证书
certificate of shortage 短缺证书
certificate of supplement 补充证书
certificate of unsealing 启封证书
certificate of valuation 估价鉴定书，价值鉴定证书
certified laboratory 认证实验室
certify 证明，证据
certifying authority 发证机构，出证机构
Chambers cotton 康伯棉（美国产）
Champion Cluster cotton 香品大铃棉（美国产）
character 特性，性质，特征
Charara cotton 恰拉拉棉（埃及产）
charge 费用，负责
charge to an account 记账
charring 炭化
check 支票，检查，核对
check and accept 验收
check payable at sight 见票即付
check variety 对照品种
chemical fiber 化学纤维

cheque 支票
chief inspection 主任检验员
chief surveyor 主任鉴定员
China Commodity Inspection Bureau [CCIB] 中国商品检验局
China National Cotton Reserves Corporation [CNCRC] 中国储备棉管理总公司
China National Standard （中国）国家标准
China National Textiles Import And Export Corporation 中国纺织品进出口总公司（简称中纺进出口公司）
Chinese cotton 中国棉花，国产棉
Chinese Export Commodities Fair [CECF] 中国出口商品交易会
chromatics 比色法，色彩学
chromometer 比色计
circumstance 情况
CL(control limit) 控制极限
claim 索赔
claim settlement 理赔
clamping length 夹钳距离，夹持长度
class 等级，种类
classer's staple 手扯长度，分级长度
classification 分类，类别，分级
classify 归类
classing room 分级室
clause 条款
clean bill 光票，清洁汇票
clean bill of health （船只的）无疫证书，船内安全报告
clean bill of lading 不带附件的提单，清洁提单
clean cargo 清洁货物
clean credit 光票信用证，无条件信用证
clean report of findings [CRF] 清洁检验报告，合格报告
cleaned cotton 净棉
cleanliness 清洁度，净度
cleanness 洁白，清洁
clear 清算，结关，清楚的
clearance rate 清除率，清关率
clearer board 绒板
Cleveland cotton 克利夫兰棉（美国产）
client 顾客，客户，委托人
close a transaction 成交，达成交易
close down 查封，关闭
closing price 闭市价
clothing 服装
coalition 结合，联盟
coasting trade 沿海贸易
coastline 海岸线
Cobweb cotton 蛛网棉（美国产）
Cochran cotton 科克伦棉（美国产）
coding 编码
coefficient 系数
coefficient of maturity 成熟系数
Coker cotton 柯克棉，柯字棉（美国产）
collecting and distributing center 集散地
collection bag 采集袋

color 颜色，色泽
color change 变色
color developing agent 显色剂
color difference 色差
color grade [C Grade] 色泽等级
color groups of cotton 棉花色泽类型
color measurement 颜色测量
color temperature 色温（度）
color ununiform 颜色不均匀，颜色不一致
colored cotton 彩棉，天然有色棉
colorimeter 色度计，比色计
colorimetric determination 比色测定
colorimetric tube 比色管
colorimetry in color card 色卡比色法
Colthorp Pride cotton 科托谱棉（美国产）
Columbia cotton 哥伦比亚棉（美国产）
comb sorter 梳片法，梳片式纤维长度分析仪
combed 精梳，梳，精梳机
combed cotton yarn 精梳棉纱
combination 组合，联合
combined declaration 联合声明
combined transport 联合运输
combustible 可燃，可燃物，易燃的
commerce 商业，贸易
commercial invoice 商业发票
commercial moisture regain 商业回潮率
commercial weight 商业重量
commission 佣金，委托
commit a crime 犯罪
commit arson 放火，纵火
commitment 承诺，（商业上的）约定
committee 委员会
commodity 商品
commodity code 商品编码
commodity inspection 商品检验
commodity inspection mark 商检标志
commodity of damage 残损证书，验残证书
commodity of origin 原产地证书
company 公司
comparative test 比对试验
compendium 概要，纲要
compensable 可补偿的
compensate 补偿，赔偿，抵消
compete 竞争，对抗
competent 胜任的，能干的
complete 完成
complex 复杂的，复合体
compliance 遵从，承诺
complicated 难懂的，复杂的
comply with 符合，遵守
composition 组成，成分
composition analysis 成分分析
compound leaf 复叶
compress 打包机，压缩
compressed bale 机压棉包，紧压棉包
computation 估计，计算，计量

concentration 浓度
concern 涉及，影响，关系
conciliation 调解，调停
conclude 决定，议定，订立
conclusion 结论
condition 条件，状态
conditioned atmosphere 公定温湿度
conditioned lint percentage of seed cotton 籽棉公定衣分率
conditioned moisture regain 公定回潮率
conditioned weight [C.W.] 公定重量
conditioned weight inspection 公量检验
conditioning 调节，调湿
conditioning house （纺织材料的）水分检验站
conduct 实施，进行，处理
conference 讨论，协商，会议
configuration 结构，构造
confirmation 确认（证），证实
confirmed credit 保兑信用证
confirming bank 保兑银行
conform to 符合，与…一致
conformity 符合性，一致
conformity assessment 符合性评定，合格评定
conformity assessment procedure 符合性评定程序
conquer 克服，征服
consensus 共识
considering 鉴于，考虑到
consign 托运，委托，交付

consignee 收货人，受托人
consigner 发货人，委托人
consignment 托付物，委托，托运
consist of 由…组成，由…构成，包括
consolidation of arbitrations 合并仲裁
constant load 固定负荷
constant rate of elongation 等速伸长
constant temperature oven 恒温烘箱
constant weight 恒重，定重
constant weight system 定重制（纤维或纱线用）
consultation 咨询，磋商
consume 耗损，消耗
consumer 消费者，用户
consumption 消费，消耗
contain 包含，含有
container 货柜，容器，集装箱
container base 集装箱基地
container by container 逐个集装箱
container freight station [CFS] 集装箱货运车站
container shi 集装箱运货船，货柜船
container yard [CY] 集装箱站或场地，集装箱堆场
contamination 污染，玷污，污染物
content 含量，容量，内容物，目录
contract 合同，契约，合约
contracted regain 合约回潮率
control 控制，支配，鉴定
control console 控制台，操纵台
control sample 对照样品

controller 监管方，监督员
conventional 惯例的，常规的，协定的
conventional lint percentage of seed cotton 籽棉准重衣分率
conventional weight 准重
conversion 换算，转化
conversion table 换算表，对照表
conveyance 运输，财产让与
convolutions in cotton 棉纤维天然转曲
Cook cotton 库克棉（美国产）
coolant 冷却剂
cooler 冷却器，散热器
Coomptah cotton 库姆塔棉（印度产）
co-operation 合作，协作
copartner 合伙人，合作者
copy 副本
corolla color 花冠色
corporation 公司
correct 矫正，纠正
corrected result 修正结果
correction coefficient 校正系数，修正系数
correction factor 校正因数，修正因子
corrective action 纠正措施
correlated color temperature 相关色温
corrugated carton 瓦楞纸箱
cost and freight [C&F] 成本加运费价
cost insurance and freight [CIF] 到岸价

cost sheet 成本单，成本报表
Cotlook 棉花展望
cotton 棉花
cotton anglyser 棉花杂质分析机
cotton aphid 棉蚜
cotton bale 棉包
cotton belt 种棉地带，产棉区
cotton boll 棉桃，棉铃
cotton boll rot 棉铃红腐病
cotton boll stage 棉花结铃期
cotton bollworm 棉铃虫
cotton breaker 棉花松包机
cotton cellulose 棉纤维素，棉浆粕
cotton classer 原棉分级员，棉检员
cotton classification 原棉分级
cotton cloth 棉布
cotton color 棉花色泽
cotton colorimeter 棉花测色仪
cotton compressing 棉花打包
cotton consumption 用棉量
cotton count 棉纱支数（英制）
cotton diseases 棉花病虫害
cotton fiber 棉纤维
cotton flock 棉束
cotton fluffer 棉花分离机（弹花机）
cotton futures 棉花期货
cotton grade 棉花品级
cotton grower 棉农，植棉者
cotton leaf curl virus 棉花曲叶病毒
cotton lint percentage 皮棉率
cotton lump 棉团，棉块
cotton mill 棉纺织厂
cotton mixing 混棉

cotton of card strips 抄斩花
cotton of card waste 车肚花
cotton picker 采棉人
cotton picking 摘棉机
cotton plant 棉株
cotton pod 棉荚
cotton press 原棉打包机
cotton property 原棉性能
cotton pulp 棉浆粕
cotton reserve 棉花储备，储备棉
cotton rot root 棉花黑根腐病
cotton rust 棉锈病
cotton sampling 原棉扦样
cotton seed 棉籽
cotton seed cake 棉籽饼
cotton seed crushers 棉籽榨油机
cotton seed oil 棉籽油
cotton seed shell 棉籽壳
cotton spinner 棉纺工人
cotton stainer 棉椿象
cotton textile 棉纺织品
cotton thread 棉线
cotton velvet 棉绒
cotton wadding 棉絮，棉胎
cotton waste 废棉，回花，回丝
cotton wax 棉蜡
cotton wilt 棉枯萎病
cotton wilt virus 棉花枯萎病毒
cotton yarn 棉纱
cotton yield before frost 霜前产棉率
Cottoneer 棉蛾（英国名称）
cotton-padded coat 棉衣，棉服
cotton-padded quilt 棉被

cottonseed huller 棉籽脱壳机
cottonseed hulls 棉籽皮
cottonseed husk 壳，棉籽皮
cottonseed meal 棉籽粉，柏子粉
cotton-spun 棉纺的
count 计算，计数，支数
counter scale 案秤，平衡锤
countermand 取消，撤回（已发出的订货单）
countervailing duty 反补贴税，反倾销税
counterweight 平衡力，平衡物
country of origin 原产国，原产地
court 法庭
court of justice 法院
cover 覆盖，包括
Cox Royal Arch cotton 喀莱阿棉（美国产）
cracking severity 撕裂强度
Crauford-cotton 克劳福德棉（美国产）
credit 信用，信贷
crimp 卷曲
criterion （判断）标准，准则，依据
critical defect 临界缺陷
crop year 作物年度，收成年度
Cross Land cotton 克罗斯兰棉（美国产）
crude fiber 粗纤维
crush 压碎，榨
crusher 轧碎机，榨油机，榨油者
cubic 立方
cubic feet 立方英尺

cultivate 耕种，种植
curling （使）卷曲
currency 货币
currency adjustment factor [CAF] 货币调整费
currency value 币值
current price 现行价格，时价
cushion 垫子，靠垫，垫层
custody 保管
customary 惯例的，通常的，习惯的
customary examination 常规检查
customer 用户，顾客
customs 海关
customs bond 海关担保，海关保税
customs clearance 报关，出口放行
customs declaration agent 报关员
customs declaration form 报关单
customs detention 海关扣留
customs duty 关税
customs entry 进口报关
customs formalities 海关手续
customs officer 海关官员
customs seal 海关加封，海关封条
customs supervision 海关监管
customs tariff 关税率，关税税则
customs valuation 海关估价
cut down 削减，缩短
cut linters 棉籽绒
cycocel 矮壮素
cylinder 滚筒，汽缸

D

····days after sight [DS] 见票···天后付款
Dacca cotton 达卡棉（孟加拉国产）
damage 损害，损失，损毁
damage cargo 货物残损
damage survey 货损检验，货损鉴定
damp 潮湿，湿气
damp-proof packing 防潮包装
date of completion of discharge 卸毕日期
date of completion of inspection 验讫日期
date of departure 离港日期，启航日期
date of dispatch 发货日期
date of shipment 装船日期，装运期
datum 数据，资料
daylight lamp 日光灯
daylighting 采光，日光照明
dead cotton 僵瓣棉
dead fiber 死纤维
dead weight 固定负载，静负载
deal sample 成交小样
Dear cotton 迪安棉（美国产）
debit and credit 借贷
debit note 借记单，借项通知单，账单

debonding 松解
debt 欠款，债务
December [Dec] 十二月
decibel 分贝
decide 决定，裁决
decided 明白的，明确的，明显的，清楚的
decimal 小数，十进位的
decimetre 分米
decision 决定，决议
declaration 申报，声明
declaration by the exporter 出口商声明
declaration by the importer 进口商声明
declaration for inspection 报检
declaration form 报检单
declaration form for entry of goods 入出境货物报检单
declaration form for exit of goods 出境货物报检单
declaration of origin 原产地声明
declaration units 报检（申报）单位
declare 声明，申报
decorative material 装饰材料
decrease 减少，减小，降低
deduct 扣除，减去，抵扣
deem 认为
defect 瑕疵，缺陷，疵点
defective bale 缺陷棉包
defective goods 次品，不合格品
defend 辩护
defendant 被告

deficiency 缺陷，不足之处
definition 释义，定义
deformation 变形，走样
degrade 降级
degree of sugar 含糖程度
delay 延期，滞期
delay of shipment 延迟装运
delayed delivery fee 滞装费
delegate 委任，委派…为代表
delete 删除
delinter 剥绒机
delinting seed 剥除短绒的棉籽
deliver 递交，交付，传送
deliver the goods 交货
delivered weight [D/W] 交货重量
delivery 交割，交付
delivery order [D/O] 交货单
delivery time [D.T] 交货期
Delta cotton 三角州棉（美国产）
Deltapine cotton 岱塔派棉（美国产）
demand 需求，要求，需要
demand draft [D/D] 即期汇票
demarcate 标定，划分
Demerara cotton 德梅拉拉棉（圭亚那产）
demurrage 滞留费
Dendara 丹达拉棉（埃及产）
density 密度，浓度
department 部门
departure date 离港日期，离开日期
depend on 依靠，按照
depreciate 贬值，降价
depth 深度，深处

depth of color 颜色深度
deputy 代理人，代表
deratization 除鼠
deratization by means of steam sterilization 蒸熏除鼠
description 货名，描述
description of goods 货物名称
design 设计，图案
design density 设计密度（棉花种植）
designate 指示，指定，指明
designated holds 预定的船舱
destination port 目的地，目的港
destroyed 销毁
detail 细节，细目，详情
determine 测定，测量
devalue 贬值
devanning 集装箱拆箱
developed country 发达国家
developing country 发展中国家
deviation 偏离，偏差
Dharwar cotton 达瓦尔棉（印度产）
diameter 直径
Diamond cotton 钻石棉（美国产）
difference 差异，区别，差数
differential treatment 区别对待
dim 暗淡
dimension 尺寸，大小，面积
dimensional change 尺寸变化
diplodia boll rot 棉铃黑果病
direct consignment 直接运输
direct cost 直接成本
direct fee 直接费用
direct users 直接用户

dirty spot 污点，污斑
disagreement 不一致，不符
discharge 卸货，解除，排除
discharging port 卸货港
disease 病害，疾病
disease index 病情指数
diseased plant rate 病株率
disinfection 消毒
disinfection passage 消毒通道
disinfection treatment 消毒处理
disinsectization 除虫
dispatch 调度，（迅速）办理
dispense 免除，豁免
dispose 处理，转让，解决，安排
disposition 处理，处置
dispute 争端
dispute settlement body 争端解决机构
distinct 有差别的，清楚的
distinguish 区分，识别
distributing centre 集散地
distribution 销售，分配，分布
distribution channel 销售渠道
divisible L/C 可分割信用证
Dixie cotton 迪克西棉（美国产）
document 单据，单证，证件，文件
document against acceptance [D/A] 承兑交单
document against payment [D/P] 付款交单
document pay [D/P] 凭单托收
documentary bill 跟单汇票
documentary credit 跟单信用证

documentary draft 跟单汇票
documentary evidence 书面证明，证明文件
documentary secretary 单证员
Domains afifi cotton 阿菲菲领域棉（埃及产）
domestic subsidy 国内补贴
downgrade 降级
dozen 一打，十二个
draft 汇票，草案，起草
draft at 90 days 90天的汇票
Drake Cluster cotton 德雷克棉（美国产）
Drake redspot cotton 德雷克红斑棉（美国产）
droppings 落棉
drought 干旱
drug-resistant 抗药性
dull 暗淡
dump good 倾销货物
dumping profit margin 倾销差价，倾销幅度
duplicate 副本，复件，一式两份，一式两份中的一份
durable 坚固的，耐用的
duration 持续（时间），期间
dust prevention 防尘
dustproof 防尘的
duty-free 免税
dyeing 染色，染色工艺
dying fire 暗火

E

Early Carolina cotton 卡罗来纳早熟棉（美国产）
earnest money 保证金，赔偿金
earthquake 地震
ease of ignition 易燃性
East Improved Georgia cotton 乔治亚改良棉（美国产）
ecological textile 生态纺织品
ecological textile 生态纺织品
Edisto cotton 埃迪斯托棉（美国产）
Egyptian cotton 埃及棉（埃及产）
elasticity 弹性，弹力
elasticity performance 弹性
electronic scales 电子天平
electrostatic spinning 静电纺纱
eletro-discharge 放电
eliminate 消除，排除
eliminate the breach 排除违约
Ellsworth cotton 埃尔斯沃恩棉（美国产）
elongation 伸长，延长
elongation at break 断裂伸长，断裂伸长率
elongation percent 伸长率
embargo 禁运，停止通航
emergeing date 出苗期

emphysema pulmonum　肺气肿
employ　雇佣
enable　使能够，授予权力或方法
enclose sample　随附样品
enclosed　封闭，附件
enclosure　附件，附上
end user　最终用户
endorse　签署，批注，开发票
enforce　实施，执行，强制
ensure　保证，担保
enter a country　入境
enterprise　企事业，事业
entitle　赋予权利
entrust　委托，委任，信托
entrusted inspection　委托检验
environmental labelling　环境标志
environmental management system　环境管理体系
environmental pollution　环境污染
environmental protection　环境保护
environmental science　环境科学
epidemic prevention　防疫
epidemic situation　疫情
equipment　设备
equipment testing　仪器测试
equivalent　当量，相当的，等效的
error　误差
escape clause　例外条款，免责条款
estate　财产，不动产
estimate　估计，预计
estimated arrival date　预计到港日期
EU scheme　欧盟计划
Eureka cotton　尤里卡棉（美国产）

evenness　均匀度，一致
every 8 working hours　每8小时工作时间
evidence　证据，证明，凭单
evidence of conformity　符合性证据
ex ship　船上交货（价）
examination　查验，检查
examine　检查，检验，调查
exceed　超出，超过，溢出
exceed the time limit　过期，逾期
excess　超过，过量，超额量
exchange　交流，交换，交易所
exchange rate　汇率
exclusion　拒绝，排除在外
execution　履行，执行，实施
exempt　豁免，免除，免税者
exemption from duty　免税
ex-factory price　出厂价格，工厂交货价
exhibit　展览，展品，展出
exogenous gene　外源基因
exothermic　放热的，加热的
expansion　扩张，伸展，膨胀
expense　费用，支出，花费
experiment　试验，实验
expire　到期，失效
expiry　期满，期限终止
expiry date　有效期，终止日期
export　出口，输出
export carton　出口箱
export contract　出口合同
export packing　出口包装
export quota　出口配额

export subsidy 出口补贴
exporter 出口商
exposure 暴露，曝光
express 快递，快件
extend 延期，延长，伸展
extension 伸长，展期
extension of certificate 证书的延期
extent 程度，范围，广度
exterior 外部，外观，外面
exterior package 外包装

external 国外的，对外的，外部的
extra 特别的，额外的，外加的
extra long staple 特长绒棉，特长纤维
extra white 特白（棉花外观描述）
extra white cotton 特白棉
extra-bract nectar 苞外蜜腺
extract 抽出，提取
extra-long fiber 超长纤维
extraneous matter 杂质，异物

F

fabric 织物
factors of cotton grade 棉花品级因素
factory 工厂
faint 暗淡的（指色泽）
fair 公平的，交易会
false declaration 虚假声明
false packed bale 掺次包装棉包，欺诈棉包
favorable balance of trade 贸易顺差
February [Feb] 二月
feet 英尺
fertile 肥沃
fertilizer 化肥，肥料
fiber 纤维
fiber bundle 纤维束
fiber bundle strength tester 束纤维强力测试仪
fiber content 纤维含量

fiber drag 棉籽上分离纤维的阻力，纤维间分离阻力
fiber end 纤维端
fiber identification 纤维鉴别
fiber length 纤维长度
fibering off 落棉，落绒
fibrograph length 跨距长度
fibrous 含纤维的，纤维状的
figure 数字，图表
filament 细丝，细绒，单纤维
file 归档
file a claim 提出索赔
fill 装满，填充
final product 最终产品
finance 金融
financial liability 金融负债
fine 优良的，好的，处罚，罚金
fine cotton yarn 细支棉纱

fine gauze 薄纱,纱布,轻纱
fineness 细度,纤度
fire behavior 着火性能
fire division wall 隔火墙,防火墙
fire hazards 火灾危险
fire prevention 防火
fireproof 防火,耐火的,消防员
first-cut linter 一类棉短绒,头道棉短绒
fix 固定
fixed assets 固定资产
fixed exchange rate 固定汇率
flammable 易燃
flashlight 手电筒
flexible package 软包装
float 浮动
floating dust 浮尘
floating pier 浮码头,浮桥
flower shape 花形
flowering date 开花期
fluctuate in line with market conditions 随行就市
fluorescent lamp 日光灯,荧光灯
flying 飞花,飞毛
force majeure 人力不可抗拒的
force of law 法律效力
foregoing 上述,前述,前面的
foreign 外来的,外国的
foreign currency 外币
foreign exchange 外汇
foreign fiber 异性纤维
foreign impurities 外来杂质,夹杂物
foreign matter 夹杂物,杂质,异物

foreign matters test 附着物检验,杂质检验
foreign trade 对外贸易
forfeit 罚金
forklift 铲车
form 格式,表格
formal 有效的,正式的,正规的
formation of image 成像
forum 论坛,讨论会
forward 期货,转交
Foster cotton 福斯特棉(美国产)
foul smell 恶臭,臭味
fragile 易碎的
frass 虫粪,幼虫的粪便
free from 不受…影响,没有…的,不含,免于
free of all average [FAA] 一切海损均不予赔偿
free of charge 免费
free on board [FOB] 离岸价,船上交货价
free trade 自由贸易
free trade zone 自由贸易区
freight 运输,货运
freight car 货车,运货车厢
freight charge 运费
freight forwarder 货运代理
freight paid 运费付讫
freight prepaid 运费预付讫
freight rates 运费率
freight yard 货场
freighter 货船,承运人
Friday 星期五

frontier trade 边境贸易
fuel adjustment factor [FAF] 燃料调整费
full container load [FCL] 整箱
fully-pressed 紧缩，高密度的
fumigation 熏蒸
fumigation processing 熏蒸处理
fungicide 杀菌剂
furnish 提供
fusarium wilt 枯萎病
fuzz 短绒，细绒，绒毛
fuzz color 短绒色

G

Gallini cotton 加利尼棉（埃及产）
Gallipoli cotton 加利波利棉（意大利产）
gallon 加仑
garments 服装
general color rendering index 一般显色指数
general terms 一般条款
generalized system of preferences [GSP] 普惠制
generally 一般地，通常地
genetic stocks 遗传材料
Georgia cotton 佐治亚棉（美国产）
germicide 杀菌剂
germinate 发芽，生长
get damp 返潮
gin fall （轧花）落棉
gin-cut cotton 轧伤棉（轧花机轧伤的纤维）
ginned cotton 皮棉，净棉
ginner 加工者，轧花商
ginnery 轧棉厂，轧花厂
ginning 轧棉，轧花，加工方式
gins 轧花机
Giza 吉萨棉（埃及产）
gloss 光泽
go mouldy 发霉
Good Middling [G.M] 上级（美棉分级标准等级）
Good Middling Lighter Spotted [G.M Lt.Sp] 上级淡点污棉（美棉分级标准等级）
Good Middling Spotted [G.M Sp] 上级点污棉（美国分级标准等级）
Good Ordinary [G.O] 平级（美棉分级标准等级）
Good Ordinary Plus [G.O.P] 平级加（美棉分级标准等级）
goods 商品
goods van 货车
goods yard 货场
gossypol 棉子醇
gossypol content in seed kernel 种仁棉酚含量

grade 等级，品级
grain 格令（重量单位）
grams per tex [G/T 或 GPT] 克/特克斯
grant 准许，同意，承认
grass seed 草籽
gray cotton 灰白棉
Griftin cotton 格里芬棉（美国产）

gross profits 毛利，总利润
gross weight 毛重
grow cotton 种植棉花
growing vigour 生长势头
growth period 生育期
guarantee 担保，保证
gunny bag 麻袋，麻包

H

H.S coding (Harmonized System) 海关编码（编码协调制度的简称）
Hagari cotton 哈加里棉（印度产）
hand feeling 手感
handing 搬运，操控，手感
harbor 港口，海湾
hard 硬的，牢固
Hard cotton 粗硬棉（巴西与秘鲁产）
harden 板结，变硬
harmful foreign matter 有害外来物，有害杂质
harmful organisms 有害生物
harmonic mean 调和平均数
harmonized commodity description and coding system [HS] 商品名称及编码协调制度
harvest 丰收
Hawkins cotton 霍金氏棉（美国产）
Hay's china cotton 海氏棉（美国产）
Hayti cotton 海地棉（海地产）

hazard 公害，危险性
hazardous material 危险品
health 健康，卫生
hectare 公顷
herbicide 除草剂
hereby 特此
hereinabove 以上，在上文
hereinafter 以下，在下文
hereinafter called 以下简称
hereinbefore 在上文中，以上
hereinbelow 以下，在下文
hereto 至此，本协议
hessian 打包麻布
high 高
high price 高价
high seas 公海
high temperature 高温
High tower cotton 高塔棉（美国产）
high volume instrument [HVI] 大容量棉花测试仪

high yield 高产
Hilliard cotton 喜勒棉（美国产）
Hinghanghat cotton 兴罕哈特棉（印度产）
hire 雇佣
histological structure 组织结构
hoist 起重机，升起
hold in pledge 抵押
hoop 箍，箍状物
Hopi Acala cotton 霍比阿卡拉棉

（印度产）
Howell cotton 霍威尔棉（美国产）
humidity 湿度
hybrid cotton 杂交棉
Hyderabad gaorani cotton 海得拉巴棉（印度产）
hygrometer 湿度计
hygroscopicity 吸湿性，吸水性
Hyphantria cunea 美国白蛾
hypothecate 抵押，担保

I

identical 同一的，完全相同的
identification 识别，鉴定
identification card 标识卡，身份证
identification code 识别码
ignite 燃点，点燃
ignition source 点火源
ill-defined 不明显，不清楚，不明确
illegal 非法的，违规的
illuminance meter 照度仪
illumination 照明，光照，照度
imaging 成像
immediate payment 即期付款，即时付款
impact 影响，效果
impartiality 公正性
implement 实施
import 进口
import content 进口商品内容

import licence 进口许可证
import quota 进口配额
import subsidy 进口补贴
import surcharge 进口附加税（费）
import variable duties 进口差价税
importer 进口商
importing country 进口国
impose 强加
improved cultivar 选育品种
improved prolific cotton 改良丰产棉
improved upland cotton 改良种陆地棉
improvement 提高，改进
impurity 杂质，不纯
in accordance with 符合，与…一致，根据…
in balance 平衡，总而言之
in debt 负债

inch 英寸
incorrect 不正确
increase 增长，增加
incubation period 潜伏期
incubator 培养箱
incur 发生，招致，导致
incur great expense 带来很大费用（花费）
indelible 不褪色的，不可磨灭的
indemnify 赔偿，补偿，保障，保护
indemonstrable 无法证明的,无法表明的
indenture 契约，双联合同，凭单
independence 独立性
index 指标，指数，索引
indicate 表明，指出，注明
indicative price 参考（指示性）价格
indirect cost 间接成本
indirect fee 间接费用
indispensable 必不可少的，必需的
indisputable 无可争辩的，无可置疑的
individual 单独的，个别的
indivisible L/C 不可分割信用证
industry 工业
infect 感染
infected plant 病株
infection 感染
infectious disease 传染病
inferior 劣等的，次的
inferior cotton 劣质棉花
inferior to 低于，次于

informal 非正式
ingrain 原料染色
inherent quality 固有性质
Inillo cotton 印尼罗棉（菲律宾产）
initial 开始的，最初的
injurious pest 有害生物
injury 损害，侵害
inner quality 内在质量
insect 昆虫
insecticide 杀虫剂
inspection 检验
inspection by attributes （品质）项目检验
inspection by sensory 感官检验
inspection certificate 检验证书
inspection certificate of packing 包装检验证书
inspection certificate of quality 品质检验证书
inspection certificate of quantity 数量检验证书
inspection certificate on damaged cargo 货物残损检验证书
inspection charges 检验费
inspection declaration agent 报检员
inspection fee 检验费
inspector 检验员
install 安设，安装
installment 分期付款，分批交货
Institute Cargo Clauses 协会货运保险条款
Institute Commodity Trades Clauses

协会商品贸易保险条款
instruction 说明
instrument and equipment 仪器设备
insurance 保险
insurance agent 保险代理人
insurance certificate 保险凭证，小保单
insurance policy 保险单，大保单
insurance premium 保险费
insured 被保险人
intact 未受损伤的，原封不动的，完好的
integral 不可分割的，整体
integration process 一体化进程
intellectual property right 知识产权
intensity 强度
interlining 夹层
interlock machine 棉毛机
intermodal transport 协调联运
internal 内部的，国内的
internal audit 内部审核
international 国际的
international calibration cotton standards [ICCS] 国际校验棉花标准
international express mail service 国际特快专递
international market price 国际市场价格
international practice 国际惯例
international railway through transport bill 国际铁路联运运单
International Standards Organization [ISO] 国际标准化组织
International System of Units [ISU] 国际单位制
international usage 国际惯例
internode length of sympodial branch 果枝节间长度
internodes length of stem 主茎节间长度
interpret 翻译
interrupt 中止，中断
interrupt contract 中止合同
interruption 中断
invalid 无效的
inverse ratio 反比，反比例
investigate and verify 查证
invoice 发票
invoice date 发票日
iron band 钢带，铁箍
iron scraps 铁屑，铁渣
iron wire 铁丝，钢丝
irradiance 照度
irregular 不规则的，无规律的，不整齐的
irregularity 不规则，不匀率
irrevocable 不可撤销的
irrevocable L/C 不可撤销的信用证
Ishan cotton 伊尚棉（尼日利日产）
isolation 隔离
issue 颁布，签证，签发

issuing authority 签证当局
issuing bank 开证银行

issuing date 签证日期
item 条款

J

Jamaica cotton 牙买加棉（西印度产）
January [Jan] 一月
jointing 拔节
Jones heblong cotton 琼斯赫朗棉（美国产）
Jones improved cotton 琼斯改良棉（美国产）
judge 评价，鉴定，判断

July [Jul] 七月
Jumbo cotton 琼博棉（美国产）
Jumel cotton 埃及马科棉（埃及产）
June [Jun] 六月
junk 垃圾，假货，冒充物
juridical person 法人
justice 公正，公平，正当，合法
justify 证明…是正当的

K

kaal finish 卡艾整理（棉毛交织物丝光整理）
Kahnami cotton 卡纳米棉（巴西及印度产）
Kaki cotton 金黄色埃及棉（埃及产）
Kalyan 卡尔杨棉（印度产）
Karachi cotton 卡拉奇棉（巴基斯坦产）
Karanak cotton 卡纳克棉（埃及产）
Karunganni cotton 卡伦甘尼棉（印度马德拉斯产）
keep 保存，保持
keep accounts 记账

keep dry 保持干燥
keep goods 仓储
keratin 角蛋白，角质
Khandesh cotton 肯地斯棉（印度产）
Khardesh roseam cotton 肯地斯混种棉（印度产）
kilogram [KG] 千克，公斤
kindling point 燃点
kink 缠绕，扭结，环结
Kirkagatch cotton 柯卡加奇棉（小亚细亚产）
Kurrachee cotton 库拉奇棉（孟加拉产）

L

lab quarantine 实验室检疫
label 标签，标记
laboratory 实验室
Lagos cotton 拉各斯棉（非洲产）
landed quality 到岸品质
landed weight 到岸重量
landing charges 卸货费
landrace 地方品种（种系）
late delivery 延期交货
lateral bud 侧芽
lateral root 侧根
law 法律
law court 法院
lawful inspection 法定检验
laws and decrees 法令
lawsuit 官司，诉讼
leaf 叶屑
leaf area 叶片面积
leaf base spot 叶基斑
leaf color 叶色
leaf form 叶型
leaf grade 叶屑等级
leaf lobe 叶裂刻深浅
leaf lobe number 叶裂片数
leaf nectar 叶蜜腺
leaf pubescence 叶茸毛
leaf shape 叶形
leaf thickness 叶片厚度
least-developed country 最不发达国家
leftover waste-cotton 下脚棉
legal 合法的，法定的，法律的
legal action 法律诉讼
legal inspection 法定检验
legal system 法制
legality 合法性
legitimate 合法的，正当的，合理的
length 长度
length uniformity 长度整齐度
length variation 长度偏差
less than 少于，低于
less than container load 拼箱，拼箱货
less weight 短重，亏重
less weight rate 短重率
letter of credit [L/C] 信用证
letter of guarantee 保函，保证书
levelness 均匀性，水平度
Lexmi cotton 勒克斯米棉（印度产）
liabilities 负债
liability 责任
liable 有责任的，有义务的
licence 许可证，执照，特许
licensee 获证方
light 光线
Light Gray cotton [Lg.G] 淡灰棉
Light Spotted cotton [Lg.Sp] 淡点污棉
lighting 照明
limit 极限

limit count　极限计数，可纺支数
limit size　极限尺寸，极限量
limited company [Co.,Ltd]　有限公司
limited corporation [Co.,Ltd]　有限公司
linear density　线密度
lint　皮棉
lint index　衣分指数
lint length　皮棉（纤维）长度
lint percentage　衣分率，皮棉百分率
lint yield before frost　霜前皮棉
linter　棉短绒
linters standard　棉短绒标准
linters test　棉短绒检验
liquidate　清理，清算
list　清单，价目表
load　载重，装载，装载量，负荷
loading list　装载清单
loading port　装船港
loading weight　装载重量
location　定位，位置
locules per boll　铃室数

lolium temulentum　毒麦
long fiber　长纤维
long-staple cotton　长绒棉
loose　松散
loss　损耗，亏率
lot　批
Low Middling [L.M]　下级（美棉分级标准等级）
Low Middling Lighter Spotted [L.M Lt.Sp]　下级淡点污棉（美棉分级标准等级）
Low Middling Plus [L.M.P]　下级加（美棉分级标准等级）
Low Middling Spotted [L.M Sp]　下级点污棉（美棉分级标准等级）
Low Middling Tinged [L.M T]　下级淡黄染棉（美棉分级标准等级）
lubricant　润滑剂
luminometer　光度计，照度计
lung disease　肺病
lustrous　光泽

M

Magrader cotton　马格鲁德棉（美国产）
Mahava cotton　马胡瓦棉（印度产）
Mahlaing cotton　马来因棉（缅甸产）
majeure　受不可抗力事件影响的
make an inventory goods in a warehouse　盘库

make up for　弥补，补偿
male parent　父本
Malwa cotton　马尔瓦棉（印度产）
Mammoth cotton　马默恩棉（美国产）
management review　管理评审
Mandypta cotton　曼地塔棉（巴拉主产）

Mandyu cotton　曼地乌棉（巴拉圭产）
manifest　仓单，货单
man-made fiber　人造纤维
manual　手册
manufacturer　制造商
manure　肥料，粪肥
March [Mar]　三月
margin　差额，利润
marine accident report　海事报告
marine cargo insurance　海上运输保险
mark　标记，唛头
mark of conformity　符合性标志
market　市场
market share　市场份额
marks of origin　原产国标记
material　原料，材料
mature fiber　成熟纤维
maturity [Mat]　成熟度
maturity coefficient　成熟系数
maturity index　成熟指数
maturity ratio　成熟度比率
maximum [Max]　最大量，最大值，上限
May　五月
mean　平均
mean length　平均长度
mean tare weight　平均皮重
means　方法，手段，工具
means of conveyance　运输工具
means of transport　运输方式
mean-square deviation　均方差
measure　尺寸，计量，措施

measurement　测量，量度，丈量
measurement list　尺码单
measuring cylinder　量筒
mediation of dispute　商业纠纷调解
medium cotton　细绒棉
medium cotton yarn　中支棉纱
medium fine　中细的
medium length fibre　中长纤维
medium staple　中绒，中长纤维
medium tenacity　中等强度
meeting　会议
member　成员
memo　码单，备忘录
merchandise　商品，货物，商业，交易
merchant　批发商
metamorphic　变质
meter　米
method　方法
metric　公制
metric count　公制支数
metric number　公支，公制支数
metric ton　公吨
microbes　微生物，细菌
micronaire [Mic]　马克隆值
middle crop　中期棉（第二次采摘的原棉）
middle part cat and weigh method　中断切断称重法（用于纤维试验）
Middling [M]　中级（美棉分级标准等级）
Middling Gray [M.G]　中级灰棉（美棉分级标准等级）
Middling Lighter Spotted [M Lt.Sp]

中级淡点污棉（美棉分级标准等级）
Middling Plus [M.P]　中级加（美棉分级标准等级）
mildew　霉变
mildew spot cotton　霉斑棉
mineral cotton　石棉，矿棉
minimum [Min]　最小值，最小量，最小的，下限，底限
ministry of agriculture　农业部
missing　遗漏的，短缺的，丢失的
mission　代表团，使命
mission statement　职责声明，宗旨
miss-planted rate　缺株率
mix　混合，掺和，调制
mixed packed bale　混杂棉包
mixed-mark　混唛，唛头混乱的
mixed-marked bales　唛头不清的棉包
mixture　混合物
modal grade　主体品级
modal length　主体长度，手扯长度
modal micronaire　主体马克隆值
mode of trade　贸易方式
model for calligraphy or painting　范本
module testing　模块测试
Moho cotton　莫霍棉（非洲产）
moisture　水分，湿气，湿度
moisture content　含水量，含湿量
moisture percentage　含水率
moisture proof packing　防潮包装
moisture regain　回潮率
moisture resistant　防潮
moisture test　水分检验

Molinos cotton　莫利诺棉（墨西哥产）
molting　蜕皮
Monday　星期一
money　货币
money for buying　货款
monopodial branch number　叶枝数
monosaccharide　单糖
more favorable treatment　更优惠待遇
mortgage　抵押
mossy cotton　多绒棉（含有短的未成熟纤维）
most-favored nation　最惠国
most-favored nation tariff　最惠国税率
most-favored nation treatment　最惠国待遇
moth-killing lamp　诱虫灯
moting　（轧棉机的）除尘作用
mouldy　发霉，发霉的
mouldy cotton　霉变棉
muck　粪肥
Mugnlai cotton　穆格莱棉（印度产）
Multan cotton　木尔坦棉（印度产）
multi-fiber agreement　多种纤维协定
Multiflora cotton　多花棉（美国产）
multilateral trade system　多边贸易体制
multimodal transport　多式联运
multiply　繁殖
musty　发霉的
mutiny　兵变
mutual　互相的，彼此的

N

Nadam cotton 纳丹棉（印度产）
nail 钉子
naked light 明火
name and address of consignee 收货人名称及地址
name and address of consignor 发货人名称及地址
name and No. of conveyance 运输工具名称及号码
name cloth 标记带
nasosinusitis 鼻窦炎
nation 国度，国家
national treatment 国民待遇
nationality 国籍
natural fiber 天然纤维
natural lighting 自然采光，日光照明
navigation 导航，引导
NCL(no control limit) 不控制界限
neatness 洁净，净度
negate 否定，无效
negotiate 谈判，协商，议定
negotiation bank 议付银行
neps 棉结
net price 净价
net weight 净重
Nickerson-Hunter cotton colorimeter 尼克森-亨特棉花色泽仪
night shift 夜班
Nikerie cotton 尼克里棉（南美洲产）

nil 无，零
Ninety days cotton 九十日棉（美国产）
no leakage 无渗漏
no spinning value cotton 无纺用价值棉花
Noba cotton 努巴棉（苏丹产）
nominal number 公称支数，名义支数
nominate 指定，任命
non-breaching party 非违约方，守约方
non-chinese cotton 非国产棉
non-conformity 不合格，不符合，不一致
non-cotton substance 非棉物质
non-lint content （皮棉）含杂率
non-lint tester 杂质分析机
non-originating 非原产的
non-restrained item 非受限制项目
non-tariff barrier 非关税壁垒
non-tariff concession 非关税减让
non-tariff measures 非关税措施
non-textile 无纺布
non-wooden packing certificate 非木质包装证
norm 规范
normal 正常的，通常的
normal condition 正常状态，基准状态

normal temperature 常温，正常温度
normative document 标准性文件
north light roof 北向采光屋顶
north skylights 北向天然昼光
Northern star 北极星棉（美国产）
note 票据，纸币，注解
notice 通知单
notice of readiness 备装通知
November [Nov] 十一月

Novo Paulista cotton 诺沃波利塔棉（巴西产）
null 零位，零
null and void 无效，作废
number 号码，编号
number and type of packages 包装种类及数量
Nurma cotton 努尔马棉（印度产）

O

objection 异议，缺陷
objective measurement 客观测定
obligation 义务，责任，约束
observer 观察员
ocean bill of lading 海运提单
ocean vessel 远洋船舶
October [Oct] 十月
odor 气味
Oekotech 环保工业用纺织品
Oeko-tex standard 纺织品生态标准
off grade 等外品，次品
off size 尺寸不符
off standard 不符合标准
off-color 色差
offer 发价，报盘
official 正式的，官方的，官员
official stamp 印章，官方印章
offset 截距
Ohollera cotton 杜来拉棉（印度产）

oil content in seed kernel 种仁脂肪含量
oil spot 油污迹
oil stain 油污，油渍
oil stained cotton 油污棉
oilseed 油籽
Old Bess cotton 老贝斯低级棉（西印度群岛产）
oligosaccharide 低聚糖
on behalf of 代表…，以…名义
on schedule 按时
on the basis of 在…基础上，根据
on time 按时
on-board bill of lading 已装运提单
on-site inspection 现场检验
on-the-spot inspection 现场检查
open fire 明火
open-end rotor 气流纺加捻杯，气流纺纱杯

open-end spinning 自由端纺纱，气流纺（纱）
opening bank 开户银行，开票银行
operate 操作，经营，运转
option 选项，选择
order 订（定）单，规则
order form 订货单，订单
organ 机构，器官
organic 有机的
organic cotton 有机棉花
organizational structure 组织机构
organoleptic evaluation 感官评定，感官检验
Orieans cotton 奥尔良棉（美国产）
origin 产地
origin receiving charge[ORC]初始接收费

origin standard 原产地标准
original 正本，原本，原始的
Orissa cotton 奥里萨棉（印度产）
oscillator 振荡器
ounce 盎司
outcome 结果
outer package 外包装
oven 烘箱
overdue 过期
overlanded & shortlanded cargo list 货物溢缺单
overlap 重叠
overleaf 在下面（背面，反面，下页）
overrule 否决
oversea 海外的，国外的
overweight 超重
owing to 由于

P

package 包（裹，装），捆，件
package intact 包装完整
package sound 包装完好
package type 包装类型
packaged machine 打包机
packaging 打包，包装
packed in pressed bales 机压包
packing 包装，打包
packing factory 打包厂
packing instruction 包装说明
packing list [P/L] 装箱单

packing mark 包装标记，唛头
packing material 包装材料
packing method 包装方法
packing sheet 包装布
packing sound 包装良好
pack 包装，打包
pallet 托盘
pallid 苍白的，暗淡的
parliament 国会
partial 一部分，分批的
partial shipment 分批装运

particular 个别项目,细节,特别的
particular terms 特殊条款
part 部分,部件,零件
party 一方,当事人,诉讼关系人
pay 支付,付款
pay a formal visit 拜谒
pay in full 一次付清,付清
pay off 付清
pay on delivery [POD] 货到付款
payable 应支付的,可支付的
payee 收款人
payer 付款人
payment 付款
payment against arrival 货到付款
payment against document 凭单付款
payment at sight 即期付款,见票即付
payment by draft 凭汇票付款
payment for goods 货款
payment method 付款方式
payment of interest 付息
payment terms 付款条件
pedigree 系谱
penalty 处罚,罚款
pencil 铅笔
per 按照,经,由,每
per steamship [Per S.S.] 由轮船装运
percent error 百分比误差
percentage content 含量百分率,百分含量
perform 履行,执行,完成,实施
performance test 性能试验
period 期限,期间,周期
permanent 永久的,持久的

permission 允许,许可,同意
permissive 非约束性条款
permit 许可,许可证,准许证
pesticide 杀虫剂,农药
pest-resistant 抗虫害的
pest 害虫,有害植物,鼠疫
petal gossypol 花瓣棉酚含量
petal size 花冠长度
petal tannin 花瓣单宁含量
pharyngitis 咽炎
photophobia 畏光
physical index 物理指标
physical property 物理性能
physical standard for cotton grade 棉花品级实物标准
physical standard for cotton length 棉花长度实物标准
physical standard for cotton micronaire 棉花马克隆值实物标准
physical standard for cotton strength 棉花强度实物标准
physical standard 实物标准
physical test 物理试验,物理测试
phytosanitary certificate 植物检疫证书
phytosanitary certificate for re-export 植物转口检疫证书
pick 采摘,挑拣
pick cotton 采棉
pier 码头
pierage 码头费
pile 堆,堆积,大量
Pima cotton 比马棉
pine wood nematode 松材线虫

pink boll rot	棉铃红粉病
pink bollworm	棉红铃虫
place of arbitration	仲裁地
place of arrival	到货地点
place of issue	签证地点
place of origin	原产地
place of receipt	收货地
place on file	归档
plan	规划，设计，进度
plant	植物
plant diseases and insect pests	植物病虫害
plant height	株高
plant residue	植物残体
plant type	株型
planted density	种植密度
plastic	塑料
plasticbags	塑料袋，塑胶袋
plastic drum	塑料桶
plastic foam	泡沫塑料
plastic sacks	塑料袋
plastic woven bags	塑编袋
plasticity	可塑性，塑性
platform balance	台秤
platform scale	台秤，磅秤
plead	辩护
pliers	钳子
plough	耕地
plume	羽毛
plywood case	胶合板箱
pneumoconiosis	尘肺，尘肺病
pneumonia	肺炎
pod	荚，壳皮
point	点
point of origin	发货港
points off	跌若干点
points on	涨若干点
policy	保险单，政策
pollination	授粉
polyester	涤纶，聚酯纤维
polypropylene fiber	聚丙烯纤维，丙纶
polysaccharide	多糖
poor feeling	手感不良
Poor Mains cotton	佃农棉（美国产）
population distribution	总体分布
population mean	总体均值
population variance	总方差
port	港口
port dues	港口税
port of arrival	到货港
port of call	途径港
port of delivery	交货港
port of departure	始发港
port of destination	目的港
port of discharge	目的港，卸货港
port of dispatch	发货港
port of embarkation	登船港，启运港
port of entry	进口港
port of loading	装货港
port of shipment	装货港，起运地
port of transshipment	转运港
port of unloading	卸货港
portion in excess	超过部分，过量
portion	部分，一份，一部分
Porto Rico cotton	波多黎各棉（西印度群岛产）
possess	具有，持有，占有

post-quota era 后配额时代
poultry feather 家禽羽毛
pound 磅
practice 惯例，实践
precise 精确的，准确的
precision 精确，精密度
precision balance 精密天平
preconditioning 预调湿，预处理
preference-giving country 给惠国
preference-receiving country 受惠国
preferential tariff 优惠关税
preferential treatment 优惠待遇
preliminary inspection 预检验，预验
premium 保险费，溢价，佣金
prepaid 预付的
preparation 制备
prescribe 规定，指定，指示
presentation 提出，呈递，描述
present 呈递，提出，现在的
preserve 保存，保护
pre-shipment 装船前
pre-shipment inspection 装船前检验
press 按压
Pressley strength 卜氏强力
pressure 压力
pre-tension 预拉伸，预加张力
prevail 盛行，流行
prevailing price 时价，现行价格
prevailing 流行的，通行的
prevent and kill off 防除
prevention 预防，防止，阻止
prevention and cure 防治
preventive action 预防措施
previous 预先的，先前的

price 价格
price difference 差价
price including commission 含佣价
price list 价目表
price term 价格条款
Pride of Georgia cotton 佐治亚大铃棉（美国产）
primary layer 初生层
primary wall 初生胞壁
prime 初期，最好的，主要的
principle 原理，原则，要素
principle of non-reciprocal treatment 非互惠待遇原则
principle of reciprocal treatment 互惠待遇原则
printing and dyeing 印染
privilege 特权，优惠
problematic bale 问题棉包
procedure 程序，过程，步骤，手续
processing 加工，处理
processing and quarantine 加工过程检疫
processing methods 加工方式
processing standard 加工标准
processing with importedmaterials 进料加工
processing with suppliedmaterials 来料加工
process 工序，环节，过程
produce 生产，产品，制造
product 产品，生产，成果
production and marketing 产销
production permit 生产许可（证）

production trial　生产试验（棉花种植）
production　产品，生产
professional standard　行业标准
profit　利润，益处，得益
profits tax　利得税，利润税
proforma invoice　形式发票
program　方案，程序，计划
prohibit　禁止，阻止
project　项目，方案，工程
promptly　迅速地，立即地
promulgate　颁布，公布
proof　证据，证明
proper　恰当的，适当的，规矩的
property　性能，特征，财产
property insurance　财产保险
proportion　比例，比率
protection　防护，保护，保护装置，保护措施
protective price　保护价
protein content in seed kernel　种仁蛋白含量
protocol　协定书，草约
ponderation　过磅
prove　证明，证实
provide　提供，供给

provided　假如，如果，以…为条件
providing　假如，如果，以…为条件
provision　规定，条款
provisional issuance　临时签发
publish　发表
Puerto Cabello cotton　卡贝略港棉（委内瑞拉产）
Pugliese cotton　普格里斯棉（意大利产）
pull length　（棉纤维）手扯长度
Pullnot cotton　普尔诺特棉（美国产）
pull of sampling　手扯棉样
pulmonary tuberculosis [TB]　肺病
punishment　处罚，惩罚
Punjab Deshi cotton　旁遮普棉（印度、巴基斯坦产）
Punjad-American cotton　旁遮普美种陆地棉（印度产）
purchase　采购，购买，进货
purchase contract　购货合同
purchase order　订货单，订单
purchaser　买方，需方
purpose　目的，用途，效用
put in storage　入库
put on sell　发售

Q

qualification　资格，合格（证明），合格鉴定
qualified　有资格的，（鉴定）合格的
qualified disposition　合格处理
qualitative analysis　定性分析
quality　质量，品质
quality assurance　品质保证
quality brand　名牌

quality certificate mark 质量标签
quality control 质量控制
quality control system 质量控制系统
quality cotton 优质棉
quality dispute 品质争议
quality evaluation 品质鉴定
quality inspection 质量检验
quality length 品质长度
quality level 质量水平
quality management system 质量管理系统
quality objective 质量目标
quality of certificate 质量证书，品质证书
quality policy 质量方针
quality standard 质量标准
quality supervision 质量监督
quality surveillance 质量监管
quality symbol 品质标记
quality system 质量体系
quantitative 定量

quantitative analysis 定量分析
quantitative method 定量法
quantitative restriction 数量限制
quantity 数量
quantity declared 报检（申报）数量
quantity of certificate 数量证书，重量证书
quarantine 检疫
quarantine and authenticate report 检疫鉴定报告
quarantine certificate for conveyance 运输工具检疫证书
quarantine pest 害虫检疫
quarter 季度
Queen cotton 皇后棉（美国产）
Queensland cotton 昆士兰棉（澳大利亚产）
query 疑问，查询
quick check 快速核查
quota 配额
quota-free product 无配额产品
quotation 报价单，引证，定价

R

race 地理种系
Radya cotton 拉德耶棉（印度产）
railway bill of lading 铁路提单
Rajputana cotton 拉普塔纳棉（孟加拉，印度产）
Rameses cotton 拉梅斯棉（美国产）
random 随机

random sampling method 随机取样法
range 范围，区域，限度
range of shades 色泽分布范围
Rangoon cotton 仰光棉（缅甸产）
Rasi cotton 拉希棉（印度产）
rate 比率，速度，等级

rate of damage	残损率
rate of dead cotton	僵瓣率
rate of duty	（关税）税率
rate of infecund seeds	不孕籽率
rate of progress	进度
rated	额定
ratification	许可，批准
ratio	比率
raw	生的，未加工的
raw cotton	原棉
raw data	原始数据
raw material	原材料，原料
raw stock	原料，未加工纤维
reach	达到
reach an agreement on	就…达成协议，就…达成共识
reading	读数
ready	准备
ready-made	现成的
reagent	试剂
real tare	实际皮重
reason	原因
reasonable	合理的，公道的，适当的
recall	罢免
receipt	收据，收到
receive	收到，接受
reciprocal	相互的，互惠的，反商，倒数
reciprocity	相互作用，互惠
recognition	承认，识别，认可
recommendation	推荐
record	记录，记载
recover	恢复，补偿

recoverability	恢复性能
recovery	回收率，回收，恢复
recycling waste cotton	回收棉
Red Peruvian cotton	红秘鲁棉
reduce	减少，缩减，还原
reduce the price	减价
reducing sugar	还原糖
reduction of a fraction	约分
reduction of output	减产
reference	参考
reflect light	反光
reflectivity	反射率
reform	改革
refund	退款，偿还，归还
refuse	拒绝，废物，垃圾
regain	回潮，吸湿（率，性）
regain percentage	回潮率
regain standard	标准回潮（率）
regard	认为，当作，考虑
regenerated fiber	再生纤维
reginned cotton	再轧棉
regional trial	区域试验
register	登记，注册
regular	正规的，正常的，有规律的
regularity	规则性，整齐
regulation	法规，条例，规则，规定
reimbursement	偿还，还款
re-inspection	复验，复检
reject	拒绝，拒收，不受理
related	有联系的，有关的
relative humidity	相对湿度
release	放行
release notice	放行单
releasing year	育成年份

relevant 有关的，相关的，相应的
reliability 可信度，可靠性
relief 免除，减免
remain 剩余，保留，保持
remark 备注，附注，注意，意见
remedial 补救的
remedial measures 补救措施
remittance 汇款
remittance with order 订货时即支付货款
remove 除去，取出，移动
render 交纳，呈递，作出（判定等）
renew 更新，修补，展望
replace 代替，替代，更换
reply 答复，回复
report 报告
report of findings 调查报告
report of inspection 检验报告
representative 代理人，代表
representative sample 代表性样品
request 请求，要求
request for arbitration 仲裁申请
requirement 要求，必要的条件，规定
reservation clause 保留条款
reserve 保留，储存
resistance to disease 抗病性
resistance to drought 抗旱性
resistance to pests 抗虫性
resolve 决定，裁决
respectively 分别地，各自地
respirator 口罩
response 答复，反映，响应
response to query 查询回复
responsibility 责任，义务，职责

restore 恢复
restrained exports 受限制出口产品
restraint of trade 贸易管制
restricted textile 受限纺织品
restriction 限制
restrictive measures investigation 设限调查
result 结果，效果
resume growth 返青
retail price 零售价
retail trade 零售业
retentivity 保持性
return 退还，返还，归还
reverse a verdict 翻案
reversible 可逆的，可撤销的
review 评论，评述
revise 修改，修正，校订
revision 更正，修改
revocable credit 可撤销信用证
rework 返工
Rex cotton 雷克斯棉（西班牙、澳大利亚和美国产）
rhinitis 鼻炎
rich 肥沃
rid 摆脱
Rifty cotton 里夫提棉（地中海东部沿岸地区产）
riot 暴乱
risk 风险，危险
Risty cotton 里士提棉（地中海地区产）
Roanoke cotton 罗阿诺克棉（美国产）
Roe cotton 罗埃棉（美国产）
roll 卷，辊，滚动
roller 辊，滚筒

rolling test 试轧（衣分测试）
roller ginned cotton [RG] 皮辊棉
roller lap waste cotton 皮辊花
Rongony cotton 龙戈尼棉（马达加斯加产）
root cap 根冠
root hair 根毛
root rot 根腐病
rotor spinning 气流纺纱
rotor spun yarn 气流纱
rough yarn 粗纱，羽毛纱

route 航线，路线
Rowden cotton 罗登棉（美国产）
ruler 规则
ruler of origin 原产地规则
rules applicable 适用规则
Rum cotton 鲁姆棉（维尔京群岛产）
rupture 断裂，破裂，破坏
rust stain 锈渍，锈斑
rust-corrosion 锈蚀
rust-proof 防锈
rusty cotton 锈斑棉

S

sack （麻，纸）袋
safeguard 保护，防护，保证条款，防护设施
safeguard measures 保障措施
safety 安全（性）
safety certification mark 安全标志
safety factor 安全系数
safety line 安全线
Saint Louis cotton 圣路易斯棉（美国产）
Saint Uincent cotton 圣文森特棉（西印度群岛产）
Sakel cotton 萨克尔棉（苏丹、索马里产）
Sakellarides cotton 萨克拉里德斯棉（埃及产）
sale 销售，出售
sales catalogue 销售目录

sales confirmation 销售确认书
sales contract 销售（货）合同
saline and alkaline land 盐碱地
sample 取样，样品
sample card 样品卡
sample of cotton 棉样
sample percentage 抽样比例
sample requirement 样品要求
sample room 样品间
sampling 取样，扦样，抽样
sampling error 抽样误差
San Martha cotton 圣马撒棉（哥伦比亚产）
San Martin cotton 圣马丁棉（西印度群岛产）
sanitary certificate for conveyance 交通工具卫生证书
sanitary inspection certificate 卫生

检验证书
sanitary treatment 卫生处理
Santos cotton 圣托斯棉（巴西产）
Sarasses cotton 萨拉瑟棉（印度产）
saturate 饱和
Saturday 星期六
save 保存
saw ginned cotton 锯齿棉
scab 病斑
scale 刻度，标尺，规模
scale car 磅秤车
schedule 计划，安排，时间表，进度
scheme under GSP 普惠制方案
sclerotium 菌核
scope 范围
scraps 渣，屑
scurviness 恶劣
sea 海
sea island cotton 海岛棉
sea transportation 海运
seal 密封，印条，印章
seal number 封条号，铅封号
seal up 查封
seal up for safekeeping 封存
seaport 海港
seaworthy 适航
secondary layer 次生层
secondary wall 次生（细胞壁）
second-cut linter 二类棉短绒
section 部分
security 安全，可靠，保证
seed 棉籽
seed cotton 籽棉

seed fuzz 短绒
seed hair length 种毛长度
seed index 棉籽指数
seed number per locule 铃室种子数
seed pigment gland 种子腺体
seedling 幼苗
seedling anthracnose 棉苗炭疽病
seedy cotton 含籽棉
Selected Triumph cotton 凯旋棉（美国产）
selective examination 抽查
sell 发售
seller 卖方
seller's agent 卖方代理
semi-finished product 半成品
send back 退回
senses 感觉
sensory evaluation 感官评价
sensory examination 感官检验
sensory test 灵敏度试验，感官测试
sepal 萼片，花萼
separate 分离，分开
separation discharge 分开卸货
September [Sept] 九月
serial 序列，连续的
serial number 序号，编号
series 连续，系列，套
serious 严重的，严肃的
seriplane 黑绒板
set on fire 焚烧
set sail 起航
settle 结算，处理，解决
settlement 结算
setup 设置，机构，调整

Sevilla cotton 塞维利亚棉（西班牙产）	Shirley analyzer 锡莱杂质分析仪
sextuplicate 一式六份	shockproof packing 防震包装
Shap cotton 夏普棉（缅甸产）	shopper 购物者，顾客
shape 形状，形态	short fiber percentage 短纤维率
share 份额，股份	short staple cotton 短绒棉
sharp 锋利的	shortage in weight 短重
shell cotton 多籽屑棉	short-term 短期的
Shem Parutti cotton 森帕鲁提棉（印度产）	short-weight 短重
Shembanon cotton 盛巴农棉（缅甸产）	show 显示
shipment 载货，装船	shrink 缩水
shipper 托运人，发货人，货主	side 边
shippers load and count 托运人自行装货点件	side effect 副作用
shipping 航运	sight draft 即期汇票
shipping advice 装船通知	sight L/C 即期信用证
shipping agent 发货代理人，装运代理人，运货代理商	sign 签名，签署，标记
shipping carton 出口箱，装运箱	signatory 签名人，签署者，签约国
shipping date 装运期，装船日期	signature 签名，签字
shipping documents 装运单据	signature of the buyer 买方签字
shipping invoice 装运单，装运发票	signature of the seller 卖方签字
shipping line 航线，船运公司	silk noil 落棉
shipping mark 唛头，标记	Sind-American cotton 信德美种棉（巴基斯坦信德产）
shipping note [S/N 或 S.N.] 装货通知单	Sind-Deshi cotton 信德孟加拉种棉（巴基斯坦信德产）
shipping order [S.O.] 装货（通知）单	single fiber strength 单纤维强度
shipping package 运输包装，外包装	sit fire to 放火
shipping schedule 船期表	site 现场
shipping space 舱位	size label 尺码标志
shipping weight [S/W 或 S.W.] 出运重量，离岸重量	size specification 尺码规格
	skid 垫木
	sledded cotton 机摘棉
	sliver 棉条
	slope 斜率
	Smith standard cotton 史密斯棉（美

国产）
Smyrna cotton 士麦拿棉（小亚细亚产）
snapped cotton 剥桃棉
soil 土壤
solenopsis invicta 红火蚁
soot 烟灰
sore shin 立枯病
sorghum halepense 假高粱
sort 分类，整理，拣选
sort out 归类
sorting the mixed mark bales 整理混唛棉包
sound 完整，完善，声音
sound goods 完好货物
South American cotton 南美洲棉（巴西及秘鲁产）
South Carolina cotton 南卡罗来纳棉（美国产）
Southern Hope cotton 南望棉（美国产）
Southern Queen cotton 阿肯色棉（美国产）
Southern Regional Research Laboratory 美国南方地区研究试验所
sow seeds 播种
sowing 播种
sowing date 播种期
span length 跨距长度
spandex 氨纶
special color rendering index 特殊显色指数
special condition 特殊条件
special defects 特殊疵点

special preferences 特惠关税
special safeguard measures 特殊保障措施
special tariff treatment 特殊关税待遇
special terms 特殊条款
specific 特定的，明确的，有特效的
specification 规格，说明书，技术要求
specified 额定
specify 详细说明，指明
specimen 样本，试样，样品
spectral distribution 光谱分布
spectrophotometer 分光光度计
spider 蜘蛛
spider mite 棉叶螨
spinning consistency index [SCI] 纺纱一致性指数
spontaneous ignition 自燃
spontaneous ignition temperature 自燃温度，自发着火点
spot 斑点，污迹
spot price 现货价格
spot transaction 现货交易
spotted cotton [sp] 点污棉
spot-check 抽查，现场检查
spring 春季
sprout 发芽
stability 稳定性
stack mixing 棉堆混棉
stain 污点，瑕疵
stain proof 防污
stained cotton 变色棉，污渍棉
stamen 雄蕊

stamp 印章，盖章，标记
stamp duty 印花税
standard 标准，规范
standard atmosphere for testing 试验用标准大气
standard export packing 标准出口包装
Standard of American Upland Cotton 美国陆地棉标准
standard stock solution 标准储备溶液
standard temperature and pressure [STP] 标准温度与压力
standard white plaque 标准白瓷板（测白度用）
standard working solution 标准工作溶液
staple analysis 纤维长度分析
staple array 纤维长度排列图
staple goods 主要产品，大路货
staple length 手扯长度
stapling 扯棉样，纤维长度分级
state 国度，州
statement 陈述，说明，声明
status 状态
statute 法律，条例
steel 钢
steel belts 钢带
steel belts hoop 钢带箍
steel seal 钢印
steel strip 钢带
steel strip hoop 钢条箍，钢带箍
Stelometer strength 斯特罗强力
stem 主茎，茎秆

stem color 茎色
stem hardness 主茎硬度
stem pubescence amount 茎毛量
stem pubescence length 茎毛长度
stench 恶臭，臭气
stencil 模板，模具
sticky 粘性，黏性
stigma length 花柱长度
stipulate 规定，约定，条款
stock 存货，库存量
stock house 库房
storage 储存，库存量，仓库
storage place quarantine 存放场所检疫
storage shelves 储存架
stow 堆垛，堆装，装载
stowage 堆装物，装载物，储藏物
stowed in 装进，装入
straight steel ruler 钢直尺
strength [Str] 强度
strength of ring spun yarn 环缕纱强度
Strict Good Ordinary [S.G.O] 次下级（美棉分级标准等级）
Strict Good Ordinary Plus [S.G.O.P] 次下级加（美棉分级标准等级）
Strict Low Middling [S.L.M] 次中级（美棉分级标准等级）
Strict Low Middling Light Spotted [S.L.M Lt.Sp] 次中级淡点污棉（美棉分级标准等级）
Strict Low Middling Plus [S.L.M.P] 次中级加（美棉分级标准等级）
Strict Low Middling Spotted [S.L.M Sp]

次中级点污棉（美棉分级标准等级）
Strict Low Middling Tinged [S.L.M Tg] 次中级淡黄点污棉（美棉分级标准等级）
Strict Middling [SM] 次上级（美棉分级标准等级）
Strict Middling Spotted [S.M Sp] 次上级点污棉（美棉分级标准等级）
Strict Middling Tinged [S.M Tg] 次上级淡黄染棉（美棉分级标准等级）
Strict Middling Yellow Stained [S.M Y.S] 次上级黄染棉（美棉分级标准等级）
strike 罢工
strippings 抄针花（包括斩刀花）
strong 强有力的，
structure 结构，构造，组织
style 式样，款式
subcontracting 分包
subject to 服从，受…支配，以…为条件
subjective inspection 主观检验，感官检验
submit 呈递，递交，提交，呈送
sub-sample 子样品，二次抽样
subsequent 随后的，后来的，继…之后的
subsequent change of vessel 后来船名的更改
subsidy 补贴
substance 物质，财物，实质
substandard 不合标准的，不合格的
substantial transformation 实质性改变
substitute 代用品，替代品，代替
subtract 扣除，减去

sufficient 足够的，充足的
sugary 含糖的
summer 夏季
Sunday 星期日
superfine fiber 超细纤维
superior to 优于，好于
supersede 取代
supervise 监督，管理
supervision and administration 监督管理
supervisor 监管人，主管
supplement 补充，增补，附加
supplementary articles 附则
supplementary contract 补充合同
supplier 供方，供应商
supply 提供，供给，补充
support value duty 保价关税
Surat cotton 苏拉特棉（印度产）
surcharge 超载，附加费
Surinam cotton 苏里南棉（圭亚那产）
surplus 剩余的，过剩，顺差，盈余
survey 鉴定，检查，测量
survey of damage 残损鉴定
survey of damaged cargo 残损货物鉴定
survey of quantity 数量鉴定
survey of weight 重量鉴定
surveyor 鉴定人，鉴定员
surveyor's report on weight 重量鉴定报告
suspend 推迟，延缓
sustain 维持，承受住
Sutton cotton 萨顿棉（美国产）
Suzani cotton 苏扎尼棉（土耳其斯

坦产）
symbol 符号，象征，标记
sympodial branch node 第一果枝

（着生）节位
sympodial branch number 果枝数
sympodial branch type 果枝类型

T

tackle 用具，装备
tactile appraisal 手感评定
tag 标签，标价牌
take care of 保管
take precautions against natural calamities 防灾
tally 符合，理货，清点
tally with 符合，与…一致
Tanguis cotton 坦圭斯棉（美国阿拉巴马产）
tare weight 皮重
tariff barrier 关税壁垒
tariff concession 关税减让
tariff quota 关税配额
tariff rate 关税税率
tatted cotton cloth 梭织棉布
tax rebate 退税
taxation 课税，税收，征税
Taxili cotton 塔斯里棉（马其顿产）
Tayiba cotton 太巴棉（苏丹产）
Taylor cotton 泰勒棉（美国阿拉巴马及南卡罗来纳产）
Tchesma cotton 彻斯马棉（马其顿产）
tear strength 撕破强度，撕裂强度
technical 技术

Technical Barriers to Trade [TBT] 技术性贸易壁垒
technical dispute 技术辩论
technical exchange 技术交流
technical regulation 技术法规
technical specification 技术规范，技术说明
technical terminology 专业用语，技术术语
technician 技术人员，技师
technology 技术，工艺
telegram 电报
telegraphic transfer [T/T] 电汇
telex 电传，用户电报
temperature 温度
temporary export licence 临时出口许可证
temporary export management 临时出口管理
tenacity 断裂强度，强力
tenacity and elongation 拉伸强力
Tennessee cotton 田纳西棉（美国产）
Tennessee Gold Dust cotton 田纳西砂金棉（美国产）
term 条款，术语，期限
term of trade 贸易条件，进出口交换

比率
terminal 末端，终端机
terminal handling charge [THC] 终点操作费
terminal receiving charge [TRC] 终点接收费
terminate 终止，解雇
terminate an agreement 解约
terms and conditions 条款和条件
terms of payment 付款方式
territory 国境
test 检测
test beard 试验须丛
test report 测试报告
testing item 检测项目
tex 特克斯
Texas [TEX] 得克萨斯（美国主要产棉区）
Texas storm proof cotton 德克萨斯抗风棉（美国产）
text 文本，正本
textile 纺织品
textile dispute 纺织品争端
textile fabric 织物，纺织布料
textile fiber 纺织纤维
textile industry 纺织行业
textile material 纺织材料
textile raw material 纺织原料
textiles safety 纺织品安全性
the average level 平均水平
the filed quarantine 田间检疫
the goods quarantine 货物检疫
the number of effective harvest plants 收获株数（实收株数）

the official plant quarantine certificate 官方植物检疫证书
the preparatory work 准备工作
the sample label 样品标签
the scene quarantine 现场检疫
thereafter 此后，之后
therefrom 所得，所产生，由此产生的，从此
therein 在其中，在那点上
thereinafter 在下一部分中，在下文中
thereinbefore 在前一部分中，在上文中
thereto 此外
thereunder 在其下，据此
thereupon 在其上
thermostat water bath 恒温水浴锅
thick 厚
thimble tube 指形管，套管
Thinawa cotton 锡那华棉（缅甸产）
thin-walled fibers 薄壁纤维
third-cut linter 三类棉短绒
this is to certify that …兹证明…
three copies of the duplications 一式三份
through bill of lading 全程联运提单
Thursday 星期四
tighten 紧固，捆扎
tighten with 用…紧固，用…捆扎
till 耕种
time limit 期限
Tinged cotton [Tg] 淡黄染棉
tinged linter content 黄根率
Tinnevelly cotton 丁内未利棉（印度产）

to be continued 未完（待续）
to order 订购，待指定
tolerance 公差，允许，容许
tolerance to salinity 耐盐性
tolerance to waterlogging 耐涝性
top 顶部，上端
total 合计，共计
total amount 总价
total lint yield 皮棉（总）产量
total quality control [TQC] 全面质量控制
total tare weight 总皮重
total value 总值
total yield 总产量
tough 坚硬的，坚韧的
tracheitis 气管炎
trade 贸易
trade consultation 贸易磋商
trade deficit 贸易逆差，贸易赤字
trade liberalization 贸易自由化
trade mark 商标
trade surplus 贸易顺差，贸易盈余
trade union 工会
trading country 贸易国
traditional 惯例的，传统的
traditional cultivar 地方品种
traffic 交易，货运，交通
transaction 交易，办理，学报
transfer 转让，传送，让与
transferable 可转让的
transform 改造
transgenic 转基因
transgenic cotton 转基因棉花

transit insurance 运送保险
transit trade 转口贸易，过境贸易
transit transportation 中转运输
translate 翻译
transparent package 透明包装
transport 运输，运送
transport goods for sale 贩运
transport package 运输包装，外包装
transport quarantine 运输工具检疫
transship 转船运输
transshipment 转运，转船
trash 杂质
trash area [Tr Area] 杂质面积
trash count [Tr Cnt] 杂质粒数
trash grade [Tr Grade] 杂质等级
treatment 处理
treaty 条约，协议
tribunal 法庭
triplicate 一式三份，一式三份中的一份
Trogoderma 斑皮蠹
Trogodermagranarium 谷斑皮蠹
troy weight 金衡制
Truitt cotton 杜鲁伊特棉（美国产）
trust 信任
truth-value 真值
Tuesday 星期二
tuft length diagram 棉束纤维长度分布图
tweezers 镊子
type 类别，型（号）
type of packages 包装种类
typical 典型的，代表性的

U

ultimate 最后的，最终的，根本的
ultraviolet rays 紫外线
unacceptable 不能接受的，难以承受的，不合格的
unavoidable 不可避免的
unclean bill of lading 不清洁提单
unconquered 未克服的
undefined 不明确
unfading paint 不褪色的颜料
unfair competition 不公平竞争
unfavorable balance of trade 贸易逆差
unforeseeable 不可预见的
uniformity 均匀度
uniformity [Unf] 整齐度
unify 统一，使一致
unilateral 单方面的，单边的
unilateral contract 单方承担义务的契约，片面义务契约
unilateral repudiation of a treaty 单方面的废除条约
unit price 单价
universal 通用的
universal density 通用密度
universal standard 通用标准

unless otherwise agreed 除非另有约定
unload 卸货，倾销
unofficial 非官方，非正式
unpack 开箱，拆包
unpaid 未付的
unqualified disposition 不合格处理
unripe cotton 未熟棉
unseal 启封
unstable 不稳定的，不牢固的
up quarter length 上四分位长度
update 最新的
upland cotton 陆地棉，细绒棉
Uppam cotton 乌帕姆棉（印度产）
upper half mean length [UHML] 上半部平均长度
Uppers cotton 上游棉（产于尼罗河流域上游）
urea 尿素
USDA Universal Cotton Standards 美国农业部通用棉花标准
user 用户，消费者
Uster irregularity 乌斯特不匀率
Uster stapling instrument 乌斯特纤维长度测定仪

V

vacuum packaging 真空包装

Valencia cotton 巴伦西亚棉（哥伦

比亚产）
valid 有效的
validity 有效期，正当，有效
Valley cotton area 盆地棉田（指美国密西西比河流域各产棉田）
valuation 估值，估计，评价
value 估值，估计，值
variance analysis 方差分析
variation 变异，变种
variety 品种，多样，种类
vegetable fiber 植物纤维
ventilate 通风
verification 验证，核实，证实
verification of conformity 符合性验证
verify 证实，验证，核实
version 版本

verticillium dahilae （棉花）黄萎病菌
verticillium wilt 黄萎病
vessel 船名，船舶
vessel age 船龄
vessel's flag 船旗
veto 否决，禁止
viral isolation 病毒分离
virus 病毒
visual assessment 目光鉴定
visual examination 目测，外观检验
visual inspection 目光（外观）检验
visual test 目光（外观）检验
void 无效的，空的
volumetric flask 容量瓶
vote down 否决
voyage 海运，航行

W

wadding fiber 絮用纤维
Wagad cotton 瓦加德棉（印度产）
Wagale cotton 瓦加利棉（缅甸产）
Wamamaker cotton 万那梅克棉（美国产）
Wani cotton 瓦尼棉（缅甸产）
Wapyu cotton 瓦比棉（缅甸产）
warehouse 仓储
warehousing 仓库，货栈
wares 货物，商品
warranty 保证书，保单
warranty period 保质期
wastage 损耗

waste 耗损，废弃的
waste cotton 废棉
waste silk 废丝
water absorption 吸水性（率）
water damaged cotton 水渍棉
waterproof 防水
way bill 运货单
Wednesday 星期三
weed identification 杂草鉴定
weighing apparatus 衡器
weight commercial 商业重量
weight declared 报检重量
weight in shortage 短重

weight list 重量单，磅码单
weight memo 重量码单
weight tolerance ratio 溢短装率
weighting 增重，称重，加权
weight-overage 溢重
Wharf 码头，停泊处
White cotton 白棉类
white porcelain 白瓷盘
white seeds 白（棉）籽
whiteness 白（色）度
whole growth period 全生育期
wholesale 批发
width 宽度
wild 野生的，野生资源
wind damage 风灾
winter 冬季

wiriness 铁丝状，手感粗硬
withdrawing of certificate 证书的撤销
within 在…期间
Wonderful 奇异棉（美国产）
wood chips 木屑
wooden case 木箱
wooden package 木制包装
wooden pallet 木托盘
work instruction 作业指导书，操作规程
worsted cotton yarn 精梳棉纱
wrap 包，裹，捆
wrappage 包裹物
wrapped with 用…包裹，用…捆扎
written form 书面形式

Y

yard 码
yarn 纱，纱线
yellow cotton 黄棉
Yellow Stained cotton [Y.S] 黄染棉
yellow-stained cotton 黄渍棉（疵点）

Yerli cotton 耶尔利棉（土耳其产）
yield 产量
yield after frost 霜后籽棉产量
yield before frost 霜前籽棉产量
yielding ability 丰产性

Z

Zafiri cotton 扎非里棉（埃及产）
Zagora cotton 扎古拉棉（埃及尼罗河三角洲产）

Zellner cotton 泽尔那棉（美国产）
zero gauge 零隔距
zone 地带，地区，区域

汉英部分
Chinese-English

a

矮壮素　cycocel
安排　arrange; assign; schedule
安全　security
安全认证标志　safety certification mark
安全系数　safety factor
安全线　safety line
安全性　safety
安设　install
安装　setup; install
氨基酸　amino acid
氨纶　spandex
氨水　ammonium hydroxide

按钮　button
按时　on time; on schedule
按压　press
按照　according to; depend on; as per; in accordance with
案秤　counter scale
案件　case
案件受理　acceptance of a case
案例　case; circumstance
暗淡　dim; dull
暗火　dying fire; blind fire
盎司　ounce

b

八月　August [Aug]
拔节　jointing
罢工　strike
罢免　recall
白（棉）籽　white seeds
白（色）度　whiteness
白瓷盘　white porcelain
白棉　white cotton
百分数　percentage
摆脱　dispense; rid
拜谒　pay a formal visit; call to pay respects
颁布　promulgate; issue

斑点　spot
斑皮蠹　Trogoderma
搬运　carry
搬运工　carrier; remover
板结　harden
版本　version
办理　handle; manage; deal with; transect; dispatch（表示迅速的、急于办理）
半成品　semi-finished product; semi-manufactured goods
半成熟纤维　semi-mature fiber
半导体　semiconductor

半纤维素	hemicellulose
绑	bind; bundle up; bundle; ligature
磅	pound
磅秤	platform scale; platform balance
磅秤车	scale car
磅码单	weight list; weight memo
包	bale; pack; wrap
包缝	wrap seam
包裹	package; wrap
包括	consist; include
包装	pack
包装布	packing sheet
包装材料	wrapping material; packing material
包装方法	packing method
包装检验证书	inspection certificate of packing
包装类型	package type
包装说明	packing instruction; packing desrciption
包装完好	packing sound
包装完整	packing intact
包装证书	certificate of packing
包装种类	type of packages; kind of packing
包装种类及数量	number and type of packages
苞齿	bract tooth
苞外蜜腺	extra-bract nectar
苞叶	bract; subtending leaf
苞叶联合	bract ally
苞叶自落	bract withered
胞壁厚度（纤维）	cell wall thinness
饱和	saturate
保持	remain; keep
保持干燥	keep dry
保持性	retentivity
保存	save; preserve; keep
保单	warranty; insurance policy
保兑信用证	confirmed letter of credit
保兑银行	confirming bank
保管	custody; take care of
保函	letter of guarantee [L/G]
保护	protect; defend; guard; safeguard; protection
保护价	protective price; conservation price
保护装置	protector; protective equipment
保价	support value
保价关税	support value duty
保留	remain; reserve; retain
保留条款	reservation clause
保暖性	heat retention; thermal retentivity
保暖性保护	warmth retention protection
保暖性能	thermal performance; thermal insulation properties
保税库	bonded warehouse
保税区	bonded zone; tariff-free zone
保险	insurance
保险代理人	insurance agent
保险单	insurance policy
保险费	insurance premium; premium
保险凭证	insurance certificate

保障　guarantee; safeguard; security
保障措施　safeguard measure
保证　ensure; guarantee
保证金　earnest money; cash deposit
保证书　guarantee; warranty
保证条款　warranty clause
保质期　warranty period; expiration date
报告　report
报关　customs clearance
报关单　customs declaration form; bill of entry
报关员　customs declaration agent; customs declarer; declarant
报价单　quotation; price sheet
报检（申报）单位　declaration units
报检　declare
报检数量　quantity declared
报检员　declare member; inspection declaration agent; inspection staff
报检重量　weight declared
报盘　offer
暴露　exposure
北向天然昼光　north skylights
备份　back-up
备注　remark
备装通知　notice of readiness
背光　backlight
被保险人　insured; the insured; assured; insurant
被告　defendant
被子植物　angiosperm
鼻窦炎　nasosinusitis
鼻炎　rhinitis
比表面积　specific surface area
比对试验　comparative test
比例　proportion; ratio
比率　proportion; rate; ratio; percentage
比马棉　Pima cotton; PIMA
比色测定　colorimetric determination
比色法　colorimetry; colorimetric method
比色管　colorimetric tube
比色计　colorimeter; chromometer
比照　according to; in the light of
比重　proportion
彼此的　mutual
币值　currency value
必需的　necessary; indispensable
必要条件　requirement
闭市价　closing price
避免　avoidance
边　side
边境贸易　frontier trade; border trade
编号　number
编码　coding
贬值　devalue; depreciate
变化　change; variation
变色棉　stain cotton
变异系数　coefficient of variation [CV]
变质　metamorphic
辩护　defend; plead
标定　demarcate
标记　mark; label; official stamp;

sign
标记及号码 Mark & No.
标价牌 tag
标签 label; tag
标识卡 identification card
标志 sign; mark; designate; symbol
标准 standard
标准白瓷板（测白度用） standard white plaque
标准差 standard deviation [STD DEV]
标准出口包装 standard export packing
标准储备溶液 standard stock solution
标准大气条件 standard atmospheric condition
标准工作溶液 standard working solution
标准温湿度 standard temperature and humidity
标准性文件 normative document
表格 table; form; tabulation; sheet
表面积 surface area
表明 indicate
表皮层 cuticular layer
表皮细胞 epidermic cells
表示 express; show
兵变 mutiny
丙基纤维素 propyl cellulose
丙纶（聚丙烯纤维的商品名） polypropylene fiber
病斑 scab

病变 lesion
病毒 virus
病毒分离 viral isolation
病害 disease
病情指数 disease index
病株 diseased plant; infected plant
病株率 diseased plant rate
剥除短绒 delint
剥除短绒的棉籽 delinting seed
剥绒机 delinter
剥桃棉 snapped cotton
播种 sow seeds; sowing
播种期 sowing date
泊位 berth; berthage
薄壁纤维 thin-walled fibers
卜氏强力 Pressley strength
补偿 compensate; compensation; indemnify
补充 replenish; supplement; additional; complementary
补充合同 supplementary contract
补充证书 certificate of supplement
补救 remedial
补救措施 remedial measures
补贴 subsidy
不成熟纤维 immature fiber
不纯 impurity
不符合 inconformity; fall short of
不符合标准 off standard
不公平竞争 unfair competition
不规则 anomaly; irregular

不含　free from
不合标准的　sub-standard
不合格　non-conformity; unqualified; below grade; off-grade; disqualification
不合格处理　unqualified disposition
不可避免的　unavoidable
不可撤销的信用证　irrevocable L/C
不可分割的　integral; indivisible
不可分割信用证　indivisible L/C
不可抗力　force majeure
不可预见的　unforeseeable
不控制极限　no control limit [NCL]
不莱梅国际循环比对试验　Bremen International Round Trials
不莱梅纤维协会　Fiber Institute Bremen
不牢固　unstable; loose
不良状态　badness condition
不明确　ambiguity; undefined; ill-defined
不能接受的　unacceptable
不清楚　ill-defined
不清洁提单　unclean bill of lading
不受…影响　not affected by...
不受理　reject
不透气的　airproof
不稳定的　unstable
不一致　inconsistent; inconformity
不匀率　irregularity
不孕籽　sterile seed
不孕籽含棉率　cotton containing sterile seed rate
不孕籽率　rate of infecund seeds
不整齐的　irregular; scruffy; untidy
不正确　incorrect
步骤　procedure
部分　part; section; portion
部门　department

C

材料　material
财产　property; asset; estate
财产保险　property insurance
裁决　decide; decision; resolve
裁决书　arbitral award
采购　purchase
采光　daylighting
采集袋　collection bag
采棉　cotton picking
采棉人　cotton picker
采摘　pick
彩棉　colored cotton
参考　consult; refer to; read sth. for reference; reference [RE]
参考价格　indicative price
参考文献　reference
参考指标　reference index
参数　parameter

残留 remain; residual
残损 damage
残损货物单 broken and damaged cargo list
残损鉴定 survey of damage
残损率 rate of damage
仓储 keep goods; storage; warehousing
仓储者 warehouser
仓单 manifest; warehouse receipt
仓库 storehouse; warehouse; depot; repertory
苍白 pale; pallid
舱位 cabin seat; shipping space
操作 operate; handle operation; manipulation;
操作规程 work instruction
草案 draft; draught
草本学 Agrostology
草棉 G. herbaceum
草约 protocol
草籽 grass seed
侧根 lateral root
侧芽 lateral bud
测定 determination; admeasurement
测量 measure; measurement
测试报告 test report
测试窗口 test window
差别 difference; disparity; dissimilarity; distinction; contrast
差别待遇 differential treatment
差额 balance; margin
差价 price difference
差异 difference; divergence; discrepancy; diversity
查封 seal up; close down
查询 query
查询回复 response to query
查验 check; examination
查证 investigate and verify
拆（集装）箱 devanning
拆包 unpack
拆包钳 bale hoop cotter
掺次包装棉包 false packed bale
掺和 admixture; blend; admix; mix
掺假 adulterate
掺杂 mix; mingle; inclusion; adulterate
缠绕 twine; bind; enlace; wind
产地 origin
产地证 certificate of origin
产量 yield; output
产品 product; produce; goods; merchandise; manufacture
产生 generate; come into being
产销 production and marketing
铲车 forklift
长度 length
长度标准棉样 length standard cotton sample
长度范围 length range
长度分布 length distribution
长度分布图 length distribution graph
长度均匀度 length uniformity
长度偏差 length variation
长期的 permanent; long-standing
长绒棉 long-staple cotton; sea island

cotton
长纤维　long fiber
常规的　conventional
常规检查　customary examination
常衡　avoirdupois
常温　normal temperature
偿还　reimbursement; repay; pay back
抄斩花　cotton of card strips; stripping
抄针花（包括轧针花）　strippings
超出　overstep; exceeding; overranging; overtop; go beyond
超额　excess; overrun
超过　exceed; outnumber; outstrip; excess
超量　excess; surplus
超细纤维　superfine fiber
超载　overload; excess freight
超长纤维　extra-long fiber
超重　overweight; superheavy
潮湿　damp; moist; humid; humidity; aquosity
车肚花　card waste
撤回（已发出的订货单）　countermand
尘肺　pneumoconiosis
陈述说明　statement; representation
衬垫材料　cushioning material
称重　weigh; weighting
成包原棉　bale cotton
成本　cost
成本单　cost sheet
成本加运费价　cost and freight [C&F]
成分　ingredient; element; composition

成分分析　composition analysis
成分商标　content label
成果　achievement; production
成交　strike a bargain; make a bargain; close a deal; bargain on
成交小样　deal sample
成铃　boll
成熟度　maturity [Mat]
成熟度比率　maturity ratio
成熟系数　coefficient of maturity
成熟纤维　mature fiber
成熟指数　maturity index
成像　imaging; imagery; image formation
成员　member;
呈递　present; submit
承兑　accept; redemption
承兑交单　document against (on) acceptance [D/A]
承认　acknowledge; recognition; admit; accept
承受　sustain
承运人　carrier; freighter
程度　degree; level; extent
程序　procedure; program
惩罚　punishment; penalty
持久的　permanent; everlasting; enduring
持票人　bearer; bill holder
持有　possess; hold
充足的　sufficient; sufficient
虫粪　frass
抽查　selective examination

抽样　sampling
抽样比例　sample percentage
抽样误差　sampling error
出厂价格　ex-factory price
出境货物通关单　bill of the customs clearance for exit cargo
出口　export
出口包装　export packing
出口补贴　export subsidy
出口合同　export contract
出口结关　customs clearance
出口配额　export quota
出口商　exporter
出口商声明　declaration by the exporter
出口退税　tax rebate
出口箱　export (shipping) carton
出苗期　period of emergence; time of emergence; season of emergence; seeding stage; emerging date
出售　sale
出运重量　shipping weight [S/W 或 S.W]
出证机构　certifying authority
初生胞壁　initial cell wall; primary cell wall
初生层　primary layer
初始接收费　origin receiving charge [ORC]
除草剂　herbicide
除虫　disinsectization
除非另有约定　unless otherwise agreed
除去　remove; eliminate
除鼠　deratization
除杂　cleaning impurity
处罚　punishment; penalty
处理　settlement; conduct; dispose; treat
处置　disposition
储备棉　cotton reserve
储存　reserve; storage
储存架　storage shelves
储棉量　cotton stock
储气罐　air storage tank
传染病　infectious disease
传送　deliver; transfer
传统的　traditional; conventional
船舶　vessel; ship
船龄　age of vessel; vessel age
船期表　shipping schedule
船旗　vessel's flag
船上交货　ex ship
船上交货价　free on board [FOB]
船运公司　shipping company
船载货物　cargo; freight
船长　captain; shipmaster
创始　initiation; originate
春季　spring
纯利润　net profits; pure profit
疵点　defect
次等的　inferior
次上级（美棉分级标准等级）　Strict Middling [S.M]
次上级淡黄染棉（美棉分级标准等级）Strict Middling Tinged [S.M Tg]

次上级点污棉（美棉分级标准等级） Strict Middling Spotted [S.M Sp]
次上级黄染棉（美棉分级标准等级） Strict Middling Yellow Stained [S.M Y.S]
次生胞壁 secondary cell wall; secondary wall
次生层 secondary layer
次下级（美棉分级标准等级） Strict Good Ordinary [S.G.O]
次下级加（美棉分级标准等级） Strict Good Ordinary Plus [S.G.O.P]
次序 order; sequence
次中级（美棉分级标准等级） Strict Low Middling [S.L.M]
次中级淡点污棉（美棉分级标准等级） Strict Low Middling Lighter Spotted [S.L.M Lt. Sp]
次中级淡黄点污棉（美棉分级标准等级） Strict Low Middling Tinged [S.L.M Tg]
次中级点污棉（美棉分级标准等级） Strict Low Middling Spotted [S.L.M Sp]
次中级加（美棉分级标准等级） Strict Low Middling Plus [S.L.M.P]
粗绒棉 Asiatic cotton
粗纱 rough yarn
粗纤维 crude fiber; coarse fiber
催速剂 accelerant
存放 store; deposit
存放场所检疫 storage place quarantine
存货 stock
磋商 negotiation; consultation

d

达成交易 close a deal; close a transaction; conclude a transaction
答复 reply; answer; response
打包 pack; packing; baling
打包厂 packing factory
打包机 baling press; packaged machine
打包结 bale ties
大保单 insurance policy
大分子链 macromolecules chain
大量生产 bulk production; mass production; fordize
大气条件 atmospheric conditions; ambient conditions
大容量棉花测试仪 high volume instrument [HVI]
代表 deputy; delegate; representative; represent; on behalf of
代表团 delegation; deputation; mission; delegacy
代表性的 representative; typical
代表性取样 representative sampling

代表性样品　representative sample
代理权　power of attorney
代理人　agent; attorney; deputy
代替　replace; substitute
带来很大费用（花费）　incur great expense
带皮重量　gross weight [G.W]
带子　band
贷款　loan; credit
待指定　to order
担保　ensure; guarantee
单独　individual; single
单方承担义务的契约　unilateral contract
单方面的　unilateral
单方面的废除条约　unilateral repudiation of a treaty
单价　price
单据　bill; receipts
单纱　single yarn
单糖　monosaccharide
单纤维强度　single fiber strength
单纤维强力机　single fiber strength tester
单证　document
单证员　documentary secretary
单株成铃数　bolls per plant
旦尼尔（纤维细度单位）　denier
淡点污棉　light spotted cotton
淡黄染棉　tinged cotton
淡灰棉　light gray cotton
淡棕色　light brown

当局　authority
当事人　party
当作　regard
刀子　knife
导航　navigation
导致　result in; incur
到岸价　cost insurance and freight [CIF]
到岸品质　landed quality
到岸重量　landed weight
到达口岸　port of destination
到达日期　date of arrival
到货地点　place of arrival
到货港　port of arrival
到货日期　date of arrival
到货通知　advice of arrival; arrival notice [A.N]
到期　expire
得克萨斯（美国主要产棉区）　Texas [TEX]
得益　profit; benefit
登记　register; check in
等分试样　aliquot sample
等级　grade
等温吸放湿平衡曲线　isothermal humidity absorption and desorption equilibrium curve
等温吸放湿曲线　isothermal sorption curve
低聚糖　oligosaccharide
低于　lower than
抵港通知　advice of arrival; arrival notice [A.N]

抵扣　deduction
抵押　mortgage; hold in pledge; hypothecate
底　bottom
地带　zone; area
地方品种　traditional cultivar; landrace
地理种系　race
地区　district; area
地震　earthquake
地址　address
地中海实蝇　Ceratitis capitata
递交　present; submit; deliver
第一果枝（着生）节位　sympodial branch node
典型的　typical
点火源　ignition source
点燃　ignite
点污棉　spotted cotton
电报　telegram; telegraph
电测器　electric logging device
电传　telex
电工学　electrotechnology; electrotechnician; electrotechnics
电荷　electric charge; electricity
电汇　cable transfer; telegraphic transfer [T/T]
电量（物体所带电荷的多少）　quantity of electricity; quantity of electric charge
电流　electric current; electricity; electrical current
电路　circuit; electric circuit; electrocircuit; circuitry
电容　electric capacity; capacitance
电位　electric potential; potential
电位差　potential difference [PD]
电压　voltage; electric tension; electric voltage
电源　current source; electric source; electrical source
电子管　electron tube; radio tube; vacuum tube
电子天平　electronic scales
电子显微镜　electron microscope
电阻　resistance; electric resistance
玷污　contamination
垫木　skid
调度　dispatch
钉子　nail
顶部　top
订单　order; order form
订货　order; order goods
订货单　indent; purchase list; purchase order
订货时即支付货款　remittance with order
订立　conclude
定价　fix a price; make a price
定量　quantitative
定量法　quantitative method
定量分析　quantitative analysis
定位　location
定性分析　qualitative analysis

定义　definition
定重　constant weight
定重制　constant weight system
冬季　winter
动态平衡　dynamic balance; dynamic equilibrium
动物尸体　animal carcass
毒麦　Lolium temulentum
独立性　independence
独有的　unique
读数　reading
度　degree
度量　measure; measurement
短期的　short-term
短缺证书　certificate of shortage
短绒　flock; seed fuzz
短绒棉　short-staple cotton
短绒色　fuzz color
短纤维　short fiber
短纤维率　short fiber percentage
短纤维指数　short fiber index [SFI]
短重　shortage in weight; weight shortage
短重率　shortage weight rate
断裂　break
断裂负荷　breaking load
断裂强力　breaking strength
断裂伸长　breaking elongation
断裂伸长率　elongation at break
断裂试验　breaking test
断裂长度　breaking length
断言　assertion; affirm; declare; allege
堆积　pile
对…有关系　concern
对比　contrast; comparison
对比试验　contrast test
对外的　external
对外贸易　foreign trade
对照品种　check variety
对照样品　control sample
多边贸易体制　multilateral trade system
多式联运　multimodal transport
多糖　polysaccharide
多种纤维协定　multi-fiber agreement
多籽屑棉　shell cotton

e

额定　rated; specified
额外的　additional; excess; extra
恶臭　foul smell; stench
恶劣　scurviness; adverse
萼片　sepal

二次抽样　sub-sampling
二类棉短绒　second-cut linter
二手的　second-hand; used
二月　February [Feb]

f

发表　publish
发达国家　developed country
发货代理人　shipping agent
发货港　port of dispatch; point of origin
发货人　consignor; consigner; shipper
发货人名称及地址　name and address of consignor
发货日期　date of dispatch
发价　offer
发霉　go mouldy; become mildewed
发票　invoice
发票号　invoice No.
发票日　invoice date
发售　sell; put on sell
发送　dispatch ; delivery
发芽　germinate; sprout
发育　growth; develop; development
发展中国家　developing country
发证检验　certification testing
罚金　fine; forfeit
法定的　legal
法定检验　lawful inspection
法规　laws and regulations; rule of law; statute
法令　laws and decrees
法律　law; statute
法律顾问　adviser on legal
法律效力　force of law

法人　juridical person
法庭　court; tribunal
法院　court of justice; law court
法制　legal system
砝码　weight
翻案　reverse a verdict
翻译　translate; interpret
凡士林　vaseline; petrolatum; petroleum jelly
繁育　breed
繁殖　breed; multiply
反比　inverse ratio
反补贴税　countervailing duty
反光　reflect light
反倾销　anti-dumping
反射　reflect
反射角　angle of reflection
反射率　reflectivity; reflectance; Rd
反应　reaction; response; repercussion; reactivity
反映　response; reflect
返潮　get damp
返工　rework
返回　return
返青　trun green; resume growth
犯法　break the law
犯罪　commit a crime
范本　model for calligraphy or painting

范畴	category
范围	scope; range
贩运	transport goods for sale; traffic
方案	project; scheme
方差分析	variance analysis
方法	method; means; way
方针	policy
防潮	dampproof; moistureproof
防潮包装	moisture proof packing
防尘	dustproof; dust-free; dust prevention
防除	prevent and kill off
防毒面具	antigas mask
防腐	antiseptic; anticorrosive
防护	protection
防火	fire prevention; fireproof
防水	waterproof
防水垢剂	anti incrustant
防锈包装	rust proof packing
防疫	epidemic prevention
防灾	take precautions against natural calamities
防震包装	shockproof packing
防止	prevent; guard against; avoid; avert
防治	prevention and cure
纺纱	spin; spinning
纺纱工人	spinner
纺纱一致性指数	spinning consistency index [SCI]
纺织材料	textile materials
纺织工业	textile industry
纺织工艺	textile technology
纺织行业	textile industry
纺织品	drygoods; textile
纺织品安全性	texiles safety
纺织纤维	textile fiber
放大镜	magnifying glass
放电	discharge; eletro-discharge
放行	release; let … pass
放行单	notice for release; release notice; release permit
放火	sit fire to; set on fire; commit arson
放热	heat release; exothermic
放湿	moisture liberation
放湿曲线	moisture liberation curve
放湿性	desorption of moisture; moisture releasability
飞花	flying; fly
飞轮	flywheel; free wheel
非法的	illegal; illicit
非关税壁垒	non-tariff barrier
非关税措施	non-tariff measures
非关税减让	non-tariff concession
非官方	unofficial
非国产棉	non-chinese cotton
非棉物质	non-cotton substance
非木质包装证	non-wooden packing certificate
非违约方	non-breaching party
非纤维性杂物	non-fibrous impurities
非约束性条款	permissive provision
非正式	unofficial; informal
非直接接触皮肤用品	products without direct contract to skin

非洲棉	Africa cotton
肥料	fertilizer; manure
肥沃	fertile; rich
肺病	lung disease; Pulmonary tuberculosis [TB]
肺气肿	emphysema pulmonum
肺炎	pneumonia
废除	abolish; abrogate
废棉	waste cotton
废丝	waste silk
废物	garbage; waste material; trash; recrement; rubbish
废止	avoidance; abolish
费用	charge; cost; expense; fee
分包	subcontracting
分贝	decibel
分别地	separately; respectively
分布	distribution
分部	branch
分公司	filiale; branch; branch office
分光光度计	spectrophotometer
分级	classify; grade; classification
分级室	classing room; grading room
分级长度	classer's staple
分开	separate; separation
分开卸货	separation discharge
分类	classify; classification; assortment; sort
分离	separate; separation
分米	decimetre
分配	allocate; distribution
分批	batch-wise
分批交货	delivery of goods by installment; segment delivery
分批装运	partial shipment
分期付款	installment; payment by installments; hire-purchase
分区取样	partitioned sampling
分析	analyze; analysis
分析证书	certificate of analysis
分支机构	branch
分子量	formula weight; molecular weight;
分子式	molecular formula; structural formula
粪肥	muck; manure
丰产性	yielding ability
丰收	foison; harvest
风险	risk; hazard; danger
风灾	wind damage; disaster caused by a windstorm
封存	seal up for safekeeping
封条	seal
封样	sealed sample
锋利的	sharp
否定	deny; negate
否决	vote down; veto; overrule
服装	garments; clothing; apparel
浮尘	floating dust
浮动	float
浮码头	floating pier
符号	symbol; sign; mark; insignia
符合	conform to; in accordance with; tally with; comply with
符合性	conformity
符合性标志	mark of conformity

符合性评定	conformity assessment

符合性评定 conformity assessment
符合性评定程序 conformity assessment procedure
符合性验证 verification of conformity
符合性证据 evidence of conformity
腐蚀 corrosion; erosion
父本 male parent
付款 pay; payment
付款方式 terms of payment; payment method
付款交单 document against payment [D/P]
付款人 payer
付款条件 payment terms
付清 pay in full; pay off
付息 payment of interest
付现 cash payment
负荷 load
负有责任的 responsible; accountable
负债 be in deb; liabilities
附加 attach; additional; additive
附加费 surcharge; additional charge
附加税 additional tax
附加物 appendage; annexation
附件 appendix; attachment; accessory
附录 appendix; addendum
附上 attach; enclosed
附则 supplementary articles
附照 attached photo
附着物 adhered substance
附注 annotation
复本 duplicate
复验 reinspect; reinspection
复叶 compound leaf
复杂 complicated; complex
复轧棉 complex gin
复制的 replication; duplicate
复制证书 certificate of duplicate
副本 copy; carbon copy
副产品 by-product; coproduct
副作用 side effect; by-effect
赋予 give; endow; entitle

g

改革 reform; innovate
改进 improve; ameliorate
改良种陆地棉 improved upland cotton
改造 transform; remould
改正 amend; correct
钙镁磷肥 calcium magnesium phosphate
盖(印章) seal; stamp
盖板花 card strips
概率 probability
概要 summary; outline; resume; abstract; compendium
干纺 dry spinning

干旱　drought
干涸期　dry period
干燥　dry; arid; desiccation
干重　dry weight
感官　senses organ
感官检验　sensory test; organoleptic inspection
感官检验室　inspection room by sensory
感官评定　organoleptic evaluation
感染　infect
纲要　outline; sketch; compendium
钢带　steel belt; steel band
钢丝　steel wire
钢印　steel seal
钢直尺　straight steel ruler; steel gauge
港口　port; harbor
港口税　port dues; harbor dues
高　high; tall
高产　high yield
高价　high price
高温　high temperature
高于　overtop; exceed; higher than
格令（重量单位）　grain
格式　form; format
隔火墙　fire wall; fire division wall
隔距　gauge
隔离　insulate; segregate
个别的　individual; single
各级　all or different levels
各自　respective; apiece
给惠国　preference-giving country

根腐病　root rot
根冠　root cap
根据　in accordance with; on the basis of
根毛　root hair
根朽病　black leg
跟单汇票　documentary bill; documentary draft; documentary bill of exchange
跟单信用证　documentary credit
更改通知书　amendment advice
更换　change; replace
更新　update; renew; renovate; replace
更正　correct; amend; revision
更正证书　certificate of revision
耕地　plough; plowland
耕种　till; cultivate
更优惠待遇　more favorable treatment; more preferential treatment
工厂　mill; factory
工程　engineering; project
工会　trade union; labor union
工具　tool; instrument; implement
工序　process; procedure; stages of production
工业　industry
工艺　craft; technology
工作校准棉样　working standard cotton samples
公报　bulletin; communique
公差　allowance; tolerance; common difference
公称支数　nominal number

公定回潮率　conditioned moisture regain
公定温湿度　conditioned atmosphere
公定重量　conditioned weight [C.W]
公断　arbitration
公吨　tonne; metric ton [MT]
公告　announcement; notice; affiche; proclamation
公海　high seas
公斤　kilogram [KG]
公量检验　conditioned weight test
公路运输　highway transportation; carriage by road
公亩　acre; are
公平　justice; fair; equity
公顷　hectare [Ha.]
公认的技术规则　acknowledged rule of technology
公司　company; corporation
公正性　impartiality; fairness
公证检验　notarial inspection
公制　the metric system
公制支数　metric count
功率　power; rate of work; capacity factor
功率因数　power factor
功能　function
共计　total; add up to
供方　supplier
供给　supply; furnish; provide
供应　supply; provision
构造　structure; construction
购货合同　purchase contract
购买　purchase; buy
估计　estimate
估价　evaluate; value
估损　assessment of loss; appraisal of damage
估值　appraisement; value of assessment
箍　hoop; band
谷斑皮蠹　Trogoderma granarium
股份　share; stock
鼓风　blast
固定　fix; immobilize
固定汇率　fixed exchange rate
固定资产　fixed assets
顾客　customer; shopper; client
雇佣　employ; hire
关税　customs duty
关税壁垒　tariff barrier
关税措施　tariff measures
关税减让　tariff concession
关税率　customs tariff; tariff rate
关税配额　tariff quota
关税税则　customs tariff
观察员　observer
官方　official; authority
官方的　official; authoritative
官方印章　official stamp
官方植物检疫证书　official plant quarantine certificate
官司　lawsuit
管理　manage; administrate
管理评审　management review
管理者　administrator; manager

惯例　convention; tradition
惯例的　conventional; customary; traditional
光波　light wave; optical wave
光程　optical distance; optical length; optical path
光程差　optical path difference [OPD]
光电长度分析仪　photoelectric length analyzer
光度计　luminometer
光合作用　photosynthesis
光滑　smooth; glossy; slick; sleek
光密媒质　optically denser medium
光谱分布　spectral distribution
光强　intensity of light
光疏媒质　optically thinner medium
光通量　light quantity; luminous flux
光线　light; ray of light
光学　optics; photology; photics
光学显微镜　light microscope; optical microscope; photon microscope
光学性质　optical properties
光源　illuminant; light; light source; optical source; lamp-house
光泽　gloss; lustrous
光栅　raster; optical grating; raster display
光轴　optic axis; optical axis; ray axis
广口瓶　wild-mouth bottle; jar
归档　place on file; file
归还　return; refund
归类　sort out; classify
归责　attributable to; imputation

规避　circumvention; avoid
规定　stipulate; stipulation; prescribe
规范　standard; norm
规格　specification; qualification
规矩的　mannered; regular
规模　scale
规则　rule; regulation; regular
辊　roller
滚筒　cylinder; roller
国产棉　chinese cotton; domestic cotton
国度　country; state; nation
国会　parliament; congress
国籍　nationality; national domicile
国际　international
国际标准化组织　International Standards Organization [ISO]
国际单位制　International System of Units
国际惯例　international practice; international usage
国际棉花标准委员会　International Cotton Standard Committee
国际市场价格　international market price
国际特快专递　international express mail service
国际铁路联运运单　international railway through transport bill
国际校验棉花标准　International Calibration Cotton Standards [ICCS]
国家标准　national standard
国家质量监督检验检疫总局

Administration of Quality Supervision, Inspection and Quarantine [AQSIQ]
国境　frontier; territory
国民待遇　national treatment
国内补贴　domestic subsidy
国内的　internal; domestic
国外的　foreign; abroad; external; oversea
果枝节间长度　internode length of sympodial branch
果枝类型　sympodial branch type
果枝数　sympodial branch number
过磅　ponderation; weighing
过成熟纤维　over mature fiber; postmature fiber
过程　course; process; procedure
过渡性保障措施　transitional safeguard measures
过量　excess
过期　exceed the time limit; overdue
过剩　surplus; excess; overplus
过重　overweight; excess weight

h

海岸线　coastline
海岛棉　sea island cotton; Gossypium Barbadense
海港　seaport; harbor
海关　customs; customhouse
海关担保　customs guarantee; customs bond
海关估价　customs evaluation; customs valuation
海关官员　customs officer
海关加封　customs seal
海关监管　customs supervision
海关扣留　customs detention
海关手续　customs procedures; customs formalities
海关总署　General Administration of Customs [GAC]
海上运输保险　marine cargo insurance
海事报告　sea protest; marine accident report
海损　average; sea damage
海损鉴定　inspection of cargoes with respect to general or particular average
海外的　oversea; overseas; transmarine
海运　sea transportation; maritime transport
海运提单　ocean bill of lading; marine bill of lading
害虫　pest; injurious insect; vermin; insect pest
害虫检疫　pest quarantine
含量　content
含量百分率　content percentage

含水量 moisture content; water content
含水率 percentage of moisture
含水率修正功能 moisture correction algorithm
含糖程度 degree of sugar
含糖的 sugary
含纤维的 fibrous
含佣金价格 price including commission
含杂率（皮棉） trash content
行业标准 professional standard
航行 voyage; navigation
航空快运 air express
航空邮件 air mail; airmail
航空运单 air waybill
航空运费 airfreight charge
航线 route; shipping line; air line
航运 shipping
毫米 millimeter [mm]
号码 number [No.]
耗损 consume; waste
合并仲裁 consolidation of arbitrations
合法的 legal; legitimate
合法性 validity; legality
合格报告 qualified report; compliance report; clean report of findings [CRF]
合格处理 qualified disposition
合格的 eligible; qualified; up to standard
合格评定 conformity assessment
合伙 copartnership
合伙人 copartner; partner

合计 total; summation
合理的 legitimate; reasonable
合同 contract
合同回潮率 contracted moisture regain
合资 joint venture
合作 cooperation; consociation
核对 check; verify
核算 business accounting
核算单位 accounting unit
核销 cancel after verification
黑籽棉 black seed cotton
恒温恒湿实验室 constant temperature and humidity laboratory
恒温烘箱 constant temperature oven
恒温水浴锅 thermostat water bath
恒重 constant weight
横截面分析法 cross-sectional analysis
衡器 weighing apparatus
烘干重量（干重） moisture-free weight
烘箱 oven; drying oven
红火蚁 Solenopsis invicta
后来的 subsequent
后配额时代 post-quota era; after-quota era
厚度 thickness
互惠 reciprocal
花瓣单宁含量 petal tannin
花瓣棉酚含量 petal gossypol
花萼 calyx
花费 cost; expense

花粉 pollen; farina
花冠色 corolla color
花冠长度 petal size; corolla length
花丝 filament; capillament
花丝色 filament color
花形 flower shape
花药 anther; semet
花药色 anther color
花柱长度 stigma length
滑准税 sliding duties
化肥 fertilizer
化学浆料 chemical pulp
化学纤维 chemical fiber
化学性质 chemical property
还款 reimbursement; refund
还原 restore; revivification
还原糖 reducing sugar
环锭纺 ring spinning
环境 environment; surroundings; circumstance
环境保护 environmental protection
环境标志 environmental mark; environmental labelling
环境管理体系 environmental management system
环境科学 environmental science
环境空气 ambient air
环境条件 ambient conditions
环境污染 environmental pollution
环缕纱强度 strength of ring spun yarn
换算 conversion
黄度 yellowness; degree of yellowness; +b
黄根率 tinged linter content
黄棉 yellow cotton
黄染棉（美棉分级标准等级） Yellow Stained [Y.S]
黄萎病 verticillium wilt; greensickness
黄渍棉（疵点） yellow-stained cotton
灰暗 murky grey
灰白 hoary; ashen
灰尘 dust; dirt
灰分 ashes
灰分含量 ash content
灰黄 sallow; grey yellow
灰棉 gray cotton; grey cotton
恢复 recover; restore
回潮率 regain; moisture regain
回归曲线 regression curve
回花 cotton waste
回花箱 lint waste box
回收率 recovery rate
回收棉 recycling waste cotton
汇款 remittance
汇率 exchange rate
汇票 bill of exchange [B/E]; draft
汇票通知单 advice of drawing
会议 meeting; conference
混纺比 blending ratio
混纺纱线 blended yarn
混合 mix; mingle; commixture
混合物 mixture; compound; admixture
混唛 mixed-mark

混棉　cotton blending; cotton mixing
混淆　confound; mix up
混杂棉包　mixed packed bale
活性成分　active ingredients
火烧棉　burned cotton
火灾危险　fire hazards
货币　money; currency
货币调整费　currency adjustment factor [CAF]
货场　goods yard; freight yard
货车　goods van; freight car
货船　freighter; cargo ship; cargo vessel
货到付款　pay on delivery [POD]; cash on delivery [COD]; payment against arrival
货柜　counter; container
货号　article No.
货款　money for buying; payment for goods
货损检验　cargo damage survey
货损鉴定　cargo damage survey
货物残损　damaged cargo
货物残损鉴定证书　survey certificate on damaged cargo
货物检疫　cargo quarantine
货物名称　description of goods
货物溢缺单　overlanded & shortlanded cargo list
货运　freight transport; freight
货运代理　freight forwarder; shipping agency
货运单　waybill
货栈　warehouse; store
货值　amount; value
货主　consignor; consigner; shipper
获证方　licensee
豁免　exempt

j

机械　machine
机压包　press-packed bale
机摘棉　sledded cotton
基本的　basic; essential; elementary; fundamental
基础级棉（据以计价）　basis cotton
基数　cardinal number
基准　datum; base; norm
级外　Below Good Ordinary [B.G.O]; off-grade
极限　limit
即期付款　immediate payment; payment at sight
即期汇票　demand draft [D/D]; sight draft
即期信用证　sight L/C
即时的　immediate; instant
集散地　collecting and distributing

centre; distributing centre
集装箱　container
集装箱堆场　container yard
集装箱货运站　container freight station
集装箱基地　container base
集装箱运货船　container ship
计划　plan; project; schedule
计量　measure; computation
计数　count
计算　calculate; count
记录　record; keep a record of
记账　keep accounts; charge to an account
技术　technique; technology
技术标准　technical standard
技术长度　technical length
技术法规　technical regulations
技术规范　technical specification
技术交流　technical exchange
技术人员　technician
技术事故争议　technical dispute
技术性贸易壁垒　technical barriers to trade [TBT]
技术用语　technical terminology
技术指标　technical index; technique data
季度　quarter
继…之后的　subsequent
加封证书　certificate of sealing
加工　process; machining
加工过程检疫　processing and quarantine
加工者　processor; ginner
加厚期　thickening stage
加利福尼亚（美国主要产棉区）California [CAL]
加仑　gallon
加权　weighting
加权平均　weighted mean
夹持长度　clamping length
夹杂物（异物）　foreign matter
家用纺织品　home textiles; household textiles
荚（壳皮）　pod
甲基纤维素　methyl cellulose
假高粱　Sorghum halepense
价格　price
价格条款　price term
价目表　list; price list
价值鉴定证书　certificate of valuation
坚固　firm; solid; strong
间接成本　indirect cost
间接费用　indirect fee
监督　supervision
监督管理　supervision and administration
监督员　supervisor; controller
检测项目　testing item
检查　check; examine
检查员　surveyor; inspector
检验　inspect; test
检验报告　report of inspection
检验费　inspection fee; inspection charges
检验日　date of inspection

检验员　inspector
检验证书　inspection certificate
检疫　quarantine
检疫鉴定报告　quarantine and authenticate report
检疫性害虫　quarantine pest
检疫证书　certificate of quarantine
减产　reduction of output
减价　reduce the price; mark down
减轻　alleviate; relieve; mitigate
减少　subtract; reduce; decrease
见票…天后付款　…day's sight; …days after sight [DS]
见票即付支票　check payable at sight
鉴别　identify; distinguish
鉴定　authenticate; survey
鉴定员　appraiser; surveyor
僵瓣棉　dead cotton
降低　reduce; cut down; decrease
降级　degrade; downgrade
交付　consign; deliver
交割　delivery
交货　deliver the goods
交货单　delivery order [D/O]
交货港　port of delivery
交货期　delivery time [D.T]
交货重量　delivered weight [D/W]
交流　exchange; interflow
交通　traffic
交通工具卫生证书　sanitary certificate for conveyance
交易　deal; trade; transaction; merchandise
交易所　exchange; bourse
胶合板　plywood
校正系数　correction coefficient
校正因数　correction factor
校准　calibrate; correct
接受　accept; receive
洁净度　cleanliness
结构　framework; configuration; construction; structure
结果　result; outcome
结论　conclusion; verdict; peroration
结算　settle accounts
截距　offset
解除　discharge; relieve
解决　solve; dispose of
解剖　anatomy
解约　break off an engagement; terminate an agreement; cancel a contract; rescind a contract
介质　medium; dielectric
借贷　debit and credit
金衡　troy weight
金融　finance
金融负债　financial liability
紧急保障措施　emergency safeguard measure
紧压棉包　compressed bale
进度　schedule
进口　import
进口报关　customs entry
进口差价税　import variable duties
进口附加税　import surcharge

进口港　port of import
进口国　importing country
进口配额　import quota
进口日期　date of arrival; date of import
进口商　importer
进口许可证　import licence
进料加工　processing with imported materials
进展　progress; evolve
近似法　approximation method
近似值　approximation
禁用偶氮燃料　banned azo colorants
禁运　embargo
禁止　prohibit; ban; forbid
茎秆　stem
茎毛长度　stem pubescence length
茎毛数量　stem pubescence amount
茎色　stem color
经营　manage; operate; run; engage in; deal in
精白　spotlessly white; pure white
精密度　precision; degree of precision
精密天平　precision balance
精确的　precise; exact; precise; accurate; pinpoint
精确度　accuracy; precision
精梳棉纱　combed cotton yarn

净价　net price
净利　net profit
净棉　cleaned cotton
净重　net weight
竞争　compete; contend
静电纺纱　electrostatic spinning
纠正措施　corrective action
九月　September [Sept]
旧的　used; old
拒绝　reject; refuse
拒收　rejection; dishonour
具有　possess; have
锯齿棉　saw ginned cotton
锯齿轧花机　saw cotton gin
聚丙烯纤维　polypropylene fiber
卷曲　curl over; crimp; crinkle
决定　decide; resolve; make up one's mind
决议　decision
绝对干燥状态　absolute dry condition
绝对干重　absolute dry weight
绝对强度　absolute intensity
绝对湿度　absolute humidity
绝对温度　absolute temperature
均方差　mean square deviation
均匀度　regularity; uniformity
菌核　sclerotium

k

开户银行　opening bank; bank of deposit

开花期	florescence; blossom; flowering date
开始	begin; start; initiate
开松	cotton opening
开箱	unpack; opening container
开证银行	issuing bank
抗病性	resistance to disease
抗虫害	pest-resistant
抗虫性	resistance to pests
抗旱性	resistance to drought
抗药性	drug-resistant
考虑	consider; regard
可补偿的	compensable; atonable; recoverable; reparable
可撤销的	revocable; invertible; voidable
可撤销信用证	revocable credit
可得到的	available; gainable; procurable; attainable
可纺性	spinnability
可纺支数	limit count
可分割信用证	divisible L/C
可耕地	arable land
可见光	visible light
可靠性	reliability; dependability
可流通（票据等）	afloat
可逆的	reversible; invertible
可燃	combustible
可染性	dyeability
可绕性	flexible; reelable
可塑性	plasticity
可用的	available
可支付	payable
可转让的	transferable; conveyable
克/特克斯	grams per tex [G/T 或 GPT]
克	gram
克服	conquer; overcome
刻度	scale
客观测量	objective measurement
空白试验	blank test
空气	air; atmosphere
空气压缩机	air compressor
空运	air transportation
空运单	air waybill
空运港	aerial port
空运货物	airborne goods
控制	control
控制极限	control limit [CL]
控制台	control console
口罩	respirator; mask
扣除	deduct; subtract
枯萎病	fusarium wilt
库存量	stock
库房	stock house
夸脱（容量单位）	quart
跨距长度	span length; fibrograph length
快递（件）	express
快速	fast; quick
快速核查	quick check
宽度	width
宽松的	loose
款式	style
矿物棉	mineral wool
亏重	shortage in weight; weight shortage

昆虫　insect
捆扎　bind; bundle up; bundle; ligature; tie
扩张　expansion; extension; dilation

L

垃圾　rubbish; garbage; trash; litter
拉伸力　drawing force; pulling force
拉伸强度　tensile strength
拉伸强力　tenacity and elongation
拉伸试验　tensile test
来料加工　processing with supplied material
劳动成本　labor cost
勒克斯（照度单位）　lux
类别　category; sort; type
冷干机　air dryer
冷却　cooling; refrigeration
冷却剂　coolant; refrigerant
冷却器　chiller; congealer; cooler; cooler body
厘米　centimeter
离岸价　free on board [FOB]
离岸重量　shipping weight [S/W 或 S.W]
离港日期　departure date
理货　tally; cargo handling
力矩　moment; moment of force
力矩平衡　moment balance; moment equilibrium
立即　instant; immediate
立枯病　sore shin; damping off
利润　profit
利益　benefit
例外条款　escape clause; exception clause
连续　series; continuous; succession
联合　alliance; unite; union; ally
联合声明　joint statement; combined declaration
联合运输　combined transport
联合组织　combination weave
联盟　alliance
量筒　measuring cylinder
亮度　luminance; light intensity; brightness; lightness
劣等的　inferior; low-grade; inferior quality
劣质棉花　inferior cotton
临时出口管理　temporary export management
临时出口许可证　temporary export licence
临时签发　provisional issuance
铃（棉、亚麻的）　boll
铃尖突起程度　boll trip
铃期　boll stage
铃色　boll color

铃室数 locules per boll
铃室种子数 seed number per locule
铃形 boll shape
铃着生方式 boll setting type
铃重 boll weight
零 nil; zero
零担 less-than-carload freight; breakbulk
零隔距 zero gauge
零售价 retail price
零位 null
流行的 prevailing; popular; fashionable
流量 flow; flux; rate of flow
六月 June [Jun]
陆地棉 Gossypium hirsutum; upland cotton
陆空陆联运 train-air-truck through transportation
陆运 carriage by land
路线 route; course; way; line
履行 perform; fulfil; carry out
滤网 filter; strainer
轮廓 outline; contour; profile
论坛 forum
罗拉 roller
罗拉式长度分析仪 roller length analyzer
落棉 fuddles（梳理机的）; noil; fibering off

m

麻袋 sacks; gunny bags; gunny sacks
马克隆值 micronaire [Mic]
码 yard
码单 weight memo
码头 wharf; pier
码头费 dock charge; pierage
买方 buyer; purchaser
买方代理 buyer's agent
买方签字 buyer signature
卖方 seller
卖方代理 seller's agent
卖方签字 seller signature
唛头 mark; shipping mark
毛利 gross profits
毛刷 brush
毛羽 hairiness
毛重 gross weight [G.W]
贸易 trade
贸易赤字 trade deficits
贸易磋商 trade consultation
贸易方式 mode of trade
贸易管制 restraint of trade
贸易国 trading nation
贸易逆差 trade deficit; unfavorable balance of trade
贸易顺差 trade surplus; favorable

balance of trade
贸易盈余　trade surplus
贸易自由化　trade liberalization
霉斑棉　mildew spot cotton
霉变　mildew; mould
霉菌　mould; mycete
每8小时工作时间　every 8 working hours
美分　cents [USC]
美国白蛾　Hyphantria cunea
美国陆地棉通用分级标准　Universal Standard for Grade of American Upland Cotton
美国南方地区研究试验所　Southern Regional Research Laboratory
美国农业部　United States Department of Agriculture [USDA]
美国农业部通用棉花标准　USDA Universal Cotton Standards
美元　dollar [USD]
米　meter
密度　density
密封　airproof; airtight; seal
棉瓣　cotton petal; cotton flap
棉包　cotton bale
棉包管理　bale management
棉被　cotton-padded quilt
棉布　cotton fabrics; cotton cloth
棉仓（纺机）　cotton bin
棉椿象　cotton stainer
棉短绒　linters
棉短绒标准　linters standard
棉短绒检验　linters test

棉堆混棉　stack mixing
棉纺工程　cotton spinning engineering
棉纺织厂　cotton mill
棉纺织品　cotton textiles
棉杆　cotton stalk
棉杆皮　cotton stalk skin
棉红腐病　cotton red rot disease
棉红铃虫　pink bollworm; Pectinophora gossypiella
棉红蜘蛛　two-spotted spider mite; Tetranychus telarius linne; Tetranychus urticae Koch
棉花　cotton; Gossypium spp
棉花包装规格　cotton packing specifications
棉花病虫害　cotton diseases
棉花测色仪　cotton colorimeter
棉花成熟度指数　causticaire index
棉花打包　cotton compressing
棉花分析机　cotton analyzer
棉花害虫　cotton insects
棉花黑根腐病　cotton rot root
棉花黄萎病菌　cotton verticillium dahilae
棉花检验　cotton inspection
棉花结铃期　cotton boll stage
棉花枯萎病毒　cotton wilt virus
棉花类别　type of cotton
棉花品级　cotton grade
棉花品级实物标准　physical standard for cotton grade
棉花品级要素　factors of cotton grade

棉花期货　cotton futures
棉花曲叶病毒　cotton leaf curl virus
棉花色泽　cotton color
棉花色泽类型　color groups of cotton
棉花水分测定仪　cotton moisture meter
棉花松包机　cotton breaker
棉花炭疽病　cotton anthracnose
棉花杂质分析机　analyzer of impurities in cotton
棉花栽培　cotton culture
棉花长度　cotton length
棉荚　cotton pod
棉浆粕　cotton pulp
棉角斑病　angular leaf spot of cotton; cotton angular leaf spot
棉结　cotton neps
棉卷　cotton roll
棉卷叶螟　cotton leaf roller
棉枯萎病　cotton wilt
棉块　cotton lump
棉蜡　cotton wax
棉蕾　cotton bud
棉铃　cotton boll
棉铃虫　cotton bollworm
棉铃黑果病　diplodia boll rot
棉铃红粉病　pink boll rot
棉铃红腐病　cotton boll rot
棉铃炭疽病　bolls anthracnose
棉苗炭疽病　seedling anthracnose
棉农　cotton grower; cotton planter
棉区　cotton region
棉绒比例　lint percentage

棉纱　cotton yarn; spun cotton
棉纱品质指标　cotton yarn quality index
棉纱支数　yarn count
棉属　Gossypiam [G]
棉束　flock
棉束纤维长度分布图　tuft length diagram
棉胎　batt; cotton wadding
棉桃　cotton boll
棉条　sliver
棉纤维　cotton fiber
棉纤维成分　cotton fiber components
棉纤维含糖量　determination of sugar in cotton fiber
棉纤维结构　cotton fiber structure
棉纤维素　cotton cellulose
棉纤维天然转曲　convolutions in cotton
棉纤维性能　properties of cotton fiber; cotton properties
棉线　cotton thread
棉锈病　cotton rust
棉絮　batt; cotton wadding; cellucotton
棉蚜虫　cotton aphid
棉样　cotton sample
棉叶　cotton leaf
棉叶螨　spider mite
棉叶跳虫　cotton leafhopper
棉衣　cotton-padded coat
棉织（制）品　cotton fabrics; cotton goods
棉株　cotton plant

棉籽 cotton seed
棉籽饼 cottonseed cake
棉籽残绒率 cottonseed residual cashmere rate
棉籽短绒 cotton hair
棉籽粉 cottonseed meal
棉籽壳 cotton seed shell
棉籽毛头率 lint percentage of cotton seeds
棉籽脱壳机 cottonseed huller
棉籽油 cotton oil; cotton-seed oil
棉籽榨油者 cottonseed crushers
免除 exempt; relief
免费 free of charge
免税 duty-free; exemption from duty
免责条款 escape clause; exception clause; exoneration clause; non-responsibility clause
面积 area

描述 description
明火 naked light; open fire
明确的 specific; explicit
模块测试 module testing
麻布（打包用） hessian
模拟昼光 simulation daylight
墨西哥棉铃象 Anthonomus grandis Boheman
木托盘 wooden pallet
木箱 wooden case
木屑 wood chips
木制包装 wooden package
目测 visual examination
目的 purpose
目的地 destination
目的港 destination port
目光检验 visual inspection
目光鉴定 visual assessment
目镜 eyepiece; eye lens; ocular

n

耐涝性 tolerance to waterlogging
耐盐性 tolerance to salinity
难以承受的 unacceptable
内部的 internal
内部审核 internal audit
内在质量 inner quality; inherent quality
尼克森-亨特色特征图 Nickerson-Hunter color chart

尼克森-亨特棉花色泽仪 Nickerson-Hunter cotton colorimeter
泥土 soil; clay
年度 annual
年度政策制定 annual policymaking
年鉴 almanac; yearbook
黏性 sticky; viscosity
捻度 twist; degree of twist
尿素 urea; carbamide

镊子 tweezers
农产品 agricultural products
农药 pesticide
农业 agriculture
农业部 Department of Agriculture; Ministry of Agriculture
农业顾问 adviser in agriculture
农业市场服务局 Agricultural Marketing Service [AMS]
农业土壤学 agrology
浓度 concentration; density

o

欧盟方案 EU scheme
偶氮染料 azo colorants; azo dyestuff; azoic dyes
偶然误差 accidental error

p

拍卖 auction
拍卖商 auctioneer
排除 exclude; eliminate
排放 discharge
盘库 make an inventory goods in a warehouse
判定 award; judge
泡沫塑料 plastic foam
培养箱 incubator
赔偿 compensation; indemnify
配额 quota
配额内 within the quotas
配额外 beyond the quotas
配棉 cotton assorting
膨胀 expansion; swell
批次 lot
批发 wholesale
批发商 merchant; wholesaler
批准 approve; authorize; ratify
皮辊 roller
皮辊棉 roller ginned cotton [RG]
皮棉（纤维）长度 lint length
皮棉 lint; ginned lint; ginned cotton
皮棉打包机 cotton press
皮棉率 cotton lint percentage
皮棉总产量 total lint yield
皮重 tare weight
偏差 deviation; offset
偏离 deviation; diverge
瓢虫 ladybug; ladybird; lady beetle

票据 bill; note; receipt
拼箱 less than container load
频率 frequency
品级 grade
品级实物标准 physical standard for cotton grade
品级长度 grade length
品牌 brand
品系 breeding line
品质 quality
品质保证 quality assurance
品质标记 quality symbol
品质检验证书 inspection certificate of quality
品质鉴定 quality evaluation
品质长度 quality length
品质争议 quality dispute
品质证书 certificate of quality
品种 variety
平行测定 parallel determination
平行光线 parallel rays
平行试验 parallel tests; duplicable test; parallel experiment
平衡 balance; equilibrium
平衡含水率 equilibrium moisture content
平衡回潮率 equilibrium moisture regain
平衡物 counterweight
平衡吸放湿曲线 equilibrium curve of desorption and ad sorption
平级（美棉分级标准） Good Ordinary [G.O]
平级加（美棉分级标准） Good Ordinary Plus [G.O.P]
平均 average; mean
平均皮重 mean tare weight
平均水平 the average level
平均长度 average length
平束法 flat bundles
评价 appraise; value
评论 remark; review
凭单 indenture; voucher
凭单付款 payment against documents
凭单托收 document pay [D/P]
凭汇票付款 payment by draft
凭小样（型号） by type
凭证 voucher; proof; evidence; credence
破裂 breakage; split; rupture
破损 breakage; damage
破籽 broken seed
普惠制 generalized system of preferences [GSP]
普惠制方案 scheme under GSP

q

七月 July [Jul]
期货 futures; forward

期间　duration; period
期满　expiry
期限　deadline; time limit
欺诈棉包　false packed bale
歧视　discrimination
企业　enterprise; firm
企业标准　enterprise standard; manufacturer's standard
启封　unseal; break the seal
启封证书　certificate of unsealing
启事　notice; announcement
启运地区　shipment region
启运港　port of embarkation; port of shipping
启运国家　shipment country
起草　draft; draught
起航　set sail
起源　origin
起重机　hoist; crane
气管炎　tracheitis
气候　climate; clime; weather
气流　airflow; air current
气流纺　air spinning; rotor spinning
气流纺加捻杯　open-end rotor
气流量　gas flow
气流纱　rotor spun yarn
气流仪　air quantity flow meter; airflow instrument
气流杂质机　air stream cleaner
气压　air pressure; atmospheric pressure; barometric pressure
契约　indenture; contract
恰当的　proper

千克　kilogram [kg]
扦样　sampling
铅笔　pencil
铅封　seal
铅封号　seal number
签发地点　place of issue
签发机构　issuing authority
签发人　signatory
签发日期　issuing date
签发证书　issue certification
签名　sign
签署　affix; sign; signature
签字　signature
前纺工序　raw stage
钳子　pliers
潜伏期　incubation period
欠款　debt
强度　strength [Str]; intensity
强加　impose
强力　strength; brute force
强力指数　strength index
强制　enforce
切线　tangent line
轻纱　fine gauze
轻油色棉　buttery cotton
倾销　unload; dump
倾销幅度　dumping profit margin
清楚的　clear; distinct
清单　bill; list
清点　tally; make an inventory
清关税　clearance duties; clear customs
清花　blowing
清洁度　cleanliness

清洁货物　clean cargo
清洁检验报告　clean report of findings [CRF]
清洁提单　clean bill of lading
清理　clean
清棉机　scutcher; cotton cleaning machine
清算　clearing; liquidate
情况　case; situation; circumstance
请求　request
秋季　autumn
区分　distinguish; differentiate
区域　range; area; region; district; zone
区域试验　regional trial
取出　take out; remove
取代　replace; supersede

取消　cancel; countermand; abrogate
取样　sampling
取样证书　certificate of sampling
权威　authority
权威的　authoritative
全棉的　all-cotton
全面质量控制　total quality control [TQC]
全生育期　whole growth period
劝说　advice; persuade
缺陷　deficiency; flaw; defect; imperfection
缺陷棉包　defective bale
缺株率　miss-planted rate
确定的　assured; reliability; determinate; doubtless
确认　acknowledge; confirm

r

燃点　ignite
燃料调整费　fuel adjustment factor [FAF]
燃烧性能　burning behaviour
燃油调整费　bunker adjustment factor [BAF]
燃着　ignition
染料　dyestuff
染色　dyeing
染污棉　stained cotton
染疫人　a person having a quarantinable infectious disease
染疫嫌疑人　a person suspected of having a quarantinable infectious disease
让与　transfer
人造毛　artificial wool
人造棉　artificial cotton
人造丝　artificial silk; rayon
人造纤维　artificial fiber; man-made fiber; staple fiber; rayon
认可　acknowledge; recognition;

approve; accept
认可实验室　accredited laboratory
认为　deem; consider; regard as
认真的　serious; earnest
认证　certification
认证实验室　certified laboratory
任命　nominate; appoint
任选要求　optional requirement
任意　random
日光灯　fluorescent lamp; daylight lamp
日光照明　natural lighting; daylight illumination
日历天　calendar day
日照　sunshine; sunlight
绒辊花（绒板花）　clearer waste
容量　capacity; volume; cubage; dimension
容量瓶　volumetric flask
容器　container; vessel
容许误差　admissible error
如下　as follows
入港报表单　bill of entry [B/E]
入境　enter a country
入境货物报检单　declaration form for entry of goods
入境货物调离通知单　transferred notice from for entry of goods
入境货物通关单　bill of the customs clearance for entry cargo
入库　put in storage
软包装　flexible package
润滑剂　lubricant

S

三极管　dynatron; triode; audion
三角带　vee belt; cone belt; triangular belt
三类棉短绒　third-cut linter
三相（电）　three-phase; triphase
三月　March [Mar]
色彩　hue; tint; color
色度　chroma; chrominance
色度计　colorimeter
色卡比色法　colorimetry in color card
色纱　color yarn; dyed yarn
色素　pigment; pigmentum; coloring material
色特征图　color chart
色温　color temperature
色泽　color and lustre; tincture; tinct
色泽等级　color grade [C Grade]
色泽分布范围　range of shades
色泽良好（羊毛及棉花分级用语）　good color
色泽特征　color characteristics
杀虫剂　insecticide; biocide; pesticide
杀菌剂　fungicide; bactericide; germicide

杀螨剂　acaricide
杀蚜虫剂　aphicide
纱线　yarn
纱线强度　yarn strength
纱线支数　yarn count; yarn size
山粉蝶　black veined white
删除　delete
商标　brand; trademark
商检标志　commodity inspection mark
商检证书　certificate of commodity inspection
商品　commodity; goods; merchandise
商品编码　commodity code
商品检验　commodity inspection
商品名称及编码协调制度　harmonized commodity description and coding system [H.S]
商业　business; commercial
商业发票　commercial invoice
商业回潮率　commercial moisture regain
商业结算重量　commercial weight
商业纠纷调解　mediation of dispute
商业重量　commercial weight
上半部平均长度　upper half mean length [UHML]
上部　upper; upside
上级（美棉分级标准等级）　Good Middling [G.M]
上级淡点污棉（美棉分级标准等级）　Good Middling Lighter Spotted [G.M Lt.Sp]
上级点污棉（美棉分级标准等级）　Good Middling Spotted [G.M Sp]
上述　above-mentioned; aforementioned; foregoing
上四分位长度　upper quartile length [UQL]
上诉　appeal
上限　maximum [Max]
烧杯　beaker
少于　less than
设备　equipment; facility
设计密度（棉花种植）　design density
涉及　concern; refer to
摄氏温度　centigrade; celsius temperature
申报　declare; declaration
申报数量　quantity declared
申请　apply; application;
申请人　applicant
伸展　extension; stretch; outspread
伸长　elongation; extension; prolongation; extend
伸长期　elongation stage; period of elongation; jointing stage
伸直长度　contour length
深度　depth; deepness
审查　inspection; investigate; examine
审核　audit; check
生产　produce; manufacture; production
生产试验（棉花种植）　production trial
生产许可（证）　production permit
生态纺织品　ecological textile

生物催化剂　biocatalyst
生物化学　biochemistry
生育期　period of duration; growth period; growth season
生长年度　crop year
生长势头　growing vigour
声明　declaration; statement
剩余　surplus; remainder
湿度　humidity; dampness
湿度计　hygrometer
湿重　natural weight
十二月　December [Dec]
十进位的　decimal
十一月　November [Nov]
十月　October [Oct]
石棉　asbestine; asbestos
时价　current price
识别　discriminate; distinguish; recognition; identification
识别码　identification code
实（试）验　experiment; test
实际到达（货）　actual arrival
实际回潮率　actual moisture regain
实际皮重　actual tare; real tare
实际支数　actual count
实际重量　actual weight [A/W]
实践　practice
实施　implement; carry out; conduct
实物标准　material standard; physical standard
实验室　laboratory
实验室检疫　lab quarantine
使命　mission
使无效　disable; invalidate; nullify; deactivate; disannul
使用　use; employ
始发港　port of departure
市场　market
市场份额　market share
事故　accident
事实　truth; fact
试行标准　trial standards
试剂　reagent; agentia
试验须丛　test beard
试验用标准大气　standard atmosphere for testing
试样　sample; specimen
试轧（衣分测试）　rolling test
适当　appropriate; proper; reasonable
适应性　adaptability
释放　release; deliver
收到　receipt; receive
收货地　place of receive
收货人　consignee
收货人名称及地址　name and address of consignee
收获株数（实收）　the number of effective harvest plants
收据　receipt; quittance; voucher
收款　proceeds; receivable
收款人　payee; beneficiary
手册　manual; handbook
手扯尺量法　hand pull dipstick method
手扯尺量长度　hand pull scale length
手扯棉样　pull of sampling; draw of sampling

手扯长度　staple length; pull length
手电筒　flashlight
手感　hand feeling; handing
手感不良　poor feeling
手续　procedure; formalities
守约　abide by
受惠国　preference-receiving country
受益人　beneficiary
授粉　pollination
授权签字人　authorized official
授予　confer; award; grant
书面形式　written form
书面证明　documentary evidence
梳棉　cotton carding
梳棉机　carding machine
梳片法　comb sorter method
梳片式长度分析仪　comb type length analyzer
输出　export
署名　signature
鼠疫　pestis; plague
束纤维强度　bundle strength; strength of fiber bundle
束纤维强力机　fiber bundle strength tester
数（重）量检验证书　inspection certificate of quantity
数据　data; datum
数量　quantity; amount
数量鉴定　survey of quantity
数量限制　quantitative restriction
数量证书　certificate of quantity
数字　figure; digit; number

双峰分布　bimodal distribution
双联合同　indenture
双面棉包　two sided bale
霜冻　frost
霜后籽棉　yield after frost
霜后籽棉产量　yield after frost
霜黄棉　frost cotton
霜期　frosty period
霜前（棉）花　cotton before frost
霜前花（重量比）率　percentage of seed cotton yield before frost
霜前皮棉产量　lint yield before frost
霜前籽棉产量　yield before frost
水分检验　moisture test
水分检验站（纺织材料的）　conditioning house
水垢　incrustant; incrustation scale; scale deposit
水解　hydrolyze; hydrolysis
水平度　levelness
水平仪　gradienter; air level
水压　water pressure
水浴锅　water-bath
水渍棉　water damaged cotton; water stained cotton
税率（关税）　tariff rate
说明　explain; instruction
丝光处理　mercerize
斯特洛强力　Stelometer strength
死纤维　dead fiber
四月　April [Apr]
松材线虫　pine wood nematode
松解　debonding

松散　loose
诉讼　legal action; lawsuit; litigation
速度　velocity; speed
塑编袋　plastic woven bag
塑料袋　plastic sack; plastic bag
塑料桶　plastic drum
随行就市　fluctuate in line with market conditions; fluctuate along with market changes
随后的　subsequent
随机　random
随机取样　random sampling
随机取样法　random sampling method

损害　damage; injure
损耗　loss; depletion
损耗量　wastage; ullage
损坏　damage; destroy
缩短　shorten; cut down
缩减　reduce; decrease; shrinkage; curtail
缩水　shrink
索棉　curly cotton
索赔　claim
索赔依据　basis for claim
索丝　winding fiber
索引　index

t

台秤　platform scale
谈判　negotiate
弹花机　cotton fluffer
弹性　elastic performance; elasticity
炭化　charring
糖分（棉花）　cotton sugar
套汇　arbitrage
套种　interplant; under sow
特别保障措施　special safeguard measures
特别的　extraordinary; extra; especial; particular; exceptional
特此证明　hereby certify; it is hereby certified that
特定的　specific

特惠关税　special preferences; preferential tariff; preference duty
特克斯（细度单位）　tex
特权　privilege; prerogative
特殊疵点　special defect
特殊关税待遇　special tariff treatment
特殊条件　particular condition; special condition
特殊条款　particular terms; special terms
特殊显色指数　special color rendering index
特效的　specific
特性（征）　characteristic; character; performance; features

特许 licence; franchise; concession; special permission
特长纤维 extra long staple
提出 file; present; submit
提出索赔 file a claim
提单（海运） bill of lading [B/L]
提高 improve; improvement; enhance; increase
提供 supply; furnish; provide
提交 submit; submission
提前装船 advance shipment
替代品 substitute; succedaneum
天花板出风口 ceiling outlet
天平 scales; balance
天然丝 natural silk
天然纤维 natural fiber
天灾 natural disaster
田间检疫 field quarantine
填充 fill; padding
挑拣 pick
挑选 select; choose; pick
条件 condition
条款 clause; stipulation; item; terms; provision
条例 regulations; rules; ordinances; byelaw
条约 treaty; pact
调和平均数 harmonic mean
调解 conciliation; mediate; reconcile; intercession
调湿平衡 moisture equilibrium
调整 adjustment; modulation
贴附 affix; attach

贴现率（银行） bank rate
铁路提单 railway B/L
铁屑 iron scraps
停泊 anchor; berth; mooring
停止 cease; shutoff; stop
通常 customary; normal; usually; generally
通风 ventilate
通用 universal; generalduty
通用标准 universal standard
通用密度 universal density
通知 inform; notice
通知方 notify party
通知书 notices; advice note; notification
通知银行 advising bank
同盟 alliance
同一的 identical; same; selfsame
同意 agree; consent; approve; permission
统计 statistics
统一 unify
筒管 bobbin
头道棉短绒 first-cut linter
投保人 policy holder
透明包装 transparent package
图表 figure; chart; diagram; graph
土壤 soil
土壤学 agrology; pedology; soil science; edaphology
吐絮 boll opening
吐絮期 opening date of bolls; boll opening date

推迟	delay	托运人	consignor; shipper
推荐	recommend	托运人自行装货点件	shippers load and count
退还	return; refund; send back		
蜕皮	ecdysis; exuviate	脱脂棉	pledget; absorbent cotton; degreasing cotton
托运	consign		

W

瓦楞纸箱	corrugated carton	伪劣品	adulteration; shoddy goods
外包装	external packing; outer packing	委托	entrust; depute; consignation; authorize
外币	foreign currency	委托检验	entrusted inspection
外部	exterior; external	委托人	consignor; client
外观	appearance; facade	委托书	a power of attorney
外汇	foreign exchange	委员会	committee
外加的	additional; excess; extra	卫生处理	sanitary treatment
外来的	foreign	卫生检验证书	sanitary inspection certificate
外源基因	exogenous gene		
完成	accomplish; complete; fulfil; finish; achieve	未付的	unpaid; owing; unliquidated
		未加工的	raw; crude
完好	intact; undamaged	未检出	absent; not detected; no find
完全相同的	identical; selfsame	未受损伤的	intact; uninjured
完整的	whole	未梳棉	uncombed cotton; raw cotton
万用电表	multimeter	未熟棉	unripe cotton
危险	risk; hazard; danger	未完（待续）	to be continued
危险品	hazardous material	畏光	photophobia
微生物	microorganism; microbes	温度	temperature [T]
韦氏机	Webster's machine	温湿度计	hygrothermograph
违规的	illegal	温湿度调节系统（空调）	air conditioning system
违约	breach of contract		
维持	sustain; maintain	文本	text

文件　document; file
文字标准　written standard
问题棉包　problematic bale
乌斯特不匀率　Uster irregularity
乌斯特公报　Uster statics
乌斯特纤维长度测定仪　Uster stapling instrument
污迹　spot; stain; smear
污染　pollute; contamination
污渍棉　stained cotton
无法证明的　indemonstrable
无纺用价值棉花　no spinning value cotton
无光泽　lack-lustre; lustreless
无规律的　irregular; ruleless
无机化学　inorganic chemistry
无机酸　inorganic acid; mineral acid
无机物　inorganics; inorganic substance
无可置疑的　indisputable
无渗漏　no leakage
无霜期　frost-free season; frost-free growing season; frost free period
五月　May
物理试验　physical test
物理性能　physical property
物理指标　physical index
物品　article; goods
物质　substance; matter
误差　error

X

吸附水　adsorbed water; adsorption water
吸附性　adsorbability; adsorptivity
吸湿保守性　moisture conservation
吸湿等温线　moisture sorption isotherm
吸湿积分热　integral thermal absorption
吸湿平衡　moisture equilibrium at dry side
吸湿曲线　moisture absorption curve
吸湿微分热　differential thermal absorbent
吸湿性　hygroscopicity; moisture absorption; absorption of moisture
吸湿滞后　sorption hysteresis; hydroscopic hysteresis
吸收　absorb; suck up; take in; assimilate
吸收水　absorbed water
锡莱杂质分析仪　Shirley analyzer
习惯的　customary; habitual
洗净剂　abstergent
系列　series; catena
系数　coefficient; modulus
细胞　cell

细胞壁（纤维） cell wall
细胞核 nucleus; caryon; cyteblast; cell nucleus; karyon
细度 fineness
细节 details; particulars; specifics
细绒棉 medium cotton; fine staple cotton
细纱 spun yarn
下级（美棉分级标准等级） Low Middling [L.M]
下级淡点污棉（美棉分级标准等级） Low Middling Lighter Spotted [L.M Lt.Sp]
下级淡黄染棉（美棉分级标准等级） Low Middling Tinged [L.M Tg]
下级点污棉（美棉分级标准等级） Low Middling Spotted [L.M Sp]
下级加（美棉分级标准等级） Low Middling Plus [L.M.P]
下脚棉 leftover waste-cotton
下限 minimum [Min]
夏季 summer
先前的 previous
纤度 fineness
纤维 fiber
纤维饱和水分 fiber saturation moisture
纤维端 fiber end
纤维含量 fiber content
纤维几何轴向 axial fiber geometry
纤维鉴别 fiber identification
纤维束 fiber bundle
纤维束强力测试仪 fiber bundle strength tester
纤维素 cellulose
纤维素伴生物 cellulose with biological
纤维素超分子结构 super-molecular structure of cellulose
纤维素大分子 cellulose macromolecules
纤维线性分子轴 fiber linear molecular axis
纤维长度 fiber length; length of fiber; length of staple
纤维照影长度 photoelectric length
纤维轴 fiber axis
纤维状的 fibrous
纤维自然长度 natural fiber length
纤维作物 fiber crops
显出 show; exhibition
显色剂 color developing agent; chromogenic reagent; color reagent
显色指数 color rendering index [CRI]
显示 show; display
现场 site; scene
现场检疫 scene quarantine
现行价格 current price
现货 cash sale
现货价格 spot price
现货交易 spot transaction
现金收入 cash debit
现款 cash
限度 limit; bound
限制 restriction; astrict; confine

线管（纺织用） bobbin; spool
线密度 linear density
相对湿度 relative humidity [RH]
相关的 relevant; related; correlative; interrelated
相关色温 correlated color temperature
相互的 mutual
相应的 homologous; corresponding
箱装货物 cargo in case
详情 detail; particular
详细说明 specify; detailed description
响应 response
项目 project; item
消除 eliminate
消除违约 eliminate the breach
消毒 disinfect; sterilize
消毒处理 disinfection treatment
消毒检验证书 disinfection certificate
消毒通道 disinfection passage
消毒证书 certificate of disinfection
消费 consumption; expense; expenditure
消费者 consumer
销毁 destroy
销售 sale; sell
销售合同 sales contract
销售目录 sales catalogue
销售渠道 distribution channel; marketing channel
销售确认书 sales confirmation
小保单 insurance certificate
小花头 flower head

小棉枝 small cotton sticks
小数 decimal
小纤维 fibril
效果 effect; result
协定的 conventional; concerted
协会 association
协会货运保险条款 Institute Cargo Clauses
协会商品贸易保险条款 Institute Commodity Trades Clauses
协商 negotiate
协调联运 intermodal transport; combined transport
协议 protocol; agreement; treaty
协助 assistance
斜率 slope
卸毕日期 date of completion of discharge
卸货 discharge; unload
卸货港 discharging port; port of discharge
信贷 credit and loan
信汇 mail transfer
信任 trust
信托 trust; entrust; confide
信用 credit; honor
信用证 letter of credit [L/C]
星期二 Tuesday
星期六 Saturday
星期日 Sunday
星期三 Wednesday
星期四 Thursday
星期五 Friday

星期一 Monday	许可 permit; ratification
行政机关 administration	许可证 licence; permit
形式发票 proforma invoice	许诺 promise; commitment
形态 shape; form; appearance	序号 serial number
型号 type; model	序列 serial
性能 performance; property	续燃 afterflame
性能试验 performance test	絮用纤维 wadding fiber
雄蕊 stamen	选项 option
修补 repair; mending	选育单位 breeding institute
修改 revise; modify; amend; alter	选育方法 breeding methods
修约规则 rules for rounding off	选育品种 advanced cultivar; improved cultivar
修正结果 corrected result	选择 option; select; choice; choose; elect
修正系数 correction factor	
锈斑棉 rusty cotton	
锈渍 rust stain	削减 cut down
虚假声明 false declaration	熏蒸 fumigation
需方 purchaser; demander	熏蒸处理 fumigation processing
需求 demand; require	熏蒸证书 certificate of fumigation

y

压舱物 ballast	轧伤棉（轧棉机轧伤的纤维） gin-cut cotton
压力 pressure; stress	轧碎机 breaker; crusher; roll crusher
压碎 crush; squash	
蚜虫 aphid	亚洲棉 Asiatic cotton
轧工质量 quality of ginning	咽炎 pharyngitis
轧花 cotton ginning	烟灰 soot
轧花机 cotton gin	延迟装运 delay of shipment
轧花落棉 gin fall	延期 extend; delay
轧棉厂 cotton ginning factory; ginnery	延期交货 back order; late delivery
	延长 extend; lengthen; prolong

严重的　severe; serious
沿海贸易　coasting trade
盐碱地　saline and alkaline land
颜色　color
颜色测量　color measurement
颜色深度　depth of color
掩埋　bury
验残　survey of damaged cargo
验残证书　certificate of damage
验讫日期　date of completion of inspection
验收　check and accept
验证　verify
样品　sample; specimen
样品标签　the sample label
样品架　specimen holder
样品间　sample room
要求　demand; require; request
要素　element; factor
野生的　wild; feral
业务　vocational work; professional work; business
叶基斑　leaf base spot
叶裂刻深浅　leaf lobe
叶裂片数　leaf lobe number
叶蜜腺　leaf nectar
叶片　leaf; blade; lamina
叶片厚度　leaf thickness
叶片面积　leaf area
叶茸毛　leaf pubescence
叶色　leaf color
叶屑　leaf
叶屑等级　leaf grade

叶形　leaf shape
叶型　leaf form; leaf type
叶枝数　monopodial branch number
夜班　night shift
液压式打包机　hydraulic packaged machine
一般地　generally; popularly
一般条款　general terms
一般显色指数　general color rendering index
一部分　partial
一打　dozen
一方　party
一类棉短绒　first-cut linter
一切海损均不予赔偿　free of all average [FAA]
一切险（全险）　all risks [AR]
一式六份　sextuplicate
一体化进程　integration process
一月　January [Jan]
一致　agreement; conformity
一致性　uniformity; consistency
衣分　ginning outturn
衣分率　lint percentage
衣分指数　lint index
依据　as per; according to; in the light of; in accordance with; on the basis of
依靠　depend; rely
仪器　instrument
仪器保养　instrument maintenance
仪器测定　instrument measurement; objective measurement; instrument

testing
仪器设备　instrument and equipment
移动　move; shift; remove
遗传材料　genetic stocks; hereditary material
乙拌磷　disulfoton
乙醇　alcohol
乙基纤维素　ethyl cellulose
乙醚　ether
已梳棉　combed cotton
已装运提单　on-board bill of lading; shipped on board B/L
义务　obligation; responsibility
议定　conclude; negotiate
议定书　protocol
议付银行　negotiating bank
异常　anomaly; unusual
异性纤维　different fiber; foreign fiber
异议　objection; dissent
异状纤维　abnormal fiber
易货贸易　barter
易燃　flammable; combustible; ease of ignition
易碎的　fragile
疫情　epidemic situation
益处　profit; advantage; benefit
意见　advice; suggestion; view
意外　unexpected; accident
溢短装率　weight tolerance ratio
溢价　premium
溢重　weight-overage
阴红　dark red
阴黄　dark yellow

阴燃　afterglow; smolder
银行汇票　bank draft [B/D]
银行贴现　bank discount
银行信贷　bank credit
引导　guide; lead; channel off
引证　quote; reference
印花税　stamp duty
印染　printing and dyeing
印章　official stamp; seal
英尺　feet
英寸　inch
英制　british system
荧光性　fluorescence [UV]
影响　affect; impact
应付票据　bills payable; notes payable
应付账款　account payable
应收票据　notes receivable; bills receivable
应收账款　account receivable
应用　apply; use; adhibition
应支付　payable
硬的　hard
硬块棉　bump cotton
佣金　commission; brokerage
永久的　permanent; perpetual; everlasting
用过的　second-hand; used
用户　consumer; user
用棉量　cotton consumption
用途　purpose
优惠待遇　preferential treatment
优惠待遇原则　principle of preferential treatment

优惠关税　preferential tariff
优良　fine; choiceness
优良产品　quality product
优于　superior to; better than
优质棉　high quality cotton
由轮船装运　per steamship
油籽　oilseed
油渍　smear; oil stains; oil spot
有差别的　distinct; different; discriminating
有关的　related; relevant; concerned; pertinent
有规律的　regular
有害生物　pest; harmful organism
有害外来物　harmful foreign matter
有害植物　harmful plant
有机棉花　organic cotton
有利的　beneficial
有联系的　relational; related
有赔偿责任　be liable for damages
有限公司　limited company; limited corporation [Co.,Ltd]
有效　effective; valid; efficacious; availability
有效期　period of validity; expiry date
有效长度　effective length; useful length
有义务的　liable; amenable
有益于　benefit; profit; avail
有责任的　liable; responsible; accountable; answerable
有资格的　qualified
幼铃　immature boll
幼苗　seedling
诱虫灯　moth-killing lamp
羽纱　camlet
雨锈棉　rain rust cotton
育成年份　releasing year
育种　breeding
预防　prevent; precaution
预防措施　precautionary measures; prevention measures
预付的　prepaid; imprest
预付货款　advance payment
预付运费　advance freight [A/F]
预计　estimate; predict; anticipated
预计到港日期　estimated arrival date
预警信息　alert information
预调湿　preconditioning
预先　beforehand; in advance; previous
预验　preliminary inspection
原产地　place of origin
原产地标准　origin criterion
原产地规则　rules of origin
原产地声明　declaration of origin
原产地证书　certificate of origin
原产国　country of origin
原产国标记　marks of origin
原封不动的　intact
原件（正本）　original
原理　principle; theory
原料　material
原料染色　ingrain
原棉　raw cotton
原棉等级　cotton grade
原棉分级　cotton classification

原棉分级员　cotton classer
原棉扦样　cotton sampling
原棉性能　cotton property
原始记录　source record; original record
原始数据　original data; primary data; initial dat; raw data
原因　cause; reason
原则　principle; tenet
远洋船舶　ocean vessel
约定（商业上的）　promise; appoint; convention
约分　reduction of a fraction
约束　constraint; restrain
允许　permit; allow
允许误差　permissible error; allowance error
运费　freight charge; freight; carriage; transportation expense
运费付讫　freight paid
运费率　freight rates
运费预付讫　freight prepaid; advanced freight
运输　freight; transport; carriage; conveyance; traffic; transit
运输包装　transport package; shipping package
运输方式　means of transport; means of conveyance; mode of transportation; type of shipping
运输工具检疫　transport quarantine
运输工具检疫证书　quarantine certificate for conveyance
运送保险　transit insurance
运载工具　carrier; means of delivery
运转　operate; run

Z

杂草鉴定　weed identification
杂交　hybridize; cross breeding; cross-fertilize
杂交棉　hybrid cotton
杂质　impurity; foreign substance; impurity substance
杂质等级　trash grade [Tr Grade]
杂质检验　trash test
杂质粒数　trash count [Tr Cnt]
杂质面积　trash area [Tr Area]
再生纤维　regenerated fiber
再轧棉　reginned cotton
载货　shipment; carry cargo; carry freight
载货单　manifest; shipping invoice
载重　load; carrying capacity
赞同　approve of; agree with; endorse; applaud
枣蔬点病　black root rot of jujube
责任　liability; obligation;

responsibility
增加 supplement; increase; raise; add; aggrandize; augment
增长 increase; rise
榨油机 crusher; oil press
粘附 affix
斩刀花 flat strips; card strips
展览 exhibit; on display; on show
展览期 extension period
占有 possess; occupy
丈量 surveying; measure
账户 account
招致 incur
着火性能 fire behaviour
照度 illuminance; intensity of illumination
照度仪 illuminance meter
照明 lighting; illumination
针状的 needlelike; acerose; spiculate
真空 vacuum; empty space
真空包装 vacuum packaging
真实的 authentic; real; true
真值 truth-value
振荡器 oscillator
争端 dispute
争端解决 dispute settlement
争端解决机构 dispute settlement body [DSB]
征收（税款、罚款等） assess; levy; impose
蒸熏除鼠 deratization by means of steam sterilization
整理 tidy; put in order; reorganize

整批销售 bulk sale
整齐度 uniformity [Unf]
整齐度指数 uniformity index [UI]
整箱 full container load [FCL]
正本 original
正常的 normal; regular
正常状态 normal condition
正当的 valid; legitimate; allowable
正规的 formal; regular; official
正态分布 normal distribution; Gaussian distribution
证件 document; certificate
证据 evidence; testimony; proof
证明 certify; prove; evidence
证明文件 documentary evidence
证实 confirm; affirm; approve; authenticate; verify
证书的撤销 withdrawing of certificate
证书的延期 extension of certificate
政策 policy
支出 expense; expend; disburse
支付 pay
支配 control; dominate; govern
支票 check; cheque
支气管炎 bronchitis
支数 number; yarn count
知识产权 intellectual property
织物 fabric
蜘蛛 araneid; spider
执行 carry out; execute; implement; enforce
执照 licence; permit
直接成本 direct cost

直接费用　direct fee
直接接触皮肤用品　products with direct contract to skin
直接用户　direct users
直接运输　direct consignment; direct shipment; direct cargo
直径　diameter
职责　responsibility; function
植物　plant
植物病虫害　plant diseases and insect pests
植物残体　plant residue
植物检疫证书　phytosanitary certificate
植物纤维　plant fiber
植物转口检疫证书　phytosanitary certificate for re-export
纸币　paper currency; banknote
纸箱　carton; paper skin
指标　index
指定　designate; nominate; specify
指定的船次　name of nominated vessel
指南　manual; guide
指示　designate; indicate; pointing
指数　index; exponent
指形管　thimble tube
制造　produce; manufacture; fabricate; make
制造商　manufacturer
质量　quality
质量标记　quality symbol
质量标准　quality standard
质量方针　quality policy
质量管理体系　quality management system
质量监督　quality surveillance; quality supervision
质量检验　quality inspection
质量控制　quality control
质量控制系统　quality control system
质量目标　quality objective; quality target
质量水平　quality level
质量体系　quality system [QS]
质量证书　certificate of quality
质量证书标志　quality certificate mark
致损原因　cause of damage
滞留　delay; retention
滞期费　demurrage
滞装费　delayed delivery fee
中等强度　medium tenacity; moderate strength
中段切断称重法（纤维试验）　middle part cut and weigh method
中断　interrupt
中国出口商品交易会　Chinese Export Commodities Fair [CECF]
中国出入境检验检疫　China's Entry and Exit Inspection and Quarantine [CIQ]
中国储备棉管理总公司　China National Cotton Reserves Corporation [CNCRC]

中国国际经济贸易仲裁委员会 China International Economic and Trade Arbitration Commission [CIETAC]
中国棉纺织行业协会 China Cotton Textile Industry Association [CCTIA]
中国棉花协会 China Cotton Association [CCA]
中国纤维检验局 China Fiber Inspection Bureau [CFI]
中级（美棉分级标准等级） Middling [M]
中级淡点污棉（美棉分级标准等级） Middling Lighter Spotted [M Lt.Sp]
中级淡黄染棉（美棉分级标准等级） Midding Tinged [M Tg]
中级点污棉（美棉分级标准等级） Midding Spotted [M Sp]
中级黄染棉（美棉分级标准等级） Midding Yellow Stained [M Y.S]
中级加（美棉分级标准等级） Middling Plus [M.P]
中期棉（第二次采摘的原棉） middle crop
中腔 lumen; mesocoele
中长纤维 medium length fibre
中支纱 medium yarn
中止 interrupt; discontinue
中止合同 interrupt contract
终点 destination
终点操作费 terminal handling charge [THC]
终点接收费 terminal receiving charge [TRC]

终端 terminal
终止 terminate
种类 kind; category; class; type
种毛长度 seed hair length
种仁蛋白含量 protein content in seed kernel
种仁棉酚含量 gossypol content in seed kernel
种仁脂肪含量 oil content in seed kernel
种植 plant; grow
种植密度 planting density
种子腺体 seed pigment gland
仲裁 arbitrate; arbitration
仲裁程序 arbitration proceedings; arbitral procedure
仲裁地 place of arbitration
仲裁规则 arbitration rules
仲裁申请 request for arbitration
仲裁庭 arbitral tribunal
仲裁委员会 arbitration commission
仲裁文件 arbitration documents
仲裁员 arbitrator
重量 quantity; weight
重量检验证书 inspection certificate of weight
重量鉴定 survey of weight
重量鉴定说明书 surveyor's report on weight
重量证书 weight certificate; certificate of quantity
周期 period; cycle
株高 plant height

株型　plant type
逐个集装箱　container by container
逐个棉包　bale by bale
主动配额　active quota
主管　supervisor
主茎　caulis; stem
主茎节间长度　internodes length of stem
主茎硬度　stem hardness
主任检验员　chief inspector
主任鉴定员　chief surveyor
主体马克隆值　modal micronaire
主体品级　modal grade
主体长度　principal length; modal length
主体长度级　staple length
主要产品　staple goods; major product; main products; principal products
注册　register; login; enroll; registration
注明　mark out; footnote
注释　annotation; note
注意　notice; remark
专利　patent
转杯纺　rotor spinning
转船运输　trans-ship; transhipment
转基因　transgenosis
转基因棉花　transgenic cotton
转交　deliver to
转口贸易　transit trade
转曲　twist; convolution
转让　transfer the possession of
转运港　port of transshipment

装船　shipment
装船港　loading port
装船计划　shipping schedule
装船前　pre-shipment
装船前检验　pre-shipment inspection [PSI]
装船日期　date of shipment
装船通知　shipping advice
装货通知单　shipping note [S.N 或 S/N]; shipping order [S.O]
装满　fill
装饰材料　decorative material
装箱单　loading list; packing list [P/L]
装运标记　shipping mark
装运代理人　shipping agent
装运单　shipping invoice; shipping document
装运期　shipping date
装载　load; embarkation
装载重量　loading weight
状态　status; state
准备　ready; prepare
准备工作　preparatory work
准确的　precise; accurate; exact
准确度　accuracy
准许　grant; permit; allow
准则　standard; criterion; rule; principle
准重　conventional weight
咨询　consult; advisory
咨询委员会　advisory committee
资产　bankroll; property
资格　qualification

资料　datum; information
子房　ovary; oophoron; germen
子样　subsample
子叶　cotyledon
籽棉　seed cotton; unginned cotton
籽棉公定衣分率　conditioned lint percentage of seed cotton
籽棉准重衣分率　conventional lint percentage of seed cotton
籽指　seed index
紫外线　ultraviolet rays
自动温度湿度控制系统　automatic control system of temperature and humidity
自然采光　natural lighting
自然光　natural light
自然长度　natural length; free length
自燃　self-ignition; pyrophoricity; breeding-fire; autoignition; self-ignite
自燃温度　spontaneous ignition temperature
自由贸易　free trade
自由贸易区　free trade zone
总产量　total yield; total output
总额　total; amount
总方差　population variance; total variance; total square deviation
总计　total
总价　total amount
总利润　gross profit

总皮重　total tare weight
总数　total amount
总体分布　population distribution
总体均值　population mean
总值　total value; gross value
奏效　effectualf; take effect
足够的　enough; sufficient
阻燃剂　flame retardant
阻止　prevent; impede; keep from; hold back
组成　constitute; form; make up; compose
组织　structure; organize; tissue
组织结构　organization structure
最不发达国家　least-developed country
最初的　initial; primary; premier
最大量　maximum [Max]
最惠国　most-favored-nation
最惠国待遇　most-favored-nation-treatment
最惠国税率　most-favored nation tariff
最小量　minimum [Min]
最新的　the latest; newest
最终用户　end user; final user
遵守　observe; abide by
作物　crop
作业指导书　work instruction; operation instruction
作用　function; effect

数据 datum; information
卵巢棉铃虫种群 every cotton-bollworm germ
子样本 subsample
子叶 cotyledon
籽棉 seed cotton; unginned cotton
籽棉净衣分率 conditioned lint percentage of seed cotton
籽棉常规衣分率 conventional lint percentage of seed cotton
籽指 seed index
紫外线 ultraviolet rays
自动控制温度和湿度系统 automatic control system of temperature and humidity
自然采光 natural lighting
自然光 natural light
自然长度 natural length; free length
自燃 self-ignition; pyrophoricity; breathing-fire; pyrofrention; self-expire
自燃温度 spontaneous ignition temperature
自由贸易 free trade
自由贸易区 free trade zone
总产量 total yield; total output
总量 total amount
总体方差 population variance; total variance; total square deviation
总计 total
总数 total amount
总利润 gross profit

总重量 total net weight
总额 total amount
总体分布 population distribution
总体均数 population mean
总值 total value; gross value
租金 the paid rate of rent
足够的 enough; sufficient
阻滞剂 drag-retardant
阻拦 prevent; impede; keep from; hold back
组成 constitute; form; make up; compose
组织 structure; organize; tissue
组织化程度 organization structure
最不发达国家 least-developed country
最初的 initial; primary; premier
最大值 maximum (Max)
最惠国 most-favored-nation
最惠国待遇 most-favored-nation treatment
最惠国关税 most-favored-nation tariff
最小值 minimum (Min)
最新的 the latest; newest
最终用户 end user; final user
遵守 observe; abide by
作物 crop
作业指导书 work instruction; operation instruction
(台用) function; effect

实用文件
Practical File

❶ 中纺棉花进出口公司购棉合同

1 Cotton Purchase Contract of China National Textile Import & Export Co.

购棉合同条款
Cotton Purchase Contract

1. 数量

1. Quantity

按公吨或按包装计算总重量,允许多交或少交 1%。

Calculated in metric tons or in bales with an allowance of $\pm 1\%$.

2. 单价

2. Unit Price

重量按每磅计算。

Weight calculated in English pound.

FOB STOWED:将棉花装入船舱内并整理好。

FOB STOWED: Cotton delivered and Stowed into ship's holds including stowage charges.

FAS:将棉花运至船的吊钩下;如 FOB 集装箱条款,须把箱运到集装箱站或场地。

FAS: Delivered to ship's tackle, case of break-bulk shipment; FOB container freight station (CFS) or container yard (CY), in case of container shipment.

CANDF:班轮条款,运费已定不变。

CANDF: Liner terms freight final.

3. 规格

3. Specification

按美国陆地棉统一标准或凭样。

as per unified grade standard of American Upland Cotton and/or by types.

4. 包装

4. Packing

适合于海运的紧缩机出口包装,外裹麻布、棉布或其他包皮,用牢固的打包铁皮或粗铁丝箍紧。每包净重约***磅(公斤),每公吨毛量平均体积***立方尺。如 FOB、FOBSTOWED、FAS VESSEL，每吨毛重超过规定的体积，其超过部分的运费由卖方负担。棉包内不准混有麻丝、破布、木屑、铁屑及铁钉等特殊杂物。如有发现，买方可根据中国商品检验局(以下简称 CCIB)或用户在使用过程中提供的证明及有关的唛头，向卖方提出对使用影响的索赔。

In export standard fully-pressed seaworthy packing, wrapped with hessian or cotton cloth or other materials, tightened with strong steel baling hoops or thick iron wires. Each bale weight about***lbs(kgs) net. The average measurement per ton in gross weight is***cubic feet. In case the measurement of cotton shipped on bases of FOB, FOBSTOWED and FAS VESSEL exceeds that specified above, any extra freight attributable thereto shall be for the seller' account. Cotton bales are not allowed to contain admixtures and extraneous matters such as iron scraps, nails wood chips, junk, rags, etc. In case these admixtures and extraneous matters are found, the Buyers shall have the right to claim against the sellers for losses arising therefrom on the basis of Inspection Certificate issued by China Commodity Inspection Bureau (thereinafter called CCIB) or of endusers' reports submitted in the course of using the cotton, as well as the shipping marks, certifying that such admixtures and extraneous matters have affected their usage.

5. 装船唛头
5. Shipping Marks

在棉包的一端和两侧的一面用不褪色的颜料按下列项目逐包刷唛。如由于卖方不刷合同号，由此而产生的混唛理货费由卖方负担。A.合同号，B.样子或等级长度，C.批号，D.包号

Shipping marks consisting of the following items shall be stenciled on one end and one side of each bale with indelible ink. In case contract numbers are not stenciled on the bales by the sellers, extra expenses incurred in tallying and sorting mixed-marked bales shall be borne by the sellers.

A. Contract No., B. Type or grade and staple, C. Lot No., D. Bale No.

6. 装运条件
6. Terms of Shipment

买方在装船前十五天左右，将船名通知卖方(允许买方以后更改船名)卖方可随时向买方了解装船时间的安排。船长在船到前 24 小时用电报通知船代理

转告卖方，说明该船预计到达目的港的时间。卖方收到船长备装通知后，应立即做好船准备。如卖方在上午收到备装通知，同天下午一时起就应开始装船。如在下午收到，应在次晨八时开始装船。等候停泊的时间不计在装船时间内。装船速度应为一船快装速度，在正常情况下，卖方向码头运货的速度，不能使指定装货舱口中断。(每八小时工作时间，卖方对每个舱口送货不少于100公吨)。

The Buyers shall inform the sellers 15 days prior to shipment of the name of nominated vessel (subsequent change of vessel by the Buyers permitted). The sellers may ask the Buyers from time to time to inform them of the shipping schedule. 24 hours before arrival of the vessel the captain shall inform the shipping agent by cable for transmission to the sellers of the EAT of the vessel. Upon receipt of captain's Notice of Readiness for Loading, the sellers should get ready for shipment promptly. If the notice of readiness is received before noon, loading shall commence at 1:00 p.m. on the same day, or commence at 8:00 a.m. next morning if received in the afternoon. Time lost in waiting for berth is not counted as loading time. Rate of loading shall be customary quick dispatch. Under normal conditions, the sellers shall keep sufficient goods at the wharf so as to enable them to be loaded to the designated hold(s) without interruption and the quantity which the sellers supply to each hold should not be less than 100 m.t within every 8 working hours.

除人力不可抗拒的原因外。如买方在合同规定的交货期内，超十天不去装货，买方需负担从超过合同规定的装船期第十一天起到船长发出备装通知之日止，每月负担货值的 1.25%迟装费。如按 C&F 成交，卖方在交货期内不能装船，卖方从超过交货期的第十一天起到船长发出备装通知之日止，付给买方每月 1.25%货值的损失。C&F 条款，在未得到买方同意的情况下，不允许转船和分运。

Except for reasons of "force majeure", the Buyers shall be liable to indemnify the sellers for carrying charges at 1.25% of the cargo value per month calculated from the 11th day after the contracted date of shipment to the day on which Captain's Notice of Readiness for Loading is issued, if the Buyers fail to ship the cargo 10 days after the period of shipment stipulated in the contract. If the business is concluded on C&F basis, the sellers shall be liable to compensate the Buyers for losses sustained as a result of late shipment at 1.25% of the cargo value per month calculated from the 11th day after the contracted date of shipment to the date of issue of the Notice of Readiness for Loading by the Captain, if the sellers fail to effect shipment 10 days after the period of shipment stipulated in the contract. Transshipment and partial shipment are not allowed without the approval of the Buyers.

如买方在合同规定的装船期内通知卖方装船,当船到后除人力不可抗拒的原因外,卖方迟交或不能交货,卖方应赔偿由此而产生的滞期和空船损失。如卖方超过合同规定的交期45天仍不能装船,买方有权决定合同是否有效。

Except for reasons of 'force majeure', the sellers shall compensate the Buyers for the demurrage and/or dead freight in case the sellers fail to make delivery of the goods within the stipulated period of shipment as notified by the Buyers. If the sellers are still unable to make delivery in 45 days after the stipulated shipment period, the Buyers shall have right to decide whether or not the contract shall continue to be in force.

7. 装船通知

7. Advice of Shipment

A. 如为FOB成交。卖方在装船后须在两个工作日内电告买方合同号、等级长度或样子、包装、净重、金额;船名、装船日期、装船口岸,并航寄装船单据副本一式三份给买方。

A. On FOB basis: Within 2 working days after completion of shipment, the sellers shall inform the Buyers by cable of the contract number, grade and staple length (types), number of bales, net weight, amount, name of carrying vessel, date of loading and port of loading and forward to the Buyers by airmail 3 copies each of the shipping documents.

B. 如为C&F成交。卖方在装船前十五天左右,将船名、船龄(老船卖方要付超龄加保费)、船旗(除南非、韩国、以色列外,其他均可)吃水、受载期、合同号、数量及目的港通知买方。船长不能超过150公尺,船宽不能超过25公尺,待买方确认后方可装船。

B. On C&F basis: the sellers shall inform the Buyers about 15 days prior to loading of the name of vessel, vessel's age (overage surcharge on insurance premium for the sellers account), flag (flags of South Africa, korea and Isreal not acceptable), draft, lay-days, contract number, quantity and port of destination. The length of the vessel shall not exceed 150 meters and the width 25 meters. Shipment can only be effected upon confirmation of the Buyers.

C. 如卖方未按上述 A 项规定电告买方，以致买方未能及时保险，由此而产生的损失由卖方负担。

C. Should the Buyers be unable to cover insurance owing to the sellers' failure in sending in time the shipping advice by cable according to Section(A), the sellers shall be liable for any and all damage and/or losses attributable to such failure.

8. 付款

8. Payment

A. 在装船前十五天左右，由买方通知北京中国银行开出以卖方为受益人的不可撤销的信用证，凭第 9 项的单据条款规定的装船单据北京见单电汇付款。信用证的金额为合同金额的 98%，其余 2%在目的港卸毕后 70 天内根据目的港的 CCIB 对该棉质量检验和重量的鉴定证书进行最后结算，多退少补。信用证的有效期、到装船后 15 天为止。

A. About 15 days prior to shipment the Buyers shall open an irrevocable L/C in favour of the sellers through the Bank of China in Beijing payable with T/T reimbursement against presentation to Bank of China, Beijing of the shipping documents as specified in Article 9. The amount of L/C covers 98% of the total contract value. The remaining 2% will be settled by airmail transfer to the sellers within 70 days after completion of discharge of cargo at port of destination subject to any necessary adjustments to be made upon examination of the quality and weight of cotton by CCIB at port of destination. The L/C shall expire 15 days after shipment.

B. 如买卖双方同意不开信用证、凭单托收时，买方凭全套装船单据由北京中国银行见单、按发票金额 98%付款，其他同 A 项。

B. If by agreement of both parties payment is made by collection against documents without opening an L/C, payment shall be effected at sight against presentation of full set of documents to Bank of China in Beijing for 98% of the total invoice value. Other terms are to go by Section(A).

9. 单据

9. Documents

在开证时，卖方需向议付行，不开证时需向北京中国银行提供下列单据：

If L/C is opened, the sellers should present the following documents to the negotiating bank and if no L/C is opened, the sellers should send the documents to the Bank of China, Beijing.

A. 汇票一式两份。

A. Draft in duplicate.

B. 商业发票一式六份(详细注明唛头)。

B. Sellers' commercial invoice, in sextuplicate, indicating detailed shipping marks.

C. 逐包过重的重量码单一式二份。

C. Weight memo, in duplicate, showing the weight of each bale.

D. 全套可转让的清洁海运提单(棉包的取样洞除外)正本一式三份。

D. Full set of negotiable Clean on Board Ocean Bill of Loading, in triplicate (holes caused by sampling on cotton bales excepted).

E. 卖方给买方的装船通知电报副本一份。提单和发票必须按每个合同分开。合同号必须标在提单上。

E. One copy of the Sellers' cable shipping advice to the Buyers. B/L and commercial invoice must be made out separately for each contract, and contract number must be indicated in the B/L.

10. 到岸品质、到岸重量

10. Landed quality and landed weight

A. 重量：由目的港的 CCIB 随机抽样 5%检验回潮率，并逐包过重，出具重量鉴定证书。回潮率为 8.5%多扣少不补。

A. Wight：Moisture Regain 8.5% not reversible. Inspection Certificate is to be issued by CCIB at port of destination on the basis of laboratory test from 5% samples drawn at random for moisture regain and on the results of weighing of every bale.

B. 皮重：CCIB 根据数量的多少、包装的差异大小。从每批货中随机抽取 3%～5%的包数，以确定平均皮重。按实际平均皮重计算到岸净重。

B. Tare：CCIB shall decide the average tare by sampling at random from each lot 3%～5% of bales depending on the quantity of bales and uniformity of packing. The net landed weight shall be calculated on the basis of the actual tare.

C. 每个合同多交或少交的到岸重量，不得超过合同数量的 1%。当装船时的纽约市价高于合同日期的纽约市价，其多交的部分按合同价计；少交的部分如超过 1%时，买方有权向卖方索取差价及空舱费的损失(指 FOB 或 FAS 条款)。如装船时的纽约市价低于合同日期的纽约市价，其多交的部分如超过 1%时，买方有权向卖方索取差价。

C. Tolerance in weight shall not exceed 1% of the contract quantity. If the prevailing New York market price at the time of loading is higher than that on the date of signing the contract, the purchase price for the portion in excess shall be calculated at the contract price and the Buyers have the right to claim with the sellers for price difference for the shortage in weight over 1% and dead freight (in case of FOB and FAS terms). The same right is also given to the Buyers to claim for price difference with the sellers for the portion in exceeding 1% in weight if the prevailing New York market price at the time of loading is lower than that on the date of signing the contract.

D. 品质：从每批货中随机抽样 10%(强力 5%)作为品质检验的依据。如交货的某些样子低于合同规定的等级，长度、强力或细度超过规定的范围时，以 CCIB 出具的检验证书为买卖双方结算的依据。

D. Quality：10% sampling for grade, staple and micronaire and 5% sampling for Pressley shall be made at random of each lot by CCIB to serve as a basis of quality inspection. The Inspection Certificate issued by CCIB shall be taken as final basis for the settlement between the Buyers and the sellers in the event that the drawn sample examined is found not up to the standard in grade, staple length, Pressley or micronaire range as specified in the contract.

11. 陆地棉品质降级差价(按合同单价的百分)

11. Quality differentials for upland cotton (based on percentage of contract unit price)

通用标准 UNIVERSAL STANDARD		小样成交 GRADE ON TYPE	
GM			
SMP	1.00%		
SM	1.25%	−1/2	1.75%
MP	1.50%	−1	3.75%
M	1.75%	−1~1/2	6.00%
SLMP	2.00%	−2	9.00%
SLM	2.50%	−2~1/2	12.50%
LMP	3.00%	−3	15.50%
LM	3.50%	−3~1/2	20.00%
SGOP	4.00%		
SGO	4.50%		

主体长度 STAPLE LENGTH		马克隆值 MICRONAIRE NCL		
1-1/16" &UP	1" &BELOW	3.5~4.9		3.8~3.9
−1/32 2.0%	1.5%	5.3&UP	2.0%	2.00%
−1/16 4.0%	3.0%	5.0-5.2	1.0%	1.00%
−3/32 7.0%	5.0%	3.5-3.7	—	1.25%
−1/8 10.0%	7.0%	3.3-3.4	1.5%	2.50%
−5/32 13.5%	10.0%	3.0-3.2	3.5%	4.50%
−3/16 17.0%		2.9&below	8.0%	9.50%

注：淡点污棉比照白棉降 1 级计算差价，即 SM. LIGHT SPOTTED 按 M 级差价计算。点污棉比照白棉降 2 级半处理，即 SM.SPOTTED 按 LMP 差价计算。商检证书内只注明 LIGHT SPOT 或 SPOT。

Remarks: Light spotted cotton shall be regarded as off-graded by 1 grade and spotted cotton as off-graded by 2-1/2 grade as per white cotton, namely:

SM.LIGHT SPOTTED　M
SM.SPOTTED　　　　LMP

Indication of L. SP. or SP. shall be made in Inspection Certificate issued by CCIB.

A. 如降等级超过 2-1/2 级、降长度超过 1/8 英寸时,买方有权按合同规定的降级差价另加 25%处理。

A. In case the cotton is degraded more than 2-1/2 and/or staple length is found shorter by 1/8", the Buyers shall have the right to impose an extra penalty at 25% in addition to price difference stipulated in the contract.

B. 1/2 级即 MP 和 M 级的差距；1 级即 SM 和 M 级的差距,余类推。

B. 1/2 grade means the difference between MP and M and 1grade means that between SM and M, and so on and so forth.

12. 检验费

12. Quality and weight inspection charges

A. 重量鉴定费由卖方负担,每公吨 50 美分(包括回潮率检验费)。

A. Weight controlling charges are to be for the Sellers' account at US＄0.50 per metric ton (including inspection of moisture regain).

B.品质检验费由买方负担,每包一美元,即每只样 10 美元。如品质降级时卖方应负担不符合同规定部分的检验费,即每只样子的等级或长度各 4.5 美元,细度 50 美分、强力 1.5 美元。

B. Quality inspection charges after landing are to be for the Buyers' account at US＄1.00 per bale, thus making US＄10.00 per sample. However, in case results of survey show that the grade does not conform to the contract stipulations, the sellers shall pay quality inspection charges proportionately according to percentage of unqualified samples at US＄4.50 for each sample of grade and staple, US＄0.50 for each sample of micronaire and US＄1.50 for each sample of Pressley.

13. 熏蒸

13. Fumigation

卖方所交之棉花,不能带有国家检疫对象的病虫害。如货到岸后有上述病虫害存在,买方将根据国家检疫规定需要熏蒸处理时,除原受载的船期损失由买方负担外,其因熏蒸而产生的各项费用,均由卖方负担。

The sellers shall guarantee that the cotton shipped is free from all pests and diseases which are subject to quarantine in the Buyers' country. In case such pests and diseases are found in the cotton upon examination at the port of destination and thus it is necessary to fumigate the cotton in accordance with the rules and regulations of the State Quarantine Organization of the Buyers' country, all expenses arising therefrom except demurrage of the carrying vessel shall be the sellers' account.

14. 人力不可抗拒

14. Force Majeure

由于一般公认的人力不可抗拒的原因而致买卖双方延迟执行合同或不能执行合同之一部分或全部，买卖双方都不承担责任。在这种情况下，买卖各方必须向订约的另一方提供政府机关或商会发给的灾害证明文件，以资证明。

Neither the sellers nor the Buyers shall be held responsible for delay in execution or non-execution of the whole or a part of the contract owing to generally recognized 'Force Majeure'. In such case, the sellers or the Buyers, as the case may be, shall deliver to the other contractual party a certificate of the accident issued by the competent government authorities or the chamber of commerce at the place where the accident occurs as evidence thereof.

15. 仲裁

15. Arbitration

本合同在执行中，如发生品质或重量以外的纠纷，应通过双方友好协商加以解决。如协商不能达成协议时，应提请仲裁。仲裁在被告国进行。如在中国，由中国国际贸易促进委员会的对外贸易仲裁委员会，根据该会的仲裁规章和程序进行。如在卖方所在国，则由该国的有关仲裁的机构进行。仲裁的裁决，须作为买卖双方最后的解决依据。仲裁费除另有规定者外，均由败诉方负。

All disputes other than those on quality and weight in connection with this contract or the execution thereof shall be settled by friendly negotiation between both parties. In case no agreement can be reached, the case in dispute shall then be submitted for arbitration which will take place in the country of the defendant. If in the Buyers' country the case will be arbitrated by the Foreign Trade Arbitration Commission of China Council for Promotion of International Trade, and if in the sellers' country it will be submitted to a competent arbitration organ for arbitration. The award by the afore-said organs shall be accepted as final and binding upon both parties. Arbitration fees shall be borne by the losing party unless otherwise awarded.

❷ 中国棉花协会棉花买卖合同
2 Cotton Purchase Contract of China Cotton Association

棉花买卖合同
COTTON PURCHASE CONTRACT

合同编号：	日期：
Contract No.:	Date:
买方：	卖方：
Buyer:	Seller:
地址：	地址：
Address:	Address:
电话：	电话：
Tel:	Tel:
传真：	传真：
Fax:	Fax:
电子邮件：	电子邮件：
E-mail:	E-mail:

本合同由买卖双方订立，根据本合同规定的条款，买方同意购买、卖方同意出售下述商品：

This Contract is made and entered into by and between the Buyer and the Seller; and in accordance with the terms and conditions of the Contract, the Buyer agrees to buy and the Seller agrees to sell the following commodity:

1 商品名称
1 Commodity
产地：
Origin:
生产年度：
Crop year:
类别：_____（细绒棉，长绒棉）

Category: _____ (upland cotton, long-staple cotton)
加工方式：□锯齿棉　□皮辊棉
Ginning: □ saw ginned　□ roller ginned

2　规格/质量
2　Specifications/Quality
级别：□　USDA 通用棉花标准
Grade: □　USDA Universal Cotton Standards
　　　　□　凭小样（小样型号）
　　　　□　by type：
长度：_____（英寸，毫米）
Staple Length: _____ (inch/mm)
马克隆值：_____NCL
Micronaire: _____NCL
断裂比强度值：最小值__克/特克斯，平均值__克/特克斯以上
Strength: minimum__grams/tex, average above__grams/tex

3　数量
3　Quantity
净重：_____（吨，磅，包）
Net Weight: _____(ton/pound/bale)
溢短装率：_____%（默认值为 1.5%）　□ 不允许多装
Weight Tolerance Ratio _____%(If not specified here, 1.5% will be applied)
□ Excess not allowed
吨与磅的换算公式：　1 吨=2204.62 磅
Conversion between ton and pound: 1 ton=2204.62 pounds

4　价格
4　Price
单价：_____（美分/磅，人民币元/吨）
Unit Price: _____(USC(cent)/pound or RMB(Yuan)/ton)
价格条件：_____（CIF，CFR，FOB，其它）
Terms: _____(CIF, CFR, FOB or others)
总价：_____（美元，人民币元）

Total Value：_____(USD/RMB)

5　付款方式
□ 信用证 □ 凭单托收 □ 其它

5　Payment Terms
□ Letter of Credit □ D/P □ Others

6　重量、质量检验
CIQ 检验证书为结算和索赔的依据。

6 Weight and Quality Inspection
CIQ Inspection Certificate shall be the basis for settlement and compensation.

7　装运/交货日期
从_____（年月日）到_____（年月日），或按月等量装运/交货（每月数量）____（吨，磅，包）。

7　Shipment/Delivery
shipment/delivery from _____ (mm/dd/yy) to ____ (mm/dd/yy) or equal monthly shipment/delivery as follows：____(ton, pound, bale).

8　目的地 _____

8　Destination _____

9　一般条款

9　General Terms
一般条款为本合同不可分割的一部分。对该条款中任何一款的修改和删除应在备注中注明。

The General Terms shall constitute an integral part of the Contract. Amendment to or deletion of any general terms shall be specified in the Remarks.

10　仲裁
凡因本合同引起的或与本合同有关的任何争议，双方同意提交：_____（□中国国际经济贸易仲裁委员会[CIETAC]；□ 国际棉花协会[ICA]；□ 其它仲裁机构），按照申请仲裁时该仲裁机构现行有效的仲裁规则进行仲裁。

10　Arbitration
Any dispute arising from or in connection with the Contract shall be referred

to＿＿（ □ CIETAC, □ICA, □ OTHERS)for arbitration in accordance with its arbitration rules effective at the time of application.

11　本合同采用书面形式，由买卖双方授权代表签字。双方在合同签订日之前以其它书面通讯方式，如信函、电报、传真或电子邮件形式达成的成交内容，须由本合同确认。

11　This Contract shall be made in written form and signed by the authorized representatives of the parties. The signed or stamped contract shall verify the terms and conditions of the contract previously agreed to at an earlier date in other written communications including mail, telegraph, fax, or e-mail.

12　备注
12　Remarks

买方签字：	卖方签字：
Signature of the Buyer:	Signature of the Seller:
日期：	日期：
Date：	Date：

一般条款
GENERAL TERMS

本一般条款是《棉花买卖合同》不可分割的一部分。
These General Terms shall be an integral part of the Cotton Purchase Contract.

1　定义
1　Definitions

在本合同中,下列词语的含义如下：
The following terms shall have the following meanings in the Contract:

- CIQ：中国出入境检验检疫机构。

CIQ：China Entry-Exit Inspection and Quarantine.

- NCL：不允许超出控制界限。

NCL：No control limit is allowed.

- USDA：美国农业部。

USDA：United States Department of Agriculture.

- 通知：采用电报、信函、传真、电子邮件等方式告知对方。

Notification: to notify the other party by telegraph, mail, fax, e-mail, or other methods.

- 皮重：棉花包装材料的重量。

Tare: the weight of cotton's packaging materials.

- 净重：总重扣除皮重后的重量。

Net Weight: the gross weight less tare.

- 非棉物质：混入棉花中对使用有严重影响的硬软杂物，如化纤丝、麻丝、破布、木屑、金属物品等。

Non-Cotton Substance: soft or hard sundries mixed in the cotton that have serious impact on the use of the same, including chemical fiber, flax, cloth, wooden chips metal articles, etc.

- 无纺用价值棉花：霉变棉、水渍棉、油污棉、火烧棉、棉花废料、棉短绒等。

No Spinning Value Cotton: mouldy cotton, water damaged cotton, oil stained cotton, burned cotton, cotton waste and linters, etc.

- 棉花废料：加工或使用棉花过程中产生的下脚回收废料等。

Cotton Waste: leftover and/or recycling waste left during the processing or use of the cotton.

- 欺诈棉包：单个棉包中：含有与棉花完全无关的非棉物质；里面含有污染棉花，但从棉包外部或可看出或看不出来；好棉花在外面，次棉花包在里面，以免在常规检查中被发现；有一定数量的无纺用价值棉花。

False Packed Bale: cotton in a single bale: containing substances entirely foreign to cotton; containing damaged cotton in the interior with or without any indication of such damage upon the exterior; composed of good cotton upon the exterior and decidedly inferior cotton in the interior, in such a manner as not to be detected by customary examination; or containing a certain amount of no spinning value cotton.

- 混杂棉包：单个棉包中含有一定数量不同品级、不同长度或不同颜色类型的棉花。

Mixed Packed Bale: a bale containing a certain amount of different grades, staples or colors of cotton.

- 溢短装率：到岸重量超出或少于合同规定重量的部分占合同总重量的百分率。

Weight Tolerance Ratio: the percentage of the part of the CIQ landed weight exceeding or shorter than the weight provided by the Contract against the total contract weight.

- 棉包密度：采用通用棉包密度，是指根据国际标准化组织——ISO 第 8115-1986（E）的规定，一个货包长度在 1060-1400 毫米，宽度 540 毫米，高度 700～950 毫米。

Bale Density: Universal Bale Density as determined by the International Standards Organization – ISO Reference No. 8115-1986 (E) is a bale with the nominal dimensions of 1060 to 1400 mm in length by 540 mm in width and 700-950 mm in height.

2 包装

2 Packing

适合于海运的紧缩机出口包装，外裹棉布或其他不能产生异性纤维的包装，捆扎牢固，包装完整。如果使用容易产生异性纤维的包装材料包装棉花，则卖方须承担全部清理异性纤维的费用。棉花须以通用密度压缩货包的形式供货。

Compressed export packing suitable for voyage, outside wrapped by cotton cloth or other packing materials that do not contain foreign matters, tightly and completely packed. If any packing materials that may easily produce foreign matters are used to pack the cotton, the Seller shall bear all the expenses for the cleaning of foreign matters. The cotton shall be supplied in forms of universal density compressed package.

3 唛头

3 Marks

除非另有约定，在棉包上挂有永久性棉包标识卡或在棉包的两侧用不褪色的颜料按下列项目逐包刷唛，其内容为：

Unless otherwise agreed, hang permanent cotton identification card onto the cotton bale or mark on both sides of each cotton bale with unfading paint the following items：

    A. 批号/包号　B. 毛重　C. 合同号

    A. Lot Number/Bale Number B. Gross Weight C. Contract Number

若唛头不清，由此而产生的混唛理货费由卖方承担。

If the marks are not clear, all the expenses arising from sorting the mixed mark bales shall be borne by the Seller.

4　装船通知

4　Shipment Notice

4.1　如为 FOB 成交：卖方应在收到船公司的装运通知后 48 小时内，通知买方合同号、品级、长度级或小样型号、包装、净重、金额；装船日期、装船口岸、目的港和预计到港日期，并航寄、传真或电子邮件的形式将装船单据副本一式三份给买方。

4.1　Under FOB terms：the Seller shall notify the Buyer by telegraph, fax or e-mail of the contract number, grade, staple or type, packing, net weight, and price; as well as shipment date, shipment port, destination port and estimated arrival date within 48 hours after notification from the shipping line and mail, fax or e-mail three copies of the duplications of the loading documents to the Buyer.

4.2　如为 CFR/CIF 成交：卖方应在收到船公司的装运通知后 48 小时内，通知买方船名、船龄(老船卖方要付超龄加保费)、船旗、装船日期、装船口岸、目的港、合同号、提单号、总金额、毛重、净重。

4.2　Under CFR/CIF terms：the Seller shall notify the Buyer of the ship name, ship age (for aged ship the Seller shall pay the over-age extra premium), ship flag, shipment date, shipment port, destination port, contract number, number of the bill of lading, total price, gross weight and net weight within 48 hours after the shipment notification from the shipping line.

4.3　如卖方未按上述 4.1、4.2 款规定通知买方，以致买方未能及时购买保险，由此而产生的损失由卖方负担。

4.3　If the Seller fails to notify the Buyer by telegraph, fax or e-mail as provided in above Article 4.1 and Article 4.2 and thus the Buyer is unable to purchase the insurance in time, all the losses arising therefrom shall be borne by the Seller.

5 单据

5 Documents

在开证时,卖方需向议付行,不开证时需向买方指定的银行提供下列单据:

The Seller shall provide the following documents for the negotiating bank in the event of issuing a letter of credit or the bank designated by the Buyer when not issuing a letter of credit:

5.1 商业发票、装箱单正本和副本各三份,详细注明信用证号、合同号。

5.1 Three originals and three copies of the original commercial invoice and packing list, specifying the letter of credit number and contract number in details;

5.2 逐包或逐集装箱过重的重量码单一式二份。

5.2 Two copies of weight memo by weighing the bales or containers one by one;

5.3 全套可转让的清洁海运提单正本和副本各三份。

5.3 Three originals and three copies of the full set of the clean bill of lading;

5.4 卖方给买方的装船通知副本一份。

5.4 One copy of the shipment notification from the Seller to the Buyer;

5.5 在 CIF 条款下,保险单正本和副本各一份。

5.5 One original and one copy of the certificate of insurance under CIF terms;

5.6 产地证明、植物检疫证明和非木质包装证明正本和副本各一份。

5.6 One original and one copy of the certificate of origin, phytosanitary certificate, and non-wooden packing certificate.

6 付款方式

6 Payment Method

6.1 在双方约定以信用证为付款方式时,在合同规定的最晚装运日前 30 天,由买方通知开户银行开出以卖方为受益人的不可撤销的信用证,凭第 5 项单据条款规定的单据电汇付款。信用证的内容应与合同规定相符。信用证的到期日为最晚装运日后的第 21 天。

6.1 In the event that the parties hereto agree to make payment by letter of credit, the Buyer shall cause the opening bank to issue an irrevocable letter of credit in favor of the Seller within 30 days prior to the latest shipment date provided by the Contract, and the payment shall be made by wire on the basis of the documents provided in Article 5-Documents. The content of the letter of credit shall be consistent with the terms and conditions of the Contract. And the expiry date of the letter of credit shall be the 21st day after the latest shipment date.

6.2 在双方约定以凭单托收为付款方式时,买方凭第 5 项单据条款规定的全套单据,由买方指定的银行按发票金额付款。

6.2 In the event that the parties hereto agree on documents against payment, the bank designated by the Buyer shall make the payment at the invoice amount on the basis of whole set of documents that are provided in Article 5-Documents.

7 到货检验

7 Inspection Upon Delivery

7.1 检验机构

7.1 Inspection Institution

货物到目的地后由 CIQ 检验,其出具的重量检验证书(包括对欺诈棉包和混杂棉包的认定)和质量检验证书(如有残破还应出具残损证书),作为买卖双方结算和索赔的依据。

CIQ will conduct inspections after the goods arrive at the destination and issue the weight inspection certificate (including the confirmation of false packed bale and mixed packed bale) and the quality inspection certificate (such as the damage certificate if the cotton is damaged), which shall be the basis for the settlement and claims between the Seller and the Buyer.

7.2 重量检验

7.2 Weight Inspection

根据实际到货情况,CIQ 将采用逐包过重或用地衡以一个集装箱为单位过重,出具重量检验证书,按实际净重结算。

In accordance with the actual delivery circumstance, CIQ will scale the weight of each bale or weigh by container with land scale, issue the weight inspection certificate and make the settlement by actual net weight.

皮重:根据数量的多少、包装类型的差异,CIQ 将从每批货中随机抽取(1~5)%的包数,以加权平均确定平均皮重。按实际平均皮重计算到岸净重。

Tare Weight: in accordance with the quantity and package type, CIQ will sample(1-5)% bales from each lot of cotton and compute the average tare weight by weighted average. And calibrate the net weight upon arrival at the port based on the actual average tare weight.

买卖双方均可派代表（须凭授权委托书）到现场察看称重、扦样过程，卖方应在货物到达目的地前通知买方，买方应予必要协助。

Both the Seller and the Buyer may assign representatives (with a power of attorney) to the site for the purposes of observing the weighing and sampling process, and the Seller shall notify the Buyer of the same prior to the goods arriving at the destination, and the Buyer shall provide necessary assistance.

7.3　质量检验

7.3　Quality Inspection

抽样：品级、长度：从每批棉包中随机抽样 10%；马克隆值和强度：从品级、长度抽取的样品中随机抽取其样品总量的一半，作为马克隆值和强力的检验样品。

Sampling: grade and staple: 10% to be sampled at random for each lot of bales; micronaire and strength: one half of the 10% is to be sampled at random for each bunch of bales.

检验方法：采用仪器测试和感官检验相结合。如双方有争议时，以感官检验为主。

Inspection method: combined equipment testing with sensory evaluation; where there is any dispute, the outcome of sensory evaluation shall prevail.

7.4　复验

7.4　Re-inspection

重量不复验。报验人对 CIQ 质量检验结果有异议的，可在收到检验结果之日起 15 日内，向做出检验结果的 CIQ 或者国家质量监督检验检疫总局申请复验。复验只对抽取留存的样品，复验结果为最终依据。

No weight re-inspections. If the applicant has objections to the CIQ quality inspection result, such applicant may apply for a re-inspection with the CIQ that renders the inspection result or with the General Administration of Quality Supervision, Inspection and Quarantine within 15 days after the receipt of the inspection result. The re-inspection will only be conducted to the sampled samples and the re-inspection result shall be final.

8　细绒棉质量降级差价(按合同单价的百分比%)

8　Price Deduction for Upland Cotton Inferior Quality Grade (% of the contract unit price)

除非双方另有约定，否则适用以下规定：

Unless otherwise agreed to by the parties, the provisions hereinafter shall apply：

8.1 品级降级差价

8.1 Price Deduction due to Inferior Quality Grade

8.1.1 按 USDA 通用棉花标准签约的：

8.1.1 In terms of Contract provided in USDA Universal Standards：

GM -, SM -3%, M -7%, SLM -12%, LM -19%, SGO -28%, GO -39%。

淡点污棉比照白棉按降 1 个级处理，即 SM LIGHT SPOTTED 按 M 级计算差价；点污棉比照白棉按降 2 个级处理，即 SM SPOTTED 按 SLM 级计算差价；淡黄染棉比照白棉按降 3 级处理，即 SM TINGED 按 LM 级计算差价；黄染棉比照白棉按降 4 级处理，即 SM YELLOW STAINED 按 SGO 计算差价。CIQ 出具的证书内只注明 LIGHT SPOTTED 或 SPOTTED 或 TINGED 或 YELLOW STAINED。

Light spotted cotton is regarded as one grade inferior with reference to white cotton, i.e., the price deduction of SM LIGHT SPOTTED shall be calculated based on M Grade; spotted cotton is regarded as two grades inferior with reference to white cotton, i.e., the price deduction of SM SPOTTED shall be calculated based on SLM Grade; light tinged cotton is regarded as three grades inferior with reference to white cotton, i.e., the price deduction of SM TINGED shall be calculated based on LM Grade; and yellow stained cotton is regarded as four grades inferior with reference to white cotton, i.e., the price deduction of SM YELLOW STAINED shall be calculated based on SGO Grade. Only LIGHT SPOTTED, SPOTTED, TINGED or YELLOW STAINED will be specified in the certificate issued by CIQ.

8.1.2 凭小样成交签约的：

8.1.2 Contract by Type

检验结果低于小样 Inspection outcome is less than the sample	扣减幅度（%） Deduction (%)
1 个级 Grade	-3.75
2 个级 Grades	-9
3 个级 Grades	-15.5

4 个级 Grades	−23

8.2 长度降级差价

8.2 Inferior Staple Deduction

8.2.1 合同长度在 1 又 1/32 英寸及以上的：

8.2.1 The Contract staple is 1-1/32 inches and longer:

检验结果低于合同长度（英寸） Inspection outcome is less than contract staple (inch)	扣减幅度（%） Deduction (%)
1/32	−2
1/16	−4
3/32	−7
1/8	−10
5/32	−13.5
3/16	−17

8.2.2 合同长度在 1 英寸及以下的：

8.2.2 The Contract staple is one inch and shorter:

检验结果低于合同长度（英寸） Inspection outcome is less than contract staple (inch)	扣减幅度（%） Deduction (%)
1/32	−1.5
1/16	−3.5
3/32	−5
1/8	−7
5/32	−10

8.3 马克隆值差价

8.3 Micronaire Variance

8.3.1 马克隆值在 3.5-4.9 范围值之间的没有折扣，超出范围值的，差价如下：

8.3.1 No deduction for a micronaire variance within the range of 3.5-4.9; and the price deduction shall be as follows for the micronaire exceeding the said range：

仪器测试值 Tested Value	扣减幅度（%） Deduction（%）
5.3 及以上 5.3 and above	−2
5.0-5.2	−1
3.3-3.4	−1.5
3.0-3.2	−3.5

| 2.9 及以下 2.9 and less | −8 |

8.3.2 对于规定了马克隆值最小值的合同,未达到此最小值的,差价如下:

8.3.2 With respect to the contract that provides the minimum micronaire, the price deduction shall be as follows if the minimum value is not met:

仪器测试值低于最小值 Value tested less than the minimum value	扣减幅度(%) Deduction (%)
0.1	−0.5
0.2	−1.0
0.3	−2.0
0.4	−3.0
0.5	−4.0
0.6	−5.0

如此类推,马克隆值每低 0.1,扣减幅度增加 1%。

Based on the foregoing, for each 0.1 micronaire value less than the minimum, the deduction shall increase 1%.

8.3.3 对于规定了马克隆值最大值的合同,超过此最大值的,差价如下:

8.3.3 With respect to the contract that provides the maximum micronaire, the price deduction shall be as follows if the maximum value is exceeded:

仪器测试值超过最大值 Value tested more than the maximum value	扣减幅度(%) Deduction (%)
0.1	−0.5
0.2	−1.0
0.3	−2.0
0.4	−3.0
0.5	−4.0
0.6	−5.0

如此类推,马克隆值每高 0.1,扣减幅度增加 1%。

Based on the foregoing, for each 0.1 micronaire value more than the maximum, the deduction shall increase 1%.

8.4 强度差价

8.4 Strength Variance

对于规定了断裂比强度最小值的合同,未达到此最小值的,差价如下:

In terms of a contract that provides the minimum strength, the price difference shall be set forth below if it fails to reach the minimum strength:

| 仪器测试值低于最小值 | 扣减幅度(%) |
Value tested less than the minimum value	Deduction (%)
1.0-2.0	-1.0
2.1-3.0	-1.5
3.1-4.0	-3.0
4.1-5.0	-5.0
5.1-6.0	-8.0

比 6.0 还低的,每低 1 克/特克斯,扣减幅度增加 4%。

If less than 6.0, for each gram/tex lowered, the deduction is 4%.

9 长绒棉质量降级差价(按合同单价的百分比%)

9 Price Deduction for Long-staple Cotton Inferior Quality Grade (% of the contract unit price)

除非双方另有约定,否则适用以下规定:

Unless otherwise agreed to by the parties, the provisions hereinafter shall apply:

9.1 品级降级差价(按 USDA 通用棉花标准)

9.1 Price Deduction due to Inferior Quality Grade (in accordance with the long staple cotton grade provided in USDA universal standards):

1 级-, 2 级-3%, 3 级-7%, 4 级-12%, 5 级-19%, 6 级-28%。

Grade1-, Grade 2 -3%, Grade 3 -7%, Grade 4 -12%, Grade 5 -19%, Grade 6 -28%。

9.2 长度降级差价

9.2 Inferior Staple Deduction

| 检验结果低于合同长度(英寸) | 扣减比例(%) |
Inspection outcome is less than contract staple (inch)	Deduction (%)
1/16	-3%
1/8	-10%
3/16	-17%

9.3 马克隆值差价

9.3 Micronaire Variance

与细绒棉相同。

As per Upland cotton.

9.4 强度差价

9.4 Strength Variance

与细绒棉相同。

As per Upland cotton.

10 违约索赔

10 Contract Settlement Differences

10.1 延迟装运

10.1 Delay of Shipment

由于卖方原因造成不能按期装运的，则卖方应从合同规定的最晚装运日的第十一天起，按照实际延迟的天数，每月付给买方货值金额 1.25% 的迟装费。

If the cotton fails to be shipped as scheduled due to the Seller's reasons, the Seller shall pay the Buyer a delayed delivery fee equivalent to 1.25% of the value of the commodity for the delay incurred in the contracted latest shipment date from the eleventh day after the month the cotton was due to be shipped.

由于买方原因造成不能按期装运的，则买方应从合同规定的最晚装运日的第十一天起，按照实际延迟的天数，每月付给卖方货值金额 1.25% 的迟装费。

If the cotton fails to be shipped as scheduled due to the Buyer's reasons, the Buyer shall compensate the Seller carrying charges equivalent to 1.25% of the value of the commodity for the delay incurred in the contracted latest shipment date from the eleventh day after the month the cotton was due to be shipped.

由于买卖双方中任何一方的原因造成超过合同规定的最晚装运日 45 天仍不能装运的，另一方有权解除合同，但违约方仍应承担违约责任。

If the cotton fails to be shipped within 45 days after the contracted latest shipment date as provided by the contract due to the reasons attributable to either Party, the other Party is entitled to terminate the contract and the breaching Party shall be liable for such termination of contract.

10.2 重量差异

10.2 Weight Differences

到岸重量与合同重量的差异在合同允差范围内的，按合同价结算。超出合同允差的多装部分买方有权拒收；超出合同允差的少装部分按其货值金额（按合同价计算）的15%赔偿给买方。

If the difference between the CIQ landed weight and the contract weight is within the contract tolerance weight, payment shall be settled at the contract price. The Buyer is entitled not to accept the part exceeding the contract tolerance weight and the Buyer shall be compensated at 15% of the value (calculated at the Contract price) of the short part.

10.3 质量不符

10.3 Quality Differences

出现下列质量不符行为，根据CIQ提供的有关证明，卖方应做如下赔偿：

The Seller shall make compensations provided as follows for any quality differences set forth below with relevant certificate provided by CIQ:

10.3.1 对合同规定品级、长度与到岸检验品级、长度相差1个级的棉包，则按上述8.1、8.2条款规定的差价率补偿。

10.3.1 In terms of the delivered bales of which the grade and staple for inspection are one grade inferior than the grade and staple provided by the Contract, compensations at the deduction rate provided in the foregoing Article 8.1 and Article 8.2 shall be made.

10.3.2 如果合同规定品级、长度与到岸检验品级、长度相差2个级的棉包数量占该批棉包总数量的比例超过5%的，除按上述8.1、8.2条款规定的差价率补偿外，还应按全部降级棉包合同金额的20%作为违约金赔偿给买方。

10.3.2 If more than 5% of the landed shipment are bales two grades inferior to the contract grade in terms of quality and staple, then in addition to price deduction compensations as in Articles 8.1 and 8.2, the Buyer shall be given extra liquidated damages equivalent to 20% of the contract value of the total defective bales.

10.3.3 对商检后棉花级别低于合同规定级别3个级及以上的棉包、欺诈棉包和混杂棉包，买方可选择：A.退货，卖方除退还全额货款外还要按这些问题棉包合同金额的50%作为违约金赔偿给买方，并承担退货费用；B. 按上述8.1、8.2条款规定的差价率补偿,并按问题棉包合同金额的50%作为违约金赔偿给买方。

10.3.3 In terms of a bale of cotton of which the quality, upon CIQ inspection, is 3 or more than 3 grades inferior to the grade agreed in the Contract, false packed bales and mixed packed bales, the Buyer may choose to: A. return the bales, in which case the Seller shall pay the Buyer liquidated damages equivalent to 50% of the contract value of the problematic bales in addition to the full refund of purchase payment as well as the expenses arising from return of bales; or B. have price deduction compensations as provided in the foregoing Articles 8.1 and 8.2, in which case the Seller shall pay the Buyer extra liquidated damages equivalent to 50% of the contract value of the problematic bales.

11　索赔期限

11　Term of Claim

除非双方另有约定，索赔期不能超过最后到港日或出保税库日后 70 天。有问题棉包的索赔期限按第 12 条款处理。

Unless otherwise agreed to by the parties, claims for noncompliance with the weight or quality provisions shall be made to the other party within 70 calendar days after the last day of landing or its release from a bonded zone. The term of claim of problematic bales shall be dealt with Article 12 below.

12　有问题棉包的处理

12　Disposal of Problematic Bales

买方须在棉花到港后 6 个月之内对欺诈棉包、混杂棉包等问题棉包提出索赔。索赔提出后棉包须另外分开存放 56 天以供卖方核查，逾期视为卖方接受索赔。卖方同意支付买方处置棉包使之恢复使用价值所需的合理费用。

The Buyer shall make claims in terms of false packed bales, mixed packed bales or other problematic bales within 6 months after the cotton arrives at the port. The bales against which the claims are made shall be stored separately for 56 days for the purposes of the Seller's inspection, and the Seller shall be deemed to accept the claims if the Seller delays the said inspection. The Seller agrees to pay the Buyer the reasonable expenses for the disposition of the bales so as to restore the use value of the same.

13　费用标准

13　Rate

13.1　检验费用

13.1　Inspection Expense

13.1.1　重量检验费由卖方负担，每包 50 美分。如果检测回潮率，卖方还应承担 80 美分的样品费和每个样品 2 美元的水分测试费。

13.1.1　The Seller shall bear the weight inspection expense at the rate of USD 50 cents per bale. In terms of moisture regain inspection, the Seller shall bear the sampling expense not approved at the rate of USD 80 cents per sample and the moisture testing expense at the rate of USD$2 per sample, in addition.

13.1.2　质量检验费由买方负担，每包 1 美元。如品级降级时，卖方应负担同样的费率。

13.1.2　The Buyer shall bear the quality inspection expense at the rate of USD1 per bale. In the event of inferior grade, the same rate shall apply to the Seller.

13.2　病虫害熏蒸费用

13.2　Expense for Insects Fumigation

卖方所交之棉花，不得带有国家检疫对象的病虫害。如棉花来自中国政府规定必须做熏蒸的国家或地区，则货到港后 CIQ 根据国家检疫规定作熏蒸处理时，其因熏蒸而产生的各项费用，均由卖方负担。

The cotton delivered by the Seller shall not contain any damage by disease or insect subject to national quarantine. When the CIQ conducts the fumigation in accordance with the national quarantine provisions after the cotton is delivered to pier, the various expenses arising from the fumigation shall be borne by the Seller.

14　合同的终止

14　Termination of the Contract

除非另有规定,本合同在下述任一种情况下终止：

Unless otherwise provided herein, the Contract may be terminated in any of the following events：

A. 通过双方共同书面协议；或

A. The Parties hereto reach an agreement in writing; or

B. 如果另一方完全因其责任在合同规定的时间内未履行其义务，程度严重，并且在收到未违约方的书面通知后 30 天内未能消除违约影响或采取补救措施，在此种情况下，非违约方有权给另一方书面通知来终止合同。

B. In the event a Party fails to perform its obligations to a material or substantial extent within any of the shipping periods provided by the Contract due to the reasons fully attributable to such Party and where such Party fails to eliminate the breach or take any remedial measures within 30 days after the receipt of the written notice from the non-breaching Party, the non-breaching Party shall be entitled to close out the Contract and subsequent thereto provide a written notice to the other Party.

合同终止不影响终止合同方的任何权利，包括但不限于因合同终止而要求损害赔偿的权利。

The termination of the Contract shall not affect any right of the Party taking such action, including, but not limited to, the right of claiming compensation for losses incurred in the termination of the Contract.

15　本合同对任何贸易术语（如 CIF、FOB、CFR）的援引都视为是对 2000 年国际商会贸易术语解释通则（INCOTERMS）的相关术语的援引。

15　Any quotation of trade terms (such as CIF, FOB and CFR) shall be deemed as the quotation of the relevant term under the 2000 International Rules for the Interpretation of Trade Terms (INCOTERMS).

16　本合同应适用《联合国国际货物销售合同公约》（CISG，即 1980 维也纳公约,the 1980 Vienna Convention on Contracts for the International Sale of Goods）。

16　This Contract is subject to the *United Nations Convention on Contracts for the International Sale of Goods* ("CISG", namely 1980 Vienna Convention - the 1980 Vienna Convention on Contracts for the International Sale of Goods).

17　不可抗力

17　Force Majeure

合同当事人因战争及严重的水灾、地震、禁运、罢工、兵变、暴乱或其他该方当事人无法控制，并在签订本合同时不能合理预见、不可避免或无法克服的事件造成其无法履行或迟延履行全部或部分合同义务，则该合同一方当事人应免责。但是，因不可抗力而影响其履约的合同一方应尽快通知另一方事件的发生，并应在事件发生后不迟于 14 天内向另一方发送由事件发生地有关政府、行业协会或当地商会出具的关于发生不可抗力事件的证明或文件。在中国，出具证明的机构为中国国际贸易促进委员会。

If either Party to the Contract is unable to perform or delays the performance of the obligations of the Contract, partially or entirely, due to reasons of war, serious fire, earthquake, embargo, strike, mutiny, riot or any other event that the parties are unable to control and is reasonably unforeseeable, unavoidable and unconquered at the time of signing the Contract, such Party hereto shall not be liable. However, the Party of which the performance is affected by the force majeure event shall notify the other Party at the time of the occurrence of the said event and shall provide evidence of such event in the form of a certificate or document issued by relevant local governmental agencies, or local trade union where the force majeure event occurs for the other Party no later than 14 days after the occurrence of the said event. In China, the institute that issues such certificate shall be the China Council for the Promotion of International Trade.

如果不可抗力事件持续超过[60]天，合同双方可协商合同的履行或终止。如果不可抗力事件发生后[90]天内双方不能达成协议，则任何一方有权终止合同。如果合同如此终止，则任一方应自行承担各自的费用，且不能对与终止合同有关的损失要求赔偿。

If a force majeure event lasts more than [60] days, the Parties hereto may negotiate whether to perform or terminate the Contract. If both Parties fail to reach an agreement within [90] days after the force majeure occurs, either Party may terminate the Contract. If the Contract is terminated, either Party shall bear the expenses on its own and shall not claim any compensations arising from the termination of the Contract against the other Party.

18　仲裁：如果双方选择由中国国际经济贸易仲裁委员会仲裁，则按照申请仲裁时该会现行有效的仲裁规则进行仲裁。

18　Arbitration: If the parties select CIETAC arbitration, the arbitration shall be conducted in accordance with the arbitration rules effective as of the time of application.

19　适用语言：本条款的中英文版本具有同等效力。

19　Applicable language: This contract is written in Chinese and English. The Chinese version and English version have equivalent legal effect.

❸ 中国国际经济贸易仲裁委员会仲裁规则
3 China International Economic and Trade Arbitration Commission (CIETAC) Arbitration Rules

第一章 总 则
Chapter I General Provisions

第一条 仲裁委员会

Article 1 The Arbitration Commission

（一）中国国际经济贸易仲裁委员会（以下简称"仲裁委员会"），原名中国国际贸易促进委员会对外贸易仲裁委员会、中国国际贸易促进委员会对外经济贸易仲裁委员会，同时使用"中国国际商会仲裁院"名称。

1. The China International Economic and Trade Arbitration Commission (hereinafter referred to as "CIETAC"), originally named the Foreign Trade Arbitration Commission of the China Council for the Promotion of International Trade and later renamed the Foreign Economic and Trade Arbitration Commission of the China Council for the Promotion of International Trade, concurrently uses as its name the "Court of Arbitration of the China Chamber of International Commerce".

（二）当事人在仲裁协议中订明由中国国际贸易促进委员会/中国国际商会仲裁，或由中国国际贸易促进委员会/中国国际商会的仲裁委员会或仲裁院仲裁的,或使用仲裁委员会原名称为仲裁机构的，均应视为同意由中国国际经济贸易仲裁委员会仲裁。

2. Where an arbitration agreement provides for arbitration by the China Council for the Promotion of International Trade/China Chamber of International Commerce, or by the Arbitration Commission or the Court of Arbitration of the China Council for the Promotion of International Trade/China Chamber of International Commerce, or refers to CIETAC's previous names, it shall be deemed that the parties have agreed to arbitration by CIETAC.

第二条 机构及职责

Article 2 The Structure and Duties

（一）仲裁委员会主任履行本规则赋予的职责。副主任根据主任的授权可以履行主任的职责。

1. The Chairman of CIETAC shall perform the functions and duties vested in him/her by these Rules while a Vice Chairman may perform the Chairman's

functions and duties with the Chairman's authorization.

（二）仲裁委员会设秘书局，在秘书长的领导下负责处理仲裁委员会的日常事务并履行本规则规定的职责。

2. CIETAC has a Secretariat, which handles its day-to-day work and performs the functions in accordance with these Rules under the direction of the Secretary General.

（三）仲裁委员会设在北京。仲裁委员会在深圳、上海、天津和重庆设有分会或中心。仲裁委员会的分会/中心是仲裁委员会的派出机构,根据仲裁委员会的授权接受仲裁申请并管理仲裁案件。

3. CIETAC is based in Beijing. It has sub-commissions or centers in Shenzhen, Shanghai, Tianjin and Chongqing. The sub-commissions/centers are CIETAC's branches, which accept arbitration applications and administer arbitration cases with CIETAC's authorization.

（四）仲裁委员会分会/中心设秘书处，在仲裁委员会分会/中心秘书长的领导下，负责处理分会/中心的日常事务并履行本规则规定由仲裁委员会秘书局履行的职责。

4. A sub-commission/center has a secretariat, which handles the day-to-day work of the sub-commission/center and performs the functions of the Secretariat of CIETAC in accordance with these Rules under the direction of the secretary general of the sub-commission/center.

（五）案件由分会/中心管理的，本规则规定由仲裁委员会秘书长履行的职责，由仲裁委员会秘书长授权的分会/中心秘书长履行。

5. Where a case is administered by a sub-commission/center, the functions and duties vested in the Secretary General of CIETAC under these Rules may, by his/her authorization, be performed by the secretary general of the relevant sub-commission/center.

（六）当事人可以约定将争议提交仲裁委员会或仲裁委员会分会/中心进行仲裁；约定由仲裁委员会进行仲裁的，由仲裁委员会秘书局接受仲裁申请并管理案件；约定由分会/中心仲裁的，由所约定的分会/中心秘书处接受仲裁申请并管理案件；约定的分会/中心不存在或约定不明的，由仲裁委员会秘书局接受仲裁申请并管理案件。如有争议，由仲裁委员会作出决定。

6. The parties may agree to submit their disputes to CIETAC or a sub-commission/center of CIETAC for arbitration. Where the parties have agreed to

arbitration by CIETAC, the Secretariat of CIETAC shall accept the arbitration application and administer the case. Where the parties have agreed to arbitration by a sub-commission/center, the secretariat of the sub-commission/center agreed upon by the parties shall accept the arbitration application and administer the case. Where the sub-commission/center agreed upon by the parties does not exist, or where the agreement is ambiguous, the Secretariat of CIETAC shall accept the arbitration application and administer the case. In the event of any dispute, a decision shall be made by CIETAC.

（七）仲裁委员会在具备条件的城市和行业设立办事处。办事处是仲裁委员会的派出机构，可以根据仲裁委员会的书面授权从事相关工作。

7. CIETAC sets up liaison offices in cities and industries where appropriate. The liaison offices are branches of CIETAC, and may perform the relevant functions in accordance with the written authorization of CIETAC.

第三条　受案范围

Article 3　Jurisdiction

（一）仲裁委员会根据当事人的约定受理契约性或非契约性的经济贸易等争议案件。

1. CIETAC accepts cases involving economic, trade and other disputes of a contractual or non-contractual nature, based on an agreement of the parties.

（二）前款所述案件包括：

1. 国际或涉外争议案件；
2. 涉及香港特别行政区、澳门特别行政区及台湾地区的争议案件；
3. 国内争议案件。

2. The cases referred to in the preceding paragraph include:

(a) International or foreign-related disputes;

(b) Disputes related to the Hong Kong Special Administrative Region, the Macao Special Administrative Region and the Taiwan region; and

(c) Domestic disputes.

第四条　规则的适用

Article 4　Scope of Application

（一）本规则统一适用于仲裁委员会及其分会/中心。

1. These Rules uniformly apply to CIETAC and its sub-commissions/centers.

（二）当事人约定将争议提交仲裁委员会仲裁的，视为同意按照本规则进行仲裁。

2. The parties shall be deemed to have agreed to arbitration in accordance with these Rules if they have agreed to arbitration by CIETAC.

（三）当事人约定将争议提交仲裁委员会仲裁但对本规则有关内容进行变更或约定适用其他仲裁规则的，从其约定，但其约定无法实施或与仲裁程序适用法强制性规定相抵触者除外。当事人约定适用其他仲裁规则的，由仲裁委员会履行相应的管理职责。

3. Where the parties agree to refer their dispute to CIETAC for arbitration but have agreed on a modification of these Rules or have agreed on the application of other arbitration rules, the parties' agreement shall prevail unless such agreement is inoperative or in conflict with a mandatory provision of the law as it applies to the arbitration proceedings. Where the parties have agreed on the application of other arbitration rules, CIETAC shall perform the relevant administrative duties.

（四）当事人约定按照本规则进行仲裁但未约定仲裁机构的，视为同意将争议提交仲裁委员会仲裁。

4. Where the parties agree to refer their dispute to arbitration under these Rules without providing the name of the arbitration institution, they shall be deemed to have agreed to refer the dispute to arbitration by CIETAC.

（五）当事人约定适用仲裁委员会制定的专业仲裁规则的，从其约定，但其争议不属于该专业仲裁规则适用范围的，适用本规则。

5. Where the parties agree to refer their disputes to arbitration under CIETAC's customized arbitration rules for a specific trade or profession, the parties' agreement shall prevail. However, if the dispute falls outside the scope of the specific rules, these Rules shall apply.

第五条 仲裁协议

Article 5　Arbitration Agreement

（一）仲裁协议系指当事人在合同中订明的仲裁条款或以其他方式达成的提交仲裁的书面协议。

1. An arbitration agreement means an arbitration clause in a contract or any other form of a written agreement concluded between the parties providing for the settlement of disputes by arbitration.

（二）仲裁协议应当采取书面形式。书面形式包括合同书、信件、电报、电传、传真、电子数据交换和电子邮件等可以有形地表现所载内容的形式。在仲裁申请书和仲裁答辩书的交换中，一方当事人声称有仲裁协议而另一方当事人不做否认表示的，视为存在书面仲裁协议。

2. The arbitration agreement shall be in writing. An arbitration agreement is in writing if it is contained in the tangible form of a document such as a contract, letter, telegram, telex, fax, EDI, or email. An arbitration agreement shall be deemed to exist where its existence is asserted by one party and not denied by the other during the exchange of the Request for Arbitration and the Statement of Defense.

（三）仲裁协议的适用法对仲裁协议的形式及效力另有规定的，从其规定。

3. Where the law as it applies to an arbitration agreement has different provisions as to the form and validity of the arbitration agreement, those provisions shall prevail.

（四）合同中的仲裁条款应视为与合同其他条款分离的、独立存在的条款，附属于合同的仲裁协议也应视为与合同其他条款分离的、独立存在的一个部分；合同的变更、解除、终止、转让、失效、无效、未生效、被撤销以及成立与否，均不影响仲裁条款或仲裁协议的效力。

4. An arbitration clause contained in a contract shall be treated as a clause independent and separate from all other clauses of the contract, and an arbitration agreement attached to a contract shall also be treated as independent and separate from all other clauses of the contract. The validity of an arbitration clause or an arbitration agreement shall not be affected by any modification, cancellation, termination, transfer, expiry, invalidity, ineffectiveness, rescission or non-existence of the contract.

第六条 对仲裁协议及/或管辖权的异议

Article 6　Objection to Arbitration Agreement and/or Jurisdiction

（一）仲裁委员会有权对仲裁协议的存在、效力以及仲裁案件的管辖权作出决定。如有必要，仲裁委员会也可以授权仲裁庭作出管辖权决定。

1. CIETAC shall have the power to determine the existence and validity of an arbitration agreement and its jurisdiction over an arbitration case. CIETAC may, where necessary, delegate such power to the arbitral tribunal.

（二）仲裁委员会依表面证据认为存在由其进行仲裁的协议的，可根据表面证据作出仲裁委员会有管辖权的决定，仲裁程序继续进行。仲裁委员会依表面证据作出的管辖权决定并不妨碍其根据仲裁庭在审理过程中发现的与表面证据不一致的事实及/或证据重新作出管辖权决定。

2. Where CIETAC is satisfied by prima facie evidence that an arbitration agreement providing for arbitration by CIETAC exists, it may make a decision based on such evidence that it has jurisdiction over the arbitration case, and the arbitration shall proceed. Such a decision shall not prevent CIETAC from making a new decision on jurisdiction based on facts and/or evidence found by the arbitral tribunal during the arbitration proceedings that are inconsistent with the prima facie evidence.

（三）仲裁庭依据仲裁委员会的授权对管辖权作出决定时，可以在仲裁程序进行中单独作出，也可以在裁决书中一并作出。

3. Where CIETAC has delegated the power to determine jurisdiction to the arbitral tribunal, the arbitral tribunal may either make a separate decision on jurisdiction during the arbitration proceedings or incorporate the decision in the final arbitral award.

（四）当事人对仲裁协议及/或仲裁案件管辖权的异议，应当在仲裁庭首次开庭前书面提出；书面审理的案件，应当在第一次实体答辩前提出。

4. An objection to an arbitration agreement and/or jurisdiction over an arbitration case shall be raised in writing before the first oral hearing is held by the arbitral tribunal. Where a case is to be decided on the basis of documents only, such an objection shall be raised before the submission of the first substantive defense.

（五）对仲裁协议及/或仲裁案件管辖权提出异议不影响仲裁程序的继续进行。

5. The arbitration shall proceed notwithstanding an objection to the arbitration agreement and/or jurisdiction over the arbitration case.

（六）上述管辖权异议及/或决定包括仲裁案件主体资格异议及/或决定。

6. The aforesaid objections to and/or decisions on jurisdiction by CIETAC shall include objections to and/or decisions on a party's standing to participate in the arbitration.

（七）仲裁委员会或经仲裁委员会授权的仲裁庭作出无管辖权决定的，应当作出撤销案件的决定。撤案决定在仲裁庭组成前由仲裁委员会秘书长作出，在仲裁庭组成后，由仲裁庭作出。

7. Where CIETAC or the authorized arbitral tribunal decides that CIETAC has no jurisdiction over an arbitration case, a decision to dismiss the case shall be made. Where a case is to be dismissed before the formation of the arbitral tribunal, the decision shall be made by the Secretary General of CIETAC. Where the case is to

be dismissed after the formation of the arbitral tribunal, the decision shall be made by the arbitral tribunal.

第七条 仲裁地

Article 7　Place of Arbitration

（一）当事人对仲裁地有约定的，从其约定。

1. Where the parties have agreed on the place of arbitration, the parties' agreement shall prevail.

（二）当事人对仲裁地未作约定或约定不明的，以管理案件的仲裁委员会或其分会/中心所在地为仲裁地；仲裁委员会也可视案件的具体情形确定其他地点为仲裁地。

2. Where the parties have not agreed on the place of arbitration or their agreement is ambiguous, the place of arbitration shall be the domicile of CIETAC or its sub-commission/center administering the case. CIETAC may also determine the place of arbitration to be another location having regard to the circumstances of the case.

（三）仲裁裁决视为在仲裁地作出。

3. The arbitral award shall be deemed as having been made at the place of arbitration.

第八条 送达及期限

Article 8　Service of Documents and Periods of Time

（一）有关仲裁的一切文书、通知、材料等均可采用当面递交、挂号信、特快专递、传真或仲裁委员会秘书局或仲裁庭认为适当的其他方式发送。

1. All documents, notices and written materials in relation to the arbitration may be delivered in person or sent by registered mail or express mail, fax, or by any other means considered proper by the Secretariat of CIETAC or the arbitral tribunal.

（二）上述第（一）款所述仲裁文件应发送当事人或其仲裁代理人自行提供的或当事人约定的地址；当事人或其仲裁代理人没有提供地址或当事人对地址没有约定的，按照对方当事人或其仲裁代理人提供的地址发送。

2. The arbitration documents referred to in the preceding Paragraph 1 shall be sent to the address provided by the party itself or by its representative(s), or to an address agreed by the parties. Where a party or its representative(s) has not provided an address or the parties have not agreed on an address, the arbitration

documents shall be sent to such party's address as provided by the other party or its representative(s).

（三）向一方当事人或其仲裁代理人发送的仲裁文件，如经当面递交收件人或发送至收件人的营业地、注册地、住所地、惯常居住地或通讯地址，或经对方当事人合理查询不能找到上述任一地点，仲裁委员会秘书局以挂号信或特快专递或能提供投递记录的其他任何手段投递给收件人最后一个为人所知的营业地、注册地、住所地、惯常居住地或通讯地址，即视为有效送达。

3. Any arbitration correspondence to a party or its representative(s) shall be deemed to have been properly served on the party if delivered to the addressee or sent to the addressee's place of business, registration, domicile, habitual residence or mailing address, or where, after reasonable inquiries by the other party, none of the aforesaid addresses can be found, the arbitration correspondence is sent by the Secretariat of CIETAC to the addressee's last known place of business, registration, domicile, habitual residence or mailing address by registered or express mail, or by any other means that can provide a record of the attempt at delivery.

（四）本规则所规定的期限，应自当事人收到或应当收到仲裁委员会秘书局向其发送的文书、通知、材料等之日的次日起计算。

4. The periods of time specified in these Rules shall begin on the day following the day when the party receives or should have received the arbitration orrespondence, notices or written materials sent by the Secretariat of CIETAC.

第九条　诚信合作

Article 9　Bona Fide Cooperation

当事人及其仲裁代理人应当诚信合作，进行仲裁程序。

The parties and their representatives shall proceed with the arbitration in bona fide cooperation.

第十条　放弃异议

Article 10　Waiver of Right to Object

一方当事人知道或理应知道本规则或仲裁协议中规定的任何条款或情事未被遵守，仍参加仲裁程序或继续进行仲裁程序而且不对此不遵守情况及时地、明示地提出书面异议的，视为放弃其提出异议的权利。

A party shall be deemed to have waived its right to object where it knows or should have known that any provision of, or requirement under, these Rules has not been complied with and yet participates in or proceeds with the arbitration proceedings without promptly and explicitly submitting its objection in writing to such non-compliance.

第二章 仲裁程序
Chapter II Arbitration Proceedings

第一节 仲裁申请、答辩、反请求
Section 1 Request for Arbitration, Defense and Counterclaim

第十一条 仲裁程序的开始

Article 11 Commencement of Arbitration

仲裁程序自仲裁委员会秘书局收到仲裁申请书之日起开始。

The arbitration proceedings shall commence on the day on which the Secretariat of CIETAC receives a Request for Arbitration.

第十二条 申请仲裁

Article 12 Application for Arbitration

当事人依据本规则申请仲裁时应：

（一）提交由申请人或申请人授权的代理人签名及/或盖章的仲裁申请书。仲裁申请书应写明。

1. 申请人和被申请人的名称和住所，包括邮政编码、电话、传真、电子邮件或其他电子通讯方式；

2. 申请仲裁所依据的仲裁协议；

3. 案情和争议要点；

4. 申请人的仲裁请求；

5. 仲裁请求所依据的事实和理由。

A party applying for arbitration under these Rules shall:

1. Submit a Request for Arbitration in writing signed and/or sealed by the Claimant or its authorized representative(s), which shall, inter alia, include:

(a) the names and addresses of the Claimant and the Respondent, including the zip code, telephone, fax, email, or any other means of electronic elecommunications;

(b) a reference to the arbitration agreement that is invoked;

(c) a statement of the facts of the case and the main issues in dispute;

(d) the claim of the Claimant;

(e) the facts and grounds on which the claim is based.

（二）在提交仲裁申请书时，附具申请人请求所依据的证据材料以及其他证明文件。

2. Attach to the Request for Arbitration the relevant documentary and other evidence on which the Claimant's claim is based.

（三）按照仲裁委员会制定的仲裁费用表的规定预缴仲裁费。

3. Pay the arbitration fee in advance to CIETAC according to its Arbitration Fee Schedule.

第十三条　案件的受理

Article 13　Acceptance of a Case

（一）仲裁委员会根据当事人在争议发生之前或在争议发生之后达成的将争议提交仲裁委员会仲裁的仲裁协议和一方当事人的书面申请，受理案件。

1. Upon a written application of a party, CIETAC shall accept a case in accordance with an arbitration agreement concluded between the parties either before or after the occurrence of the dispute, in which it is provided that disputes are to be referred to arbitration by CIETAC.

（二）仲裁委员会秘书局收到申请人的仲裁申请书及其附件后，经审查，认为申请仲裁的手续完备的，应将仲裁通知、仲裁委员会仲裁规则和仲裁员名册各一份发送给双方当事人；申请人的仲裁申请书及其附件也应同时发送给被申请人。

2. Upon receipt of a Request for Arbitration and its attachments, if after examination, the Secretariat of CIETAC finds the formalities required for arbitration application to be complete, it shall send a Notice of Arbitration to both parties together with one copy each of these Rules and CIETAC's Panel of Arbitrators. The Request for Arbitration and its attachments submitted by the Claimant shall be sent to the Respondent under the same cover.

（三）仲裁委员会秘书局经审查认为申请仲裁的手续不完备的，可以要求申请人在一定的期限内予以完备。申请人未能在规定期限内完备申请仲裁手续的，视同申请人未提出仲裁申请；申请人的仲裁申请书及其附件，仲裁委员会秘书局不予留存。

3. Where after examination the Secretariat of CIETAC finds the formalities required for the arbitration application to be incomplete, it may request the Claimant to complete them within a specified time period. The Claimant shall be deemed not to have submitted a Request for Arbitration if it fails to complete the

required formalities within the specified time period. In such a case, the Claimant's Request for Arbitration and its attachments shall not be kept on file by the Secretariat of CIETAC.

（四）仲裁委员会受理案件后，秘书局应指定一名案件秘书协助仲裁案件的程序管理工作。

4. After CIETAC accepts a case, its Secretariat shall designate a Case Manager to assist with the procedural administration of the case.

第十四条　答辩
Article 14　Statement of Defense

（一）被申请人应自收到仲裁通知后 45 天内提交答辩书。被申请人确有正当理由请求延长提交答辩期限的，由仲裁庭决定是否延长答辩期限；仲裁庭尚未组成的，由仲裁委员会秘书局作出决定。

1. The Respondent shall file a Statement of Defense in writing within forty-five (45) days from the date of receipt of the Notice of Arbitration. If the Respondent has justified reasons to request an extension of the time period, the arbitral tribunal shall decide whether to grant an extension. Where the arbitral tribunal has not yet been formed, the decision on whether to grant the extension of the time period shall be made by the Secretariat of CIETAC.

（二）答辩书由被申请人或被申请人授权的代理人签名及/或盖章，并应包括下列内容及附件：

1. 被申请人的名称和住所，包括邮政编码、电话、传真、电子邮件或其他电子通讯方式；
2. 对仲裁申请书的答辩及所依据的事实和理由；
3. 答辩所依据的证据材料以及其他证明文件。

2. The Statement of Defense shall be signed and/or sealed by the Respondent or its authorized representative(s), and shall, inter alia, include the following contents and attachments:

(a) the name and address of the Respondent, including the zip code, telephone, fax, email, or any other means of electronic telecommunications;

(b) the defense to the Request for Arbitration setting forth the facts and grounds on which the defense is based;

(c) the relevant documentary and other evidence on which the defense is based.

（三）仲裁庭有权决定是否接受逾期提交的答辩书。

3. The arbitral tribunal has the power to decide whether to accept a Statement of Defense submitted after the expiration of the above time limit.

（四）被申请人未提交答辩书，不影响仲裁程序的进行。

4. Failure by the Respondent to file a Statement of Defense shall not affect the conduct of the arbitration proceedings.

第十五条 反请求

Article 15 Counterclaim

（一）被申请人如有反请求，应自收到仲裁通知后 45 天内以书面形式提交。被申请人确有正当理由请求延长提交反请求期限的，由仲裁庭决定是否延长反请求期限；仲裁庭尚未组成的，由仲裁委员会秘书局作出决定。

1. The Respondent shall file a counterclaim, if any, in writing within forty-five (45) days from the date of receipt of the Notice of Arbitration. If the Respondent has justified reasons to request an extension of the time period, the arbitral tribunal shall decide whether to grant an extension. Where the arbitral tribunal has not yet been formed, the decision on whether to grant the extension of the time period shall be made by the Secretariat of CIETAC.

（二）被申请人提出反请求时，应在其反请求申请书中写明具体的反请求事项及其所依据的事实和理由，并附具有关的证据材料以及其他证明文件。

2. When filing its counterclaim, the Respondent shall specify its counterclaim in its Statement of Counterclaim and state the facts and grounds on which its counterclaim is based with the relevant documentary and other evidence attached thereto.

（三）被申请人提出反请求，应按照仲裁委员会制定的仲裁费用表在规定的时间内预缴仲裁费。被申请人未按期缴纳反请求仲裁费的，视同未提出反请求申请。

3. When filing its counterclaim, the Respondent shall pay an arbitration fee in advance according to the Arbitration Fee Schedule of CIETAC within a specified time period, failing which the Respondent shall be deemed not to have filed any counterclaim.

（四）仲裁委员会秘书局认为被申请人提出反请求的手续已完备的，应向双方当事人发出反请求受理通知。申请人应在收到反请求受理通知后 30 天内对被申请人的反请求提交答辩。申请人确有正当理由请求延长提交答辩期限的，由仲裁庭决定是否延长答辩期限；仲裁庭尚未组成的，由仲裁委员会秘书局作出决定。

4. Where the formalities required for filing a counterclaim are found to be complete, the Secretariat of CIETAC shall send a Notice of Acceptance of Counterclaim to the parties. The Claimant shall submit its Statement of Defense in writing within thirty (30) days from the date of receipt of the Notice. If the Claimant has justified reasons to request an extension of the time period, the arbitral tribunal shall decide whether to grant such an extension. Where the arbitral tribunal has not yet been formed, the decision on whether to grant the extension of the time period shall be made by the Secretariat of CIETAC.

（五）仲裁庭有权决定是否接受逾期提交的反请求答辩书。

5. The arbitral tribunal has the power to decide whether to accept a Statement of Defense submitted after the expiration of the above time limit.

（六）申请人对被申请人的反请求未提出书面答辩的，不影响仲裁程序的进行。

6. Failure of the Claimant to file a Statement of Defense to the Respondent's counterclaim shall not affect the conduct of the arbitration proceedings.

第十六条　变更仲裁请求或反请求

Article 16　Amendment to the Claim or Counterclaim

申请人可以申请对其仲裁请求进行更改，被申请人也可以申请对其反请求进行更改；但是仲裁庭认为其提出更改的时间过迟而影响仲裁程序正常进行的，可以拒绝其更改请求。

The Claimant may apply to amend its claim and the Respondent may apply to amend its counterclaim. However, the arbitral tribunal may not permit any such amendment if it considers that the amendment is too late and may delay the arbitration proceedings.

第十七条　合并仲裁

Article 17　Consolidation of Arbitrations

（一）经一方当事人请求并经其他各方当事人同意，或仲裁委员会认为必要并经各方当事人同意，仲裁委员会可以决定将根据本规则进行的两个或两个以上的仲裁案件合并为一个仲裁案件，进行审理。

1. At the request of a party and with the agreement of all the other parties, or where CIETAC believes it necessary and all the parties have agreed, CIETAC may

consolidate two or more arbitrations pending under these Rules into a single arbitration.

（二）根据上述第（一）款决定合并仲裁时，仲裁委员会应考虑相关仲裁案件之间的关联性，包括不同仲裁案件的请求是否依据同一仲裁协议提出，不同仲裁案件的当事人是否相同，以及不同案件的仲裁员的选定或指定情况。

2. In deciding whether to consolidate the arbitrations in accordance with the preceding Paragraph 1, CIETAC may take into account any factors it considers relevant in respect of the different arbitrations, including whether all of the claims in the different arbitrations are made under the same arbitration agreement, whether the different arbitrations are between the same parties, or whether one or more arbitrators have been nominated or appointed in the different arbitrations.

（三）除非各方当事人另有约定，合并的仲裁案件应合并于最先开始仲裁程序的仲裁案件。

3. Unless otherwise agreed by all the parties, the arbitrations shall be consolidated into the arbitration that was first commenced.

第十八条 仲裁文件的提交与交换

Article 18　Submissions and Exchange of Arbitration Documents

（一）当事人的仲裁文件应提交至仲裁委员会秘书局。

1. All arbitration documents from the parties shall be submitted to the Secretariat of CIETAC.

（二）仲裁程序中需发送或转交的仲裁文件，由仲裁委员会秘书局发送或转交仲裁庭及当事人，当事人另有约定并经仲裁庭同意或仲裁庭另有决定者除外。

2. All arbitration documents to be exchanged during the arbitration proceedings shall be exchanged among the arbitral tribunal and the parties by the Secretariat of CIETAC unless otherwise agreed by the parties and with the consent of the arbitral tribunal or otherwise decided by the arbitral tribunal.

第十九条 仲裁文件的份数

Article 19　Copies of Submissions

当事人提交的仲裁申请书、答辩书、反请求书和证据材料以及其他仲裁文件，应一式五份；多方当事人的案件，应增加相应份数；当事人提出财产保全申请或证据保全申请的，应增加相应份数；仲裁庭组成人数为一人的，应相应减少两份。

When submitting the Request for Arbitration, the Statement of Defense, the Statement of Counterclaim, evidence, and other arbitration documents, the parties shall make their submissions in quintuplicate. Where there are multiple parties, additional copies shall be provided accordingly. Where preservation of property or protection of evidence is applied for, the party shall also provide additional copies accordingly. Where the arbitral tribunal is composed of a sole arbitrator, the number of copies submitted may be reduced by two.

第二十条 仲裁代理人

Article 20 Representation

当事人可以授权中国及/或外国的仲裁代理人办理有关仲裁事项。当事人或其仲裁代理人应向仲裁委员会秘书局提交授权委托书。

A party may be represented by its authorized Chinese and/or foreign representative(s) in handling matters relating to the arbitration. In such a case, a Power of Attorney shall be forwarded to the Secretariat of CIETAC by the party or its authorized representative(s).

第二十一条 保全及临时措施

Article 21 Conservatory and Interim Measures

（一）当事人依据中国法律规定申请保全的，仲裁委员会秘书局应当依法将当事人的保全申请转交当事人指明的有管辖权的法院。

1. Where a party applies for conservatory measures pursuant to the laws of the People's Republic of China, the secretariat of CIETAC shall forward the party's application to the competent court designated by that party in accordance with the law.

（二）经一方当事人请求，仲裁庭依据所适用的法律可以决定采取其认为必要或适当的临时措施，并有权决定请求临时措施的一方提供适当的担保。仲裁庭采取临时措施的决定，可以程序令或中间裁决的方式作出。

2. At the request of a party, the arbitral tribunal may order any interim measure it deems necessary or proper in accordance with the applicable law, and may require the requesting party to provide appropriate security in connection with the measure. The order of an interim measure by the arbitral tribunal may take the form of a procedural order or an interlocutory award.

第二节 仲裁员及仲裁庭

Section 2 Arbitrators and the Arbitral Tribunal

第二十二条 仲裁员的义务

Article 22 Duties of Arbitrator

仲裁员不代表任何一方当事人,应独立于各方当事人,平等地对待各方当事人。

An arbitrator shall not represent either party, and shall be and remain independent of the parties and treat them equally.

第二十三条 仲裁庭的人数

Article 23 Number of Arbitrators

(一)仲裁庭由一名或三名仲裁员组成。

1. The arbitral tribunal shall be composed of one or three arbitrators.

(二)除非当事人另有约定或本规则另有规定,仲裁庭由三名仲裁员组成。

2. Unless otherwise agreed by the parties or provided by these Rules, the arbitral tribunal shall be composed of three arbitrators.

第二十四条 仲裁员的选定或指定

Article 24 Nomination or Appointment of Arbitrator

(一)仲裁委员会制定统一适用于仲裁委员会及其分会/中心的仲裁员名册;当事人从仲裁委员会制定的仲裁员名册中选定仲裁员。

1. CIETAC establishes a Panel of Arbitrators which uniformly applies to itself and all its sub-commissions/centers. The parties shall nominate arbitrators from the Panel of Arbitrators provided by CIETAC.

(二)当事人约定在仲裁委员会仲裁员名册之外选定仲裁员的,当事人选定的或根据当事人之间的协议指定的人士经仲裁委员会主任依法确认后可以担任仲裁员。

2. Where the parties have agreed to nominate arbitrators from outside CIETAC's Panel of Arbitrators, an arbitrator so nominated by the parties or nominated according to the agreement of the parties may act as arbitrator subject to the confirmation by the Chairman of CIETAC in accordance with the law.

第二十五条 三人仲裁庭的组成

Article 25 Three-Arbitrator Tribunal

(一)申请人和被申请人应各自在收到仲裁通知后 15 天内选定或委托仲裁委员会主任指定一名仲裁员。当事人未在上述期限内选定或委托仲裁委员会主任指定的,由仲裁委员会主任指定。

1. Within fifteen (15) days from the date of receipt of the Notice of Arbitration, the Claimant and the Respondent shall each nominate, or entrust the Chairman of CIETAC to appoint, an arbitrator, failing which the arbitrator shall be appointed by

the Chairman of CIETAC.

（二）第三名仲裁员由双方当事人在被申请人收到仲裁通知后 15 天内共同选定或共同委托仲裁委员会主任指定。第三名仲裁员为仲裁庭的首席仲裁员。

2. Within fifteen (15) days from the date of the Respondent's receipt of the Notice of Arbitration, the parties shall jointly nominate, or entrust the Chairman of CIETAC to appoint, the third arbitrator, who shall act as the presiding arbitrator.

（三）双方当事人可以各自推荐一至五名候选人作为首席仲裁员人选，并按照上述第（二）款规定的期限提交推荐名单。双方当事人的推荐名单中有一名人选相同的，该人选为双方当事人共同选定的首席仲裁员；有一名以上人选相同的，由仲裁委员会主任根据案件的具体情况在相同人选中确定一名首席仲裁员，该名首席仲裁员仍为双方共同选定的首席仲裁员；推荐名单中没有相同人选时，由仲裁委员会主任指定首席仲裁员。

3. The parties may each recommend one to five arbitrators as candidates for presiding arbitrator and shall each submit a list of recommended candidates within the time period specified in the preceding Paragraph 2. Where there is only one common candidate on the lists, such candidate shall be the presiding arbitrator jointly nominated by the parties. Where there is more than one common candidate on the lists, the Chairman of CIETAC shall choose a presiding arbitrator from among the common candidates having regard to the circumstances of the case, and he/she shall act as the presiding arbitrator jointly nominated by the parties. Where there is no common candidate on the lists, the presiding arbitrator shall be appointed by the Chairman of CIETAC.

（四）双方当事人未能按照上述规定共同选定首席仲裁员的，由仲裁委员会主任指定首席仲裁员。

4. Where the parties have failed to jointly nominate the presiding arbitrator according to the above provisions, the presiding arbitrator shall be appointed by the Chairman of CIETAC.

第二十六条　独任仲裁庭的组成

Article 26 Sole-Arbitrator Tribunal

仲裁庭由一名仲裁员组成的，按照本规则第二十五条第（二）、（三）、（四）款规定的程序，选定或指定该独任仲裁员。

Where the arbitral tribunal is composed of one arbitrator, the sole arbitrator shall be nominated pursuant to the procedures stipulated in Paragraphs 2, 3 and 4 of

Article 25 of these Rules.

第二十七条　多方当事人仲裁庭的组成

Article 27　Multiple-Party Tribunal

（一）仲裁案件有两个或两个以上申请人及/或被申请人时,申请人方及/或被申请人方应各自协商,各方共同选定或共同委托仲裁委员会主任指定一名仲裁员。

1. Where there are two or more Claimants and/or Respondents in an arbitration case, the Claimant side and/or the Respondent side, following consultations, shall each jointly nominate or jointly entrust the Chairman of CIETAC to appoint one arbitrator.

（二）首席仲裁员或独任仲裁员应按照本规则第二十五条第（二）、（三）、（四）款规定的程序选定或指定。申请人方及/或被申请人方按照本规则第二十五条第（三）款的规定选定首席仲裁员或独任仲裁员时,应各方共同协商,并提交各方共同选定的候选人名单。

2. The presiding arbitrator or the sole arbitrator shall be nominated in accordance with the procedures stipulated in Paragraphs 2, 3 and 4 of Article 25 of these Rules. When making such nomination pursuant to Paragraph 3 of Article 25 of these Rules, the Claimant side and/or the Respondent side, following consultations, shall each submit a list of their jointly agreed candidates.

（三）如果申请人方及/或被申请人方未能在收到仲裁通知后15天内各方共同选定或各方共同委托仲裁委员会主任指定一名仲裁员,则由仲裁委员会主任指定仲裁庭三名仲裁员,并从中确定一人担任首席仲裁员。

3. Where either the Claimant side or the Respondent side fails to jointly nominate or jointly entrust the Chairman of CIETAC with appointing one arbitrator within fifteen (15) days from the date of receipt of the Notice of Arbitration, the Chairman of CIETAC shall appoint all three members of the arbitral tribunal and designate one of them to act as the presiding arbitrator.

第二十八条　指定仲裁员的考虑因素

Article 28　Considerations in Appointing Arbitrators

仲裁委员会主任根据本规则的规定指定仲裁员时,应考虑争议的适用法律、仲裁地、仲裁语言、当事人国籍,以及仲裁委员会主任认为应考虑的其他因素。

When appointing arbitrators pursuant to these Rules, the Chairman of CIETAC shall take into consideration the law as it applies to the dispute, the place of arbitration, the language of arbitration, the nationalities of the parties, and any other factor(s) the Chairman considers relevant.

第二十九条 披露

Article 29 Disclosure

(一)被选定或被指定的仲裁员应签署声明书,披露可能引起对其公正性和独立性产生合理怀疑的任何事实或情况。

1. An arbitrator nominated by the parties or appointed by the Chairman of CIETAC shall sign a Declaration and disclose any facts or circumstances likely to give rise to justifiable doubts as to his/her impartiality or independence.

(二)在仲裁程序中出现应披露情形的,仲裁员应立即书面披露。

2. If circumstances that need to be disclosed arise during the arbitration proceedings, the arbitrator shall promptly disclose such circumstances in writing.

(三)仲裁员的声明书及/或披露的信息应提交仲裁委员会秘书局并由其转交各方当事人。

3. The Declaration and/or the disclosure of the arbitrator shall be submitted to the Secretariat of CIETAC and communicated to the parties by the Secretariat of CIETAC.

第三十条 仲裁员的回避

Article 30 Challenge to the Arbitrator

(一)当事人收到仲裁员的声明书及/或书面披露后,如果以仲裁员披露的事实或情况为理由要求该仲裁员回避,则应于收到仲裁员的书面披露后10天内书面提出。逾期没有申请回避的,不得以仲裁员曾经披露的事项为由申请该仲裁员回避。

1. Upon receipt of the Declaration and/or the written disclosure of an arbitrator, a party which intends to challenge the arbitrator on the grounds of the facts or circumstances disclosed by the arbitrator shall forward the challenge in writing within ten (10) days from the date of such receipt. If a party fails to file a challenge within the above time period, it may not subsequently challenge the arbitrator on the basis of the matters disclosed by the arbitrator.

(二)当事人对被选定或被指定的仲裁员的公正性和独立性产生具有正当理由的怀疑时,可以书面提出要求该仲裁员回避的请求,但应说明提出回避请求所依据的具体事实和理由,并举证。

2. A party which has justifiable doubts as to the impartiality or independence of an arbitrator may challenge that arbitrator in writing and shall state the facts and reasons on which the challenge is based with supporting evidence.

（三）对仲裁员的回避请求应在收到组庭通知后15天内以书面形式提出；在此之后得知要求回避事由的，可以在得知回避事由后15天内提出，但应不晚于最后一次开庭终结。

3. A party may challenge an arbitrator in writing within fifteen (15) days from the date it receives the Notice of Formation of the Arbitral tribunal. Where a party becomes aware of a reason for a challenge after such receipt, the party may challenge the arbitrator in writing within fifteen (15) days after such reason has become known, but no later than the conclusion of the last oral hearing.

（四）当事人的回避请求应当立即转交另一方当事人、被请求回避的仲裁员及仲裁庭其他成员。

4. The challenge by one party shall be promptly communicated to the other party, the arbitrator being challenged and the other members of the arbitral tribunal.

（五）如果一方当事人请求仲裁员回避，另一方当事人同意回避请求，或被请求回避的仲裁员主动提出不再担任该仲裁案件的仲裁员，则该仲裁员不再担任仲裁员审理本案。上述情形并不表示当事人提出回避的理由成立。

5. Where an arbitrator is challenged by one party and the other party agrees to the challenge, or the arbitrator being challenged voluntarily withdraws from his/her office, such arbitrator shall no longer be a member of the arbitral tribunal. However, in neither case shall it be implied that the reasons for the challenge are sustained.

（六）除上述第（五）款规定的情形外，仲裁员是否回避，由仲裁委员会主任作出终局决定并可以不说明理由。

6. In circumstances other than those specified in the preceding Paragraph 5, the Chairman of CIETAC shall make a final decision on the challenge with or without stating the reasons.

（七）在仲裁委员会主任就仲裁员是否回避作出决定前，被请求回避的仲裁员应继续履行职责。

7. An arbitrator who has been challenged shall continue to serve on the arbitral tribunal until a final decision on the challenge has been made by the Chairman of CIETAC.

第三十一条 仲裁员的更换
Article 31 Replacement of Arbitrator

（一）仲裁员在法律上或事实上不能履行其职责，或没有按照本规则的要求或在本规则规定的期限内履行应尽职责时，仲裁委员会主任有权决定将其更换；该仲裁员也可以主动申请不再担任仲裁员。

1. In the event that an arbitrator is prevented de jure or de facto from fulfilling his/her functions, or fails to fulfill his/her functions in accordance with the requirements of these Rules or within the time period specified in these Rules, the Chairman of CIETAC shall have the power to decide to replace the arbitrator. Such arbitrator may also voluntarily withdraw from his/her office.

（二）是否更换仲裁员，由仲裁委员会主任作出终局决定并可以不说明理由。

2. The Chairman of CIETAC shall make a final decision on whether or not an arbitrator should be replaced with or without stating the reasons.

（三）仲裁员因回避或更换不能履行职责时，应按照原选定或指定该仲裁员的方式和期限，选定或指定替代的仲裁员。当事人未按照原方式和期限选定替代仲裁员的，由仲裁委员会主任指定替代的仲裁员。

3. In the event that an arbitrator is unable to fulfill his/her functions due to being challenged or replaced, a substitute arbitrator shall be nominated according to the same procedure and time period that applied to the nomination of the arbitrator being challenged or replaced. If a party fails to nominate a substitute arbitrate or accordingly, the substitute arbitrator shall be appointed by the Chairman of CIETAC.

（四）重新选定或指定仲裁员后，由仲裁庭决定是否重新审理及重新审理的范围。

4. After the replacement of an arbitrator, the arbitral tribunal shall decide whether and to what extent the previous proceedings in the case shall be repeated.

第三十二条 多数仲裁员继续仲裁程序
Article 32 Majority Continuation of Arbitration

最后一次开庭终结后，如果三人仲裁庭中的一名仲裁员因死亡或被除名等情形而不能参加合议及/或作出裁决，另外两名仲裁员可以请求仲裁委员会主任按照第三十一条的规定更换该仲裁员；在征求双方当事人意见并经仲裁委员会主任同意后，该两名仲裁员也可以继续进行仲裁程序，作出决定或裁决。仲裁委员会秘书局应将上述情况通知双方当事人。

After the conclusion of the last oral hearing, if an arbitrator on a three-member tribunal is unable to participate in the deliberations and/or to render the award owing to his/her demise or to his/her removal from CIETAC's Panel of Arbitrators, or for any other reason, the other two arbitrators may request the Chairman of CIETAC to replace that arbitrator pursuant to Article 31 of these Rules. After consulting with the parties and upon the approval of the Chairman of CIETAC, the other two arbitrators may also continue the arbitration proceedings and make decisions, rulings, or render the award. The Secretariat of CIETAC shall notify the parties of the above circumstances.

第三节　审　理
Section 3　Hearing

第三十三条　审理方式

Article 33　Conduct of Hearing

（一）除非当事人另有约定，仲裁庭可以按照其认为适当的方式审理案件。在任何情形下，仲裁庭均应公平和公正地行事，给予双方当事人陈述与辩论的合理机会。

1. The arbitral tribunal shall examine the case in any way it deems appropriate unless otherwise agreed by the parties. Under all circumstances, the arbitral tribunal shall act impartially and fairly and shall afford a reasonable opportunity to all parties to make submissions and arguments.

（二）仲裁庭应开庭审理案件，但双方当事人约定并经仲裁庭同意或仲裁庭认为不必开庭审理并征得双方当事人同意的，可以只依据书面文件进行审理。

2. The arbitral tribunal shall hold oral hearings when examining the case. However, the arbitral tribunal may examine the case on the basis of documents only if the parties so agree and the arbitral tribunal consents or the arbitral tribunal deems that oral hearings are unnecessary and the parties so agree.

（三）除非当事人另有约定，仲裁庭可以根据案件的具体情况采用询问式或辩论式审理案件。

3. Unless otherwise agreed by the parties, the arbitral tribunal may adopt an inquisitorial or adversarial approach in hearing the case having regard to the circumstances of the case.

（四）仲裁庭可以在其认为适当的地点以其认为适当的方式进行合议。

4. The arbitral tribunal may hold deliberations at any place or in any manner that it considers appropriate.

（五）除非当事人另有约定，仲裁庭认为必要时可以发布程序令、发出问题单、制作审理范围书、举行庭前会议等。

5. Unless otherwise agreed by the parties, the arbitral tribunal may, if it considers it necessary, issue procedural orders or question lists, produce terms of reference, or hold pre-hearing conferences, etc.

第三十四条 开庭地

Article 34 Place of Oral Hearing

（一）当事人约定了开庭地点的，仲裁案件的开庭审理应当在约定的地点进行，但出现本规则第七十二条第（三）款规定的情形的除外。

1. Where the parties have agreed on the place of an oral hearing, the case shall be heard at that agreed place except in the circumstances stipulated in Paragraph 3 of Article 72 of these Rules.

（二）除非当事人另有约定，由仲裁委员会秘书局或其分会/中心秘书处管理的案件应分别在北京或分会/中心所在地开庭审理；如仲裁庭认为必要，经仲裁委员会秘书长同意，也可以在其他地点开庭审理。

2. Unless otherwise agreed by the parties, the place of oral hearings shall be in Beijing for a case administered by the Secretariat of CIETAC or at the domicile of the sub-commission/center which administers the case, or if the arbitral tribunal considers it necessary and with the approval of the Secretary-General of CIETAC, at another location.

第三十五条 开庭通知

Article 35 Notice of Oral Hearing

（一）开庭审理的案件，仲裁庭确定第一次开庭日期后，应不晚于开庭前20天将开庭日期通知双方当事人。当事人有正当理由的，可以请求延期开庭，但应于收到开庭通知后5天内提出书面延期申请；是否延期，由仲裁庭决定。

1. Where a case is to be examined by way of an oral hearing, the parties shall be notified of the date of the first oral hearing at least twenty (20) days in advance of the oral hearing. A party having justified reasons may request a postponement of the oral hearing. However, such request must be communicated in writing to the arbitral tribunal within five (5) days of the receipt of the notice of the oral hearing. The arbitral tribunal shall decide whether or not to postpone the oral hearing.

(二)当事人有正当理由未能按上述第(一)款规定的期限提出延期开庭申请的,是否接受其延期申请,由仲裁庭决定。

2. Where a party has justified reasons for failure to submit a request for a postponement of the oral hearing within the time period specified in the preceding Paragraph 1, the arbitral tribunal shall decide whether or not to accept the request.

(三)再次开庭审理的日期及延期后开庭审理日期的通知及其延期申请,不受上述第(一)款中期限的限制。

3. A notice of a subsequent oral hearing, a notice of a postponed oral hearing, as well as a request for postponement of such an oral hearing shall not be subject to the time periods specified in the preceding Paragraph 1.

第三十六条 保密

Article 36 Confidentiality

(一)仲裁庭审理案件不公开进行。双方当事人要求公开审理的,由仲裁庭决定是否公开审理。

1. Hearings shall be held in camera. Where both parties request an open hearing, the arbitral tribunal shall make a decision.

(二)不公开审理的案件,双方当事人及其仲裁代理人、仲裁员、证人、翻译、仲裁庭咨询的专家和指定的鉴定人,以及其他有关人员,均不得对外界透露案件实体和程序的有关情况。

2. For cases heard in camera, the parties and their representatives, the arbitrators, the witnesses, the interpreters, the experts consulted by the arbitral tribunal, the appraisers appointed by the arbitral tribunal and other relevant persons shall not disclose to any outsider any substantive or procedural matters relating to the case.

第三十七条 当事人缺席

Article 37 Default

(一)申请人无正当理由开庭时不到庭的,或在开庭审理时未经仲裁庭许可中途退庭的,可以视为撤回仲裁申请;被申请人提出反请求的,不影响仲裁庭就反请求进行审理,并作出裁决。

1. If the Claimant fails to appear at an oral hearing without showing sufficient cause, or withdraws from an on-going oral hearing without the permission of the arbitral tribunal, the Claimant may be deemed to have withdrawn its Request for Arbitration. In such a case, if the Respondent has filed a counterclaim, the arbitral

tribunal shall proceed with the hearing of the counterclaim and make a default award.

（二）被申请人无正当理由开庭时不到庭的，或在开庭审理时未经仲裁庭许可中途退庭的，仲裁庭可以进行缺席审理并作出裁决；被申请人提出反请求的，可以视为撤回反请求。

2. If the Respondent fails to appear at an oral hearing without showing sufficient cause, or withdraws from an on-going oral hearing without the permission of the arbitral tribunal, the arbitral tribunal may proceed with the arbitration and make a default award. In such a case, if the Respondent has filed a counterclaim, the Respondent may be deemed to have withdrawn its counterclaim.

第三十八条　庭审笔录

Article 38　Record of Oral Hearing

（一）开庭审理时，仲裁庭可以制作庭审笔录及/或影音记录。仲裁庭认为必要时，可以制作庭审要点，并要求当事人及/或其代理人、证人及/或其他有关人员在庭审笔录或庭审要点上签字或盖章。

1. The arbitral tribunal may arrange for a stenographic and/or an audio-visual record to be made of an oral hearing. The arbitral tribunal may, if it considers it necessary, take minutes of the oral hearing and request the parties and/or their representatives, witnesses and/or other persons involved to sign and/or affix their seals to the stenographic record or the minutes.

（二）庭审笔录、庭审要点和影音记录供仲裁庭查用。

2. The stenographic record, the minutes and the audio-visual record of an oral hearing shall be available for use and reference by the arbitral tribunal.

第三十九条　举证

Article 39　Evidence

（一）当事人应对其申请、答辩和反请求所依据的事实提供证据加以证明，对其主张、辩论及抗辩要点提供依据。

1. Each party shall bear the burden of proving the facts on which it relies to support its claim, defense or counterclaim and provide the basis for its opinions, arguments and counter-arguments.

（二）仲裁庭可以规定当事人提交证据的期限。当事人应在规定的期限内提交证据。逾期提交的，仲裁庭可以不予接受。当事人在举证期限内提交证据材料确有困难的，可以在期限届满前申请延长举证期限。是否延长，由仲裁庭决定。

2. The arbitral tribunal may specify a time period for the parties to produce evidence and the parties shall produce evidence within the specified time period. The arbitral tribunal may refuse to admit any evidence produced after that time period. If a party experiences difficulties in producing evidence within the specified time period, it may apply for an extension before the end of the period. The arbitral tribunal shall decide whether or not to extend the time period.

（三）当事人未能在规定的期限内提交证据，或虽提交证据但不足以证明其主张的，负有举证责任的当事人承担因此产生的后果。

3. If a party bearing the burden of proof fails to produce evidence within the specified time period, or if the produced evidence is not sufficient to support its claim or counterclaim, it shall bear the consequences thereof.

第四十条　质证

Article 40　Examination of Evidence

（一）开庭审理的案件，证据应在开庭时出示，当事人可以质证。

1. Where a case is examined by way of an oral hearing, the evidence shall be produced at the hearing and may be examined by the parties.

（二）对于书面审理的案件的证据材料，或对于开庭后提交的证据材料且当事人同意书面质证的，可以进行书面质证。书面质证时，当事人应在仲裁庭规定的期限内提交书面质证意见。

2. Where a case is to be decided on the basis of documents only, or where the evidence is submitted after the hearing and both parties have agreed to examine the evidence by means of writing, the parties may examine the evidence without an oral hearing. In such circumstances, the parties shall submit their written opinions on the evidence within the time period specified by the arbitral tribunal.

第四十一条　仲裁庭自行调查

Article 41　Investigation by the Arbitral Tribunal

（一）仲裁庭认为必要时，可以自行调查事实，收集证据。

1. The arbitral tribunal may undertake investigations and collect evidence on its own initiative as it considers necessary.

（二）仲裁庭自行调查事实、收集证据时，可以通知当事人到场。经通知，一方或双方当事人不到场的，不影响仲裁庭自行调查事实和收集证据。

2. When investigating and collecting evidence on its own initiative, the arbitral

tribunal may notify the parties to be present. In the event that one or both parties fail to be present after being notified, the investigation and collection of evidence shall proceed without being affected.

（三）仲裁庭自行调查收集的证据，应转交当事人，给予当事人提出意见的机会。

3. Evidence collected by the arbitral tribunal through its own investigation shall be forwarded to the parties for their comments.

第四十二条 专家报告及鉴定报告

Article 42　Expert's Report and Appraiser's Report

（一）仲裁庭可以就案件中的专门问题向专家咨询或指定鉴定人进行鉴定。专家和鉴定人可以是中国或外国的机构或自然人。

1. The arbitral tribunal may consult experts or appoint appraisers for clarification on specific issues of the case. Such an expert or appraiser may be a Chinese or foreign institution or natural person.

（二）仲裁庭有权要求当事人、当事人也有义务向专家或鉴定人提供或出示任何有关资料、文件或财产、货物，以供专家或鉴定人审阅、检验或鉴定。

2. The arbitral tribunal has the power to request the parties, and the parties are also obliged, to deliver or produce to the expert or appraiser any relevant materials, documents, property, or goods for checking, inspection or appraisal by the expert or appraiser.

（三）专家报告和鉴定报告的副本应转交当事人，给予当事人提出意见的机会。任何一方当事人要求专家或鉴定人参加开庭的，经仲裁庭同意，专家或鉴定人应参加开庭，并在仲裁庭认为必要时就所作出的报告进行解释。

3. Copies of the expert's report and the appraiser's report shall be communicated to the parties for their comments. At the request of either party and with the approval of the arbitral tribunal, the expert or appraiser shall participate in an oral hearing and give explanations on the report when the arbitral tribunal considers it necessary.

第四十三条 程序中止

Article 43　Suspension of the Arbitration Proceedings

（一）当事人请求中止仲裁程序，或出现其他需要中止仲裁程序的情形的，仲裁程序可以中止。

1. Where parties request a suspension of the arbitration proceedings or under

circumstances where such suspension is necessary, the arbitration proceedings may be suspended.

（二）中止程序的原因消失或中止程序期满后，仲裁程序恢复进行。

2. The arbitration proceedings shall resume as soon as the reason for the suspension disappears or the suspension period ends.

（三）仲裁程序的中止及恢复，由仲裁庭决定；仲裁庭尚未组成的，由仲裁委员会秘书长决定。

3. The arbitral tribunal shall decide whether to suspend or resume the arbitration proceedings. Where the arbitral tribunal has not yet been formed, the decision shall be made by the Secretary General of CIETAC.

第四十四条　撤回申请和撤销案件

Article 44　Withdrawal and Dismissal

（一）当事人可以撤回全部仲裁请求或全部仲裁反请求。申请人撤回全部仲裁请求的，不影响仲裁庭就被申请人的仲裁反请求进行审理和裁决。被申请人撤回全部仲裁反请求的，不影响仲裁庭就申请人的仲裁请求进行审理和裁决。

1. A party may withdraw its claim or counterclaim in its entirety. In the event that the Claimant withdraws its claim in its entirety, the arbitral tribunal shall proceed with its examination of the counterclaim and render an arbitral award thereon. In the event that the Respondent withdraws its counterclaim in its entirety, the arbitral tribunal shall proceed with the examination of the claim and render an arbitral award thereon.

（二）因当事人自身原因致使仲裁程序不能进行的，可以视为其撤回仲裁请求。

2. A party may be deemed to have withdrawn its claim or counterclaim if it becomes impossible to carry on the arbitration proceedings for reasons attributable to that party.

（三）仲裁请求和反请求全部撤回的，案件撤销。在仲裁庭组成前撤销案件的，由仲裁委员会秘书长作出撤案决定；仲裁庭组成后撤销案件的，由仲裁庭作出撤案决定。

3. A case shall be dismissed if the claim and counterclaim have been withdrawn in their entirety. Where a case is to be dismissed prior to the formation of the arbitral tribunal, the Secretary General of CIETAC shall make a decision on the dismissal. Where a case is to be dismissed after the formation of the arbitral

tribunal, the arbitral tribunal shall make the decision.

（四）上述第（三）款及本规则第六条第（七）款所述撤案决定应加盖"中国国际经济贸易仲裁委员会"印章。

4. The seal of CIETAC shall be affixed to the dismissal decision referred to in the preceding Paragraph 3 and Paragraph 7 of Article 6 of these Rules.

第四十五条　仲裁与调解相结合

Article 45　Combination of Conciliation with Arbitration

（一）双方当事人有调解愿望的，或一方当事人有调解愿望并经仲裁庭征得另一方当事人同意的，仲裁庭可以在仲裁程序进行过程中对其审理的案件进行调解。双方当事人也可以自行和解。

1. Where both parties wish to conciliate, or where one party wishes to conciliate and the other party's consent has been obtained by the arbitral tribunal, the arbitral tribunal may conciliate the case during the course of the arbitration proceedings. The parties may also settle the case by themselves.

（二）仲裁庭在征得双方当事人同意后可以按照其认为适当的方式进行调解。

2. With the consent of both parties, the arbitral tribunal may conciliate the case in a manner it considers appropriate.

（三）调解过程中，任何一方当事人提出终止调解或仲裁庭认为已无调解成功的可能时，仲裁庭应停止调解。

3. During the process of conciliation, the arbitral tribunal shall terminate the conciliation proceedings if either party so requests or if the arbitral tribunal believes that further conciliation efforts shall be futile.

（四）经仲裁庭调解达成和解或双方当事人自行和解的，双方当事人应签订和解协议。

4. Where settlement is reached through conciliation by the arbitral tribunal or by the parties themselves, the parties shall sign a settlement agreement.

（五）经调解或当事人自行达成和解协议的，当事人可以撤回仲裁请求或反请求；当事人也可以请求仲裁庭根据当事人和解协议的内容作出裁决书或制作调解书。

5. Where a settlement agreement is reached through conciliation by the arbitral tribunal or by the parties themselves, the parties may withdraw their claim or counterclaim. The parties may also request the arbitral tribunal to render an arbitral award or a conciliation statement in accordance with the terms of the settlement

agreement.

（六）当事人请求制作调解书的,调解书应当写明仲裁请求和当事人书面和解协议的内容，由仲裁员署名，并加盖"中国国际经济贸易仲裁委员会"印章，送达双方当事人。

6. Where the parties request for a conciliation statement, the conciliation statement shall clearly set forth the claims of the parties and the terms of the settlement agreement. It shall be signed by the arbitrators, sealed by CIETAC, and served upon both parties.

（七）调解不成功的，仲裁庭应当继续进行仲裁程序并作出裁决。

7. Where conciliation fails, the arbitral tribunal shall resume the arbitration proceedings and render an arbitral award.

（八）当事人有调解愿望但不愿在仲裁庭主持下进行调解的，经双方当事人同意，仲裁委员会可以协助当事人以适当的方式和程序进行调解。

8. Where the parties wish to conciliate their dispute but do not wish to have conciliation conducted by the arbitral tribunal, CIETAC may, with the consent of both parties, assist the parties to conciliate the dispute in a manner and procedure it considers appropriate.

（九）如果调解不成功，任何一方当事人均不得在其后的仲裁程序、司法程序和其他任何程序中援引对方当事人或仲裁庭在调解过程中曾发表的意见、提出的观点、作出的陈述、表示认同或否定的建议或主张作为其请求、答辩或反请求的依据。

9. Where conciliation fails, any opinion, view or statement, and any proposal or proposition expressing acceptance or opposition by either party or by the arbitral tribunal in the process of conciliation, shall not be invoked by either party as grounds for any claim, defense or counterclaim in the subsequent arbitration proceedings, judicial proceedings, or any other proceedings.

（十）当事人在仲裁程序开始之前自行达成或经调解达成和解协议的，可以依据由仲裁委员会仲裁的仲裁协议及其和解协议，请求仲裁委员会组成仲裁庭，按照和解协议的内容作出仲裁裁决。除非当事人另有约定，仲裁委员会主任指定一名独任仲裁员组成仲裁庭，按照仲裁庭认为适当的程序进行审理并作出裁决。具体程序和期限，不受本规则其他条款关于程序和期限的限制。

10. Where the parties have reached a settlement agreement by themselves through negotiation or conciliation before the commencement of an arbitration

proceeding, either party may, based on an arbitration agreement concluded between them that provides for arbitration by CIETAC and the settlement agreement, request CIETAC to constitute an arbitral tribunal to render an arbitral award in accordance with the terms of the settlement agreement. Unless otherwise agreed by the parties, the Chairman of CIETAC shall appoint a sole arbitrator to form such an arbitral tribunal, which shall examine the case in a procedure it considers appropriate and render an award in due course.The specific procedure and time limit for rendering the award shall not be subject to other provisions of these Rules.

第三章 裁 决
Chapter Ⅲ Arbitral Award

第四十六条 作出裁决的期限

Article 46 Time Limit for Rendering Award

（一）仲裁庭应在组庭后6个月内作出裁决书。

1. The arbitral tribunal shall render an arbitral award within six (6) months from the date on which the arbitral tribunal is formed.

（二）经仲裁庭请求，仲裁委员会秘书长认为确有正当理由和必要的，可以延长该期限。

2. Upon the request of the arbitral tribunal, the Secretary General of CIETAC may extend the time limit if he/she considers it truly necessary and the reasons for the extension are truly justified.

（三）程序中止的期间不计入上述第（一）款规定的裁决期限。

3. Any suspension period shall be excluded when calculating the time limit in the preceding Paragraph 1.

第四十七条 裁决的作出

Article 47 Making of Award

（一）仲裁庭应当根据事实和合同约定，依照法律规定，参考国际惯例，公平合理、独立公正地作出裁决。

1. The arbitral tribunal shall independently and impartially render a fair and reasonable arbitral award based on the facts of the case and the terms of the contract, in accordance with the law, and with reference to international practices.

（二）当事人对于案件实体适用法有约定的，从其约定。当事人没有约定或其约定与法律强制性规定相抵触的，由仲裁庭决定案件实体的法律适用。

2. Where the parties have agreed on the law as it applies to the merits of their dispute, the parties' agreement shall prevail. In the absence of such an agreement or where such agreement is in conflict with a mandatory provision of the law, the arbitral tribunal shall determine the law as it applies to the merits of the dispute.

（三）仲裁庭在其作出的裁决书中，应写明仲裁请求、争议事实、裁决理由、裁决结果、仲裁费用的承担、裁决的日期和地点。当事人协议不写明争议事实和裁决理由的，以及按照双方当事人和解协议的内容作出裁决书的，可以不写明争议事实和裁决理由。仲裁庭有权在裁决书中确定当事人履行裁决的具体期限及逾期履行所应承担的责任。

3. The arbitral tribunal shall state in the award the claims, the facts of the dispute, the reasons on which the award is based, the result of the award, the allocation of the arbitration costs, and the date on which and the place at which the award is made. The facts of the dispute and the reasons on which the award is based may not be stated in the award if the parties have so agreed, or if the award is made in accordance with the terms of a settlement agreement between the parties. The arbitral tribunal has the power to determine in the award the specific time period for the parties to carry out the award and the liabilities for failure to do so within the specified time period.

（四）裁决书应加盖"中国国际经济贸易仲裁委员会"印章。

4. The seal of CIETAC shall be affixed on the arbitral award.

（五）由三名仲裁员组成的仲裁庭审理的案件，裁决依全体仲裁员或多数仲裁员的意见作出。少数仲裁员的书面意见应附卷，并可以附在裁决书后，该书面意见不构成裁决书的组成部分。

5. Where a case is examined by an arbitral tribunal composed of three arbitrators, the award shall be rendered by all three arbitrators or a majority of the arbitrators. A written dissenting opinion shall be kept with the file and may be appended to the award. Such dissenting opinion shall not form a part of the award.

（六）仲裁庭不能形成多数意见时，裁决依首席仲裁员的意见作出。其他仲裁员的书面意见应附卷，并可以附在裁决书后，该书面意见不构成裁决书的组成部分。

6. Where the arbitral tribunal cannot reach a majority opinion, the arbitral award shall be rendered in accordance with the presiding arbitrator's opinion. The written opinions of the other arbitrators shall be kept with the file and may be

appended to the award. Such written opinions shall not form a part of the award.

（七）除非裁决依首席仲裁员意见或独任仲裁员意见作出并由其署名，裁决书应由多数仲裁员署名。持有不同意见的仲裁员可以在裁决书上署名，也可以不署名。

7. Unless the arbitral award is made in accordance with the opinion of the presiding arbitrator or the sole arbitrator and signed by the same, the arbitral award shall be signed by a majority of the arbitrators. An arbitrator who has a dissenting opinion may or may not sign his/her name on the award.

（八）作出裁决书的日期，即为裁决发生法律效力的日期。

8. The date on which the award is made shall be the date on which the award comes into legal effect.

（九）裁决是终局的，对双方当事人均有约束力。任何一方当事人均不得向法院起诉，也不得向其他任何机构提出变更仲裁裁决的请求。

9. The arbitral award is final and binding upon both parties. Neither party may bring a lawsuit before a court or make a request to any other organization for revision of the award.

第四十八条　部分裁决

Article 48　Partial Award

（一）仲裁庭认为必要或当事人提出请求经仲裁庭同意的，仲裁庭可以在作出最终裁决之前，就当事人的某些请求事项作出部分裁决。部分裁决是终局的，对双方当事人均有约束力。

1. Where the arbitral tribunal considers it necessary, or where a party so requests and the arbitral tribunal agrees, the arbitral tribunal may render a partial award on any part of the claim before rendering the final award. A partial award is final and binding upon both parties.

（二）一方当事人不履行部分裁决，不影响仲裁程序的继续进行，也不影响仲裁庭作出最终裁决。

2. Failure of either party to implement a partial award shall not affect the arbitration proceedings, nor prevent the arbitral tribunal from making the final award.

第四十九条　裁决书草案的核阅

Article 49　Scrutiny of Draft Award

仲裁庭应在签署裁决书之前将裁决书草案提交仲裁委员会核阅。在不影响仲裁庭独立裁决的情况下，仲裁委员会可以就裁决书的有关问题提请仲裁庭注意。

The arbitral tribunal shall submit its draft award to CIETAC for scrutiny before signing the award. CIETAC may bring to the attention of the arbitral tribunal issues addressed in the award on the condition that the arbitral tribunal's independence in rendering the award is not affected.

第五十条　费用承担

Article 50　Allocation of Fees

（一）仲裁庭有权在裁决书中裁定当事人最终应向仲裁委员会支付的仲裁费和其他费用。

1. The arbitral tribunal has the power to determine in the arbitral award the arbitration fees and other expenses to be paid by the parties to CIETAC.

（二）仲裁庭有权根据案件的具体情况在裁决书中裁定败诉方应补偿胜诉方因办理案件而支出的合理的费用。仲裁庭裁定败诉方补偿胜诉方因办理案件而支出的费用是否合理时，应具体考虑案件的裁决结果、复杂程度、胜诉方当事人及/或代理人的实际工作量以及案件的争议金额等因素。

2. The arbitral tribunal has the power to decide in the arbitral award, having regard to the circumstances of the case, that the losing party shall compensate the winning party for the expenses reasonably incurred by it in pursuing the case. In deciding whether or not the winning party's expenses incurred in pursuing the case are reasonable, the arbitral tribunal shall take into consideration such specific factors as the outcome and complexity of the case, the workload of the winning party and/or its representative(s), and the amount in dispute, etc.

第五十一条　裁决书的更正

Article 51　Correction of Award

（一）仲裁庭可以在发出裁决书后的合理时间内自行以书面形式对裁决书中的书写、打印、计算上的错误或其他类似性质的错误作出更正。

1. Within a reasonable time after the award is made, the arbitral tribunal may, on its own initiative, make corrections in writing of any clerical, typographical or computational errors, or any errors of a similar nature contained in the award.

（二）任何一方当事人均可以在收到裁决书后 30 天内就裁决书中的书写、打印、计算上的错误或其他类似性质的错误，书面申请仲裁庭作出更正；如确有错误，仲裁庭应在收到书面申请后 30 天内作出书面更正。

2. Within thirty (30) days from its receipt of the arbitral award, either party

may request the arbitral tribunal in writing for a correction of any clerical, typographical or computational errors, or any errors of a similar nature contained in the award. If such an error does exist in the award, the arbitral tribunal shall make a correction in writing within thirty (30) days of receipt of the written request for the correction.

（三）上述书面更正构成裁决书的组成部分，应适用本规则第四十七条第（四）至（九）款的规定。

3. The above written correction shall form a part of the arbitral award and shall be subject to the provisions in Paragraphs 4 to 9 of Article 47 of these Rules.

第五十二条 补充裁决

Article 52 Additional Award

（一）如果裁决书中有遗漏事项，仲裁庭可以在发出裁决书后的合理时间内自行作出补充裁决。

1. Where any matter which should have been decided by the arbitral tribunal was omitted from the arbitral award, the arbitral tribunal may, on its own initiative, make an additional award within a reasonable time after the award is made.

（二）任何一方当事人可以在收到裁决书后 30 天内以书面形式请求仲裁庭就裁决书中遗漏的事项作出补充裁决；如确有漏裁事项，仲裁庭应在收到上述书面申请后 30 天内作出补充裁决。

2. Either party may, within thirty (30) days from its receipt of the arbitral award, request the arbitral tribunal in writing for an additional award on any claim or counterclaim which was advanced in the arbitration proceedings but was omitted from the award. If such an omission does exist, the arbitral tribunal shall make an additional award within thirty (30) days of receipt of the written request.

（三）该补充裁决构成裁决书的一部分，应适用本规则第四十七条第（四）至（九）款的规定。

3. Such additional award shall form a part of the arbitral award and shall be subject to the provisions in Paragraphs 4 to 9 of Article 47 of these Rules.

第五十三条 裁决的履行

Article 53 Carrying out of Award

（一）当事人应依照裁决书写明的期限履行仲裁裁决；裁决书未写明履行期限的，应立即履行。

1. The parties shall automatically carry out the arbitral award within the time

period specified in the award. If no time limit is specified in the award, the parties shall carry out the award immediately.

（二）一方当事人不履行裁决的，另一方当事人可以依法向有管辖权的法院申请执行。

2. Where one party fails to carry out the award, the other party may apply to a competent court for enforcement of the award in accordance with the law.

第四章　简易程序
Chapter Ⅳ　Summary Procedure

第五十四条　简易程序的适用
Article 54　Application

（一）除非当事人另有约定，凡争议金额不超过人民币 200 万元，或争议金额超过人民币 200 万元，但经一方当事人书面申请并征得另一方当事人书面同意的，适用简易程序。

1. Unless otherwise agreed by the parties, Summary Procedure shall apply to any case where the amount in dispute does not exceed RMB 2,000,000yuan; or to any case where the amount in dispute exceeds RMB 2,000,000yuan, yet one party applies for arbitration under the Summary Procedure and the other party agrees in writing.

（二）没有争议金额或争议金额不明确的，由仲裁委员会根据案件的复杂程度、涉及利益的大小以及其他有关因素综合考虑决定是否适用简易程序。

2. Where no monetary claim is specified or the amount in dispute is not clear, CIETAC shall determine whether or not to apply the Summary Procedure after a full consideration of relevant factors, including but not limited to the complexity of the case and the interests involved.

第五十五条　仲裁通知
Article 55　Notice of Arbitration

申请人提出仲裁申请，经审查可以受理并适用简易程序的，仲裁委员会秘书局应向双方当事人发出仲裁通知。

Where a Request for Arbitration submitted by the Claimant is found to be acceptable for arbitration under the Summary Procedure, the Secretariat of CIETAC shall send a Notice of Arbitration to both parties.

第五十六条　仲裁庭的组成

Article 56　Formation of Arbitral Tribunal

除非当事人另有约定，适用简易程序的案件，依照本规则第二十六条的规定成立独任仲裁庭审理案件。

Unless otherwise agreed by the parties, a sole-arbitrator tribunal shall be formed in accordance with Article 26 of these Rules to hear a case under Summary Procedure.

第五十七条　答辩和反请求

Article 57　Defense and Counterclaim

（一）被申请人应在收到仲裁通知后 20 天内提交答辩书及证据材料以及其他证明文件；如有反请求，也应在此期限内提交反请求书及证据材料以及其他证明文件。

1. The Respondent shall submit its Statement of Defense, evidence and other supporting documents within twenty (20) days of receipt of the Notice of Arbitration; counterclaim, if any, shall also be filed with evidence and supporting documents within the time period.

（二）申请人应在收到反请求书及其附件后 20 天内对被申请人的反请求提交答辩。

2. The Claimant shall file its Statement of Defense to the Respondent's counterclaim within twenty (20) days of receipt of the counterclaim and its attachments.

（三）当事人确有正当理由请求延长上述期限的，由仲裁庭决定是否延长；仲裁庭尚未组成的，由仲裁委员会秘书局作出决定。

3. If a party has justified reasons to request an extension of the time limit, the arbitral tribunal shall decide whether to grant such extension. Where the arbitral tribunal has not yet been formed, such decision shall be made by the Secretariat of CIETAC.

第五十八条　审理方式

Article 58　Conduct of Hearing

仲裁庭可以按照其认为适当的方式审理案件；可以决定只依据当事人提交的书面材料和证据进行书面审理，也可以决定开庭审理。

The arbitral tribunal may examine the case in the manner it considers appropriate. The arbitral tribunal may decide whether to examine the case solely on the basis of the written materials and evidence submitted by the parties or to hold an oral hearing.

第五十九条　开庭通知
Article 59　Notice of Oral Hearing

（一）对于开庭审理的案件，仲裁庭确定第一次开庭日期后，应不晚于开庭前 15 天将开庭日期通知双方当事人。当事人有正当理由的，可以请求延期开庭，但应于收到开庭通知后 3 天内提出书面延期申请；是否延期，由仲裁庭决定。

1. For a case examined by way of an oral hearing, after the arbitral tribunal has fixed a date for the first oral hearing, the parties shall be notified of the date at least fifteen (15) days in advance of the oral hearing. A party having justified reasons may request a postponement of the oral hearing. However, such request must be communicated in writing to the arbitral tribunal within three days of receipt of the notice of the oral hearing. The arbitral tribunal shall decide whether or not to postpone the oral hearing.

（二）当事人有正当理由未能按上述第（一）款规定的期限提出延期开庭申请的，是否接受其延期申请，由仲裁庭决定。

2. If a party has justified reasons for failure to submit a request for a postponement of the oral hearing within the time period specified in the preceding Paragraph 1, the arbitral tribunal shall decide whether to accept such a request.

（三）再次开庭审理的日期及延期后开庭审理日期的通知及其延期申请，不受上述第（一）款中期限的限制。

3. A notice of a subsequent oral hearing, a notice of a postponed oral hearing, as well as a request for postponement of such oral hearing shall not be subject to the time limits specified in the preceding Paragraph 1.

第六十条　作出裁决的期限
Article 60　Time Limit for Rendering Award

（一）仲裁庭应在组庭后 3 个月内作出裁决书。

1. The arbitral tribunal shall render an arbitral award within three (3) months from the date on which the arbitral tribunal is formed.

（二）经仲裁庭请求，仲裁委员会秘书长认为确有正当理由和必要的，可以延长该期限。

2. Upon the request of the arbitral tribunal, the Secretary General of CIETAC may extend the time limit if he/she considers it truly necessary and the reasons for

the extension are truly justified.

（三）程序中止的期间不计入上述第（一）款规定的裁决期限。

3. Any suspension period shall be excluded when calculating the time period in the preceding Paragraph 1.

第六十一条　程序变更

Article 61　Change of Procedure

仲裁请求的变更或反请求的提出，不影响简易程序的继续进行。经变更的仲裁请求或反请求所涉争议金额分别超过人民币 200 万元的案件，除非当事人约定或仲裁庭认为有必要变更为普通程序，继续适用简易程序。

The application of Summary Procedure shall not be affected by any amendment to the claim or by the filing of a counterclaim. Where the amount in dispute of the amended claim or that of the counterclaim exceeds RMB 2,000,000 yuan, Summary Procedure shall continue to apply unless the parties agree or the arbitral tribunal decides that a change to the general procedure is necessary.

第六十二条　本规则其他条款的适用

Article 62　Context Reference

本章未规定的事项，适用本规则其他各章的有关规定。

The relevant provisions in the other Chapters of these Rules shall apply to matters not covered in this Chapter.

第五章　国内仲裁的特别规定

Chapter V　Special Provisions for Domestic Arbitration

第六十三条　本章的适用

Article 63　Application

（一）国内仲裁案件，适用本章规定。

1. The provisions of this Chapter shall apply to domestic arbitration cases.

（二）符合本规则第五十四条规定的国内仲裁案件，适用第四章简易程序的规定。

2. The provisions of Summary Procedure in Chapter IV shall apply if a domestic arbitration case falls within the scope of Article 54 of these Rules.

第六十四条　案件的受理

Article 64　Acceptance

（一）收到仲裁申请书后，仲裁委员会认为仲裁申请符合本规则第十二条规定的受理条件的，仲裁委员会秘书局应当在 5 天内通知当事人；认为不符合受理条件的，应书面通知当事人不予受理，并说明理由。

1. Upon receipt of a Request for Arbitration, if CIETAC finds the Request to meet the requirements specified in Article 12 of these Rules, the Secretariat of CIETAC shall notify the parties accordingly within five (5) days from its receipt of the Request. Where a Request for Arbitration is found not to be in conformity with the requirements, the Secretariat of CIETAC shall notify the party in writing of its refusal of acceptance with reasons stated.

（二）收到仲裁申请书后，仲裁委员会秘书局经审查认为申请仲裁的手续不符合本规则第十二条规定的，可以要求当事人在规定的期限内予以完备。

2. Upon receipt of a Request for Arbitration, if after examination, the Secretariat of CIETAC finds the Request not to be in conformity with the formality requirements specified in Article 12 of these Rules; it may request the Claimant to complete the requirements within a specified time period.

第六十五条　仲裁庭的组成

Article 65　Formation of the Arbitral Tribunal

仲裁庭应按照本规则第二十三条、第二十四条、第二十五条、第二十六条、第二十七条和第二十八条的规定组成。

The arbitral tribunal shall be formed in accordance with the provisions of Articles 23, 24, 25, 26, 27 and 28 of these Rules.

第六十六条　答辩和反请求

Article 66　Defense and Counterclaim

（一）被申请人应在收到仲裁通知后 20 天内提交答辩书及所依据的证据材料以及其他证明文件；如有反请求，也应在此期限内提交反请求书及所依据的证据材料以及其他证明文件。

1. Within twenty (20) days from the date of receipt of the Notice of Arbitration, the Respondent shall submit its Statement of Defense, evidence and other supporting documents; counterclaim, if any, shall also be filed with evidence and other supporting documents within the time period.

（二）申请人应在收到反请求书及其附件后 20 天内对被申请人的反请求提交答辩。

2. The Claimant shall file its Statement of Defense to the Respondent's counterclaim within twenty (20) days from the date of receipt of the counterclaim

and its attachments.

（三）当事人确有正当理由请求延长上述期限的，由仲裁庭决定是否延长；仲裁庭尚未组成的，由仲裁委员会秘书局作出决定。

3. If the Respondent has justified reasons to request an extension of the time limit, the arbitral tribunal shall decide whether to grant such extension. Where the arbitral tribunal has not yet been formed, such decision shall be made by the Secretariat of CIETAC.

第六十七条　开庭通知

Article 67　Notice of Oral Hearing

（一）对于开庭审理的案件，仲裁庭确定第一次开庭日期后，应不晚于开庭前 15 天将开庭日期通知双方当事人。当事人有正当理由的，可以请求延期开庭，但应于收到开庭通知后 3 天内提出书面延期申请；是否延期，由仲裁庭决定。

1. For a case examined by way of an oral hearing, after the arbitral tribunal has fixed a date for the first oral hearing, the parties shall be notified of the date at least fifteen (15) days in advance of the oral hearing. A party having justified reason may request a postponement of the oral hearing. However, such request must be communicated in writing to the arbitral tribunal within three days of receipt of the notice of the oral hearing. The arbitral tribunal shall decide whether or not to postpone the oral hearing.

（二）当事人有正当理由未能按上述第（一）款规定的期限提出延期开庭申请的，是否接受其延期申请，由仲裁庭决定。

2. If a party has justified reasons for failure to submit a request for a postponement of the oral hearing within the time period provided in the preceding Paragraph 1, the arbitral tribunal shall decide whether to accept such a request.

（三）再次开庭审理的日期及延期后开庭审理日期的通知及其延期申请，不受上述第（一）款中期限的限制。

3. A notice of a subsequent oral hearing, a notice of a postponed oral hearing, as well as a request for postponement of such oral hearing shall not be subject to the time limits specified in the preceding Paragraph 1.

第六十八条　庭审笔录

Article 68　Record of Oral Hearing

（一）仲裁庭应将开庭情况记入笔录。当事人和其他仲裁参与人认为对自己陈述的记录有遗漏或有差错的，可以申请补正；仲裁庭不同意其补正的，应将该申请记录在案。

1. The arbitral tribunal shall make a written record of the oral hearing. Any party or participant in the arbitration may apply for a correction upon finding any omission or mistake in the record regarding its own statements. If the application is refused by the arbitral tribunal, it shall nevertheless be recorded and kept with the file.

（二）庭审笔录由仲裁员、记录人员、当事人和其他仲裁参与人签名或盖章。

2. The written record shall be signed or sealed by the arbitrator(s), the recorder, the parties, and any other participant in the arbitration.

第六十九条　作出裁决的期限

Article 69　Time Limit for Rendering Award

（一）仲裁庭应在组庭后4个月内作出裁决书。

1. The arbitral tribunal shall render an arbitral award within four (4) months from the date on which the arbitral tribunal is formed.

（二）经仲裁庭请求，仲裁委员会秘书长认为确有正当理由和必要的，可以延长该期限。

2. Upon the request of the arbitral tribunal, the Secretary General of CIETAC may extend the time period if he/she considers it truly necessary and the reasons for the extension truly justified.

（三）程序中止的期间不计入上述第（一）款规定的裁决期限。

3. Any suspension period shall be excluded when calculating the time limit in the preceding Paragraph 1.

第七十条　本规则其他条款的适用

Article 70　Context Reference

本章未规定的事项，适用本规则其他各章的有关规定。

The relevant provisions in the other Chapters of these Rules shall apply to matters not covered in this Chapter.

第六章　附　则
Chapter Ⅵ　Supplementary Provisions

第七十一条　仲裁语言

Article 71　Language

（一）当事人对仲裁语言有约定的，从其约定。当事人没有约定的，仲裁程序以中文为仲裁语言，或以仲裁委员会视案件的具体情形确定的其他语言为仲裁语言。

1. Where the parties have agreed on the language of arbitration, their agreement shall prevail. In the absence of such agreement, the language of arbitration to be used in the proceedings shall be Chinese or any other language designated by CIETAC having regard to the circumstances of the case.

（二）仲裁庭开庭时，当事人或其代理人、证人需要语言翻译的，可由仲裁委员会秘书局提供译员，也可由当事人自行提供译员。

2. If a party or its representative(s) or witness (es) requires interpretation at an oral hearing, the Secretariat of CIETAC may provide an interpreter, or the party may employ its own interpreter.

（三）当事人提交的各种文书和证明材料，仲裁庭或仲裁委员会秘书局认为必要时，可以要求当事人提供相应的中文译本或其他语言译本。

3. The arbitral tribunal or the Secretariat of CIETAC may, if it considers it necessary, require the parties to submit a corresponding translation of their documents and evidence into Chinese or other languages.

第七十二条　仲裁费用及实际费用

Article 72　Arbitration Fees and Costs

（一）仲裁委员会除按照其制定的仲裁费用表向当事人收取仲裁费外,可以向当事人收取其他额外的、合理的实际费用,包括仲裁员办理案件的特殊报酬、差旅费、食宿费以及仲裁庭聘请专家、鉴定人和翻译等的费用。

1. Apart from the arbitration fees charged in accordance with its Fees Schedule, CIETAC may charge the parties any other extra and reasonable costs, including but not limited to arbitrators' special remuneration, their travel and accommodation expenses incurred in dealing with the case, as well as the costs and expenses of experts, appraisers or interpreters appointed by the arbitral tribunal.

（二）当事人未在仲裁委员会规定的期限内为其选定的需要开支差旅费、食宿费等实际费用的仲裁员预缴实际费用的，视为没有选定仲裁员。

2. Where a party has nominated an arbitrator who will incur actual costs such as travel and accommodation expenses, but fails to pay in advance a deposit for such costs within the time period specified by CIETAC, the party shall be deemed not to have nominated the arbitrator.

（三）当事人约定在仲裁委员会或其分会/中心所在地之外开庭的，应预缴因此而发生的差旅费、食宿费等实际费用。当事人未在仲裁委员会规定的期限内预缴有关实际费用的，应在仲裁委员会或其分会/中心所在地开庭。

3. Where the parties have agreed to hold an oral hearing at a place other than the domicile of CIETAC or its relevant sub-commission/center, they shall pay a deposit in advance for the actual costs such as travel and accommodation expenses incurred thereby. In the event that the parties fail to do so within the time period specified by CIETAC, the oral hearing shall be held at the domicile of CIETAC or its relevant sub-commission/center.

（四）当事人约定以两种或两种以上语言为仲裁语言的，或根据本规则第五十四条的规定适用简易程序的案件但当事人约定由三人仲裁庭审理的，仲裁委员会可以向当事人收取额外的、合理的费用。

4. Where the parties have agreed to use two or more languages as the languages of arbitration, or where the parties have agreed on a three-arbitrator tribunal in a case to which Summary Procedure shall apply in accordance with Article 54 of these Rules, CIETAC may charge the parties for the extra and reasonable costs.

第七十三条　规则的解释

Article 73　Interpretation

（一）本规则条文标题不用于解释条文含义。

1. The headings of the articles in these Rules shall not be construed as interpretations of the contents of the provisions contained therein.

（二）本规则由仲裁委员会负责解释。

2. These Rules shall be interpreted by CIETAC.

第七十四条　规则的施行

Article 74　Coming into Force

本规则自 2012 年 5 月 1 日起施行。本规则施行前仲裁委员会秘书局及其分会/中心秘书处管理的案件，仍适用受理案件时适用的仲裁规则；双方当事人同意的，也可以适用本规则。

These Rules shall be effective as from May 1, 2012. For cases administered by the Secretariat of CIETAC or the secretariat of one of its sub-commissions/centers before these Rules come into force, the Arbitration Rules effective at the time of acceptance shall apply, or where both parties agree, these Rules shall apply.

④ 国际棉花协会棉花规章与规则
4 Bylaws and Rules of the International Cotton Association Limited

Section 1 Introduction
第 1 部分 简介

Bylaws are the mandatory provisions of the Association which cannot be changed or varied by the parties.

规章为本协会强制性规定，当事方不得变更或变动。

Definitions
定义

Bylaw 100
规章 100

In our Bylaws and Rules, and in any contract made under our Bylaws and Rules, the following expressions will have the meanings given unless their context clearly shows them to have a different use:

在我方的规章及规则以及据此签署的任何合同中，除非上下文明白显示其另有所指，否则以下表述的含义如下：

Administrative terms
行政术语

1 'Approved panel' means the list of individuals, approved annually by the Board of Directors, from which the Directors will appoint the Preliminary Investigation Committee.

1 "经批准小组"指每年经董事会批准、从中任命初步调查委员会成员的人员名单。

2 'Articles' means our Articles of Association and any changes to them which are in force.

2 "章程"指我们的公司章程以及对其所作的一切有效变更。

3 'Bylaws' and 'Rules' mean all our bylaws and rules which are in force.

3 "规章"和"规则"指我们所有有效的规章和规则。

4 'Committee' means any committee elected by the Individual Members. Committee members will include anyone eligible, appointed or nominated to serve under our Articles.

4 "委员会"指任何由个人会员选举产生的委员会。委员会成员包括任何适格、接受指定或提名依据章程供职者。

5 'Director' means any of our Directors, whether Ordinary or Associate, and includes the President, First Vice-President, Second Vice-President, Treasurer and immediate past President.

5 "董事"指我们的任何董事,无论为常任董事还是联席董事,包括会长、第一副会长、第二副会长、财务主管和前一任会长。

'Associate Director' means a Director invited each year by the Directors and approved by the Members to serve the common interests of the industry.

"联席董事"指每年由董事会邀请并经会员通过为行业共同利益服务的董事。

'Ordinary Director' means a Director elected by the Individual Members. It does not include the President, First Vice-President, Second Vice-President, Treasurer or immediate Past President.

"普通董事"指由各个人会员推举产生的董事。其中不包括会长、第一副会长、第二副会长、财务主管和前一任会长。

'Immediate Past President' does not include a President who is removed pursuant to Article 86 or ceases to be a Director pursuant to Article 94.

"前一任会长"不包括依据条例第86条免除的会长或依据条例第94条终止担任董事的会长。

6 'General Meeting' means a meeting of our Individual Members called under our Articles.

6 "会员大会"指依据我们的条例召集的个人会员大会。

7 'Month' means a calendar month.

7 "月"指日历月。

8 'Observer' means a probationary arbitrator who, for training purposes, may be appointed by the Association to act as an unpaid observer on technical arbitration tribunals and technical appeal committees. The observer will not participate in, nor influence, the tribunal's decision making process.

8 "观察员"指由于培训目的,可能被协会委任为技术仲裁庭和技术申诉委员会无薪观察员的试用仲裁员。此类仲裁员不参与或影响仲裁庭的决策过程。

9 'Our' means whatever is owned by us or issued by us.

9 "我们的/我方"指任何由我方所有或由我方发布的任何东西。

10 'President' includes the First Vice-President or Second Vice-President or

anyone appointed by the Directors under our Articles to carry out the duties of an absent President.

10 "会长"包括依据章程在会长缺席期间履行会长职务的第一副会长或第二副会长或由董事会任命的任何人士。

11 'Place of business' of any Individual Member or Registered Firm means an office where the Directors consider an Individual Member or Registered Firm carries out business.

11 任何个人会员或注册公司的"营业地点"指董事会认定的个人会员或注册公司从事业务所在的办公场所。

12 The 'Rule Book' means the book in which we publish our Bylaws and Rules.

12 "规则手册"指我们在其中公布规章和规则的手册。

13 The 'Secretary' means the person the Directors have appointed to act as Secretary. An Alternate Secretary appointed by the Directors may act in place of the Secretary.

13 "秘书"指由董事会任命行使秘书职责的个人。董事会任命的代理秘书可以代理秘书履行职务。

14 'We', 'us' and 'ICA' mean The International Cotton Association Limited.

14 "我们"、"我方"和"ICA"指国际棉花协会有限公司。

15 'In writing' and 'written' include printing and other ways of reproducing words on paper or on a screen or website. Written correspondence can be delivered by post, hand, fax, e-mail and so on.

15 "书面形式"包括印刷品及其他将文字复制于纸张、屏幕或网站的方式。书面通信可以通过邮寄、专人递交、传真、电子邮件和其他方式传达。

16 'ICA List of Unfulfilled Awards' (also known as the 'ICA Default List') means the list of unfulfilled awards circulated by the Association at the request of reporting parties.

16 "ICA 未执行裁决书名单"(也称为"ICA 违约名单")指协会在报告方要求下发布的未执行裁决书清单。

Membership and registration terms
成员资格与注册术语

17 'Affiliate Industry Firm' means a firm or organisation registered as such under our Bylaws.

17　"关联行业公司"指依据我们的规章注册为关联行业公司的公司或组织。

18　'Agent Firm' means a firm or organisation registered as such under our Bylaws.

18　"代理公司"指依据我们的规章注册为代理公司的公司或组织。

19　'Firm' means any partnership, un-incorporated body or company carrying out business.

19　"公司"指任何开展经营业务的合伙、非法人组织或公司。

20　'Individual Member' means a person elected to be an Individual Member of a member firm under our Articles.

20　"个人会员"指依据我们的章程选举为成员公司个人会员的个人。

21　'Member Firm' means a Principal Firm, an Association Member Firm, an Affiliate Industry Firm, an Agent Firm or a Related Company.

21　"成员公司"指一家主公司、协会成员公司、关联行业公司、代理公司或关联公司。

22　'Non-member' means any person who is not an Individual Member of the Association.

22　"非会员"指任何不是协会个人会员的人。

23　'Non-registered firm' means any firm that is not a Registered Firm of the Association.

23　"非注册公司"指任何不是协会注册公司的公司。

24　'Principal Firm' is a Merchant, Producer or Mill and means a firm or company registered as such under our Articles and Bylaws.

24　"主公司"是一家经销商、生产商或工厂,根据我们的章程和规章注册为主公司的公司。

25　'Registered Firm' means all Principal Firms, Affiliate Industry Firms, Related Companies, Affiliated Associations, Association Member Firms and Agent Firms, details of which are entered in the Register of Registered Firms.

25　"注册公司"是指所有主公司、关联行业公司、关联公司、附属协会、协会成员公司和代理公司,其详细情况登记于注册公司的登记簿中。

26　'Registered' means registered or re-registered and 'Registration' means registering or re-registering.

26　"经注册"指经注册或经再注册,"注册"指注册或再注册。

27　For the purposes of these Bylaws and Rules, 'Register of Registered

Firms' means our list of Principal Firms, Affiliate Industry Firms, Related Companies, Affiliated Associations, Association Member Firms and Agent Firms.

27 在本章和规则中,"注册公司登记簿"指我们的主公司、关联行业公司、关联公司、附属协会、协会成员公司和代理公司清单。

28 'Registered Firm', means any firm listed in our register of Registered Firms as defined in our Articles.

28 "注册公司"指任何列入我们的章程所定义的注册公司登记簿的公司。

29 'Related Company' means a company related to a Principal Firm or an Affiliate Industry Firm.

29 "关联公司"指一家与主公司或关联行业公司存在关联关系的公司。

General trading terms
一般贸易术语

30 'American cotton' means all cotton grown anywhere within the contiguous states of the United States of America, including cotton known as Upland, Gulf or Texas cotton, but not including the Sea Island or Pima varieties.

30 "美国棉"指在美利坚合众国境内各州种植的棉花,包括所谓陆地棉、海湾棉及德克萨斯棉,但不包括海岛或皮马棉各品种。

31 'Certified laboratory' means a laboratory that is on an approved list issued by us.

31 "认证实验室"指列入我们发布的经核准清单的实验室。

32 'Combined transport', 'intermodal transport' and 'multimodal transport' mean delivering cotton from one place to another using at least two different means of transport.

32 "联合运输"、"协调联运"和"多式联运"指使用两种及以上运输方式将棉花从一地运至另一地。

33 'Combined transport document' means a bill of lading or other document of title produced by a shipping company, combined transport operator or agent covering cotton being moved by combined transport, intermodal transport or multimodal transport.

33 "联合运输单据"指涉及以联合运输、协调联运或多式联运方式运送棉花的船运公司、联运承运人或代理所开具的提单或其他物权文件。

34 'Combined transport operator' means a person or firm which produces a combined transport document.

34 "联运承运人"指提供联运文件的个人或公司。

35 'Container freight station', 'CFS' and 'container base' mean a place where the carrier or his agent loads or unloads containers under their control.

35 "集装箱货运站"、"CFS"和"集装箱基地"指承运人或其代理装载或卸载其控制下的集装箱的地点。

36 'Container yard' and 'CY' mean a place where containers can be parked, picked up or delivered, full or empty. A container yard or CY may also be a place where containers are loaded (or stuffed) or unloaded (or de-vanned).

36 "集装箱堆场"和"CY"指满载或空载集装箱停放、提货或交货的地点。集装箱堆场或 CY 可以同时为集装箱装载（或填货）或卸载（或开箱）的地点。

37 'Control limit' means the variation in readings taken on different instruments, using the same cotton.

37 "控制界限"指用不同仪器测量同一棉纤维时读数的变化。

38 'Cotton waste' will be treated as cotton if it had been included in contracts which are subject to our Bylaws and Rules.

38 "废棉"如被包含在受我们的规章和规则规范的合同中，则作为棉花处理。

39 'Country damage' is the damage or deterioration of the fibre caused by the absorption of excessive moisture, dust or sand from the exterior because it has been:

39 "产地污损"是指因以下原因从外界吸收过量湿气、灰尘或沙粒而导致的纤维损坏或变质：

exposed to the weather; or
暴露在空气中；或

stored on wet or contaminated surfaces, prior to loading to trucks/containers or the vessel.

储存在潮湿或污损的地面；以上发生在装载入卡车/集装箱或船只前。

Country damage does not include:
产地污损不包括：

any internal damage; or
任何内部损坏；或

any other contamination; or
任何其他污损；或

any damage which takes place after loading to trucks/containers or the vessel.

任何在装载入卡车/集装箱或船只后发生的损坏。

40 'Date of arrival' will, depending on the context, have one of the following meanings:

40 "到达日" 依上下文取下列涵义之一：

For break bulk shipments, it will mean the date the vessel arrives in the port of destination named in the bill of lading.But, if the vessel is diverted or the cotton is moved to another ship, it will be the date the cotton arrives in the port stated in the bill of lading or in another port acceptable to the buyer.

在散货运输情况下，指船只到达提单指明目的港的日期。但若船只改道或棉花移至另一船只，则为棉花到达提单所述港口或买方可以接受的其他港口的日期。

For cotton shipped in containers, it will be the date the cotton arrives in the port of destination named in the bill of lading or the combined transport document. But, if the carrying vessel is diverted or the containers are moved to another ship, it will be the date the containers arrive in the port stated in the bill of lading or in another port acceptable to the buyer.

在棉花以集装箱运输的情况下，为棉花到达提单或联合运输单据所述目的港的日期。但若载运船只改道或集装箱移至另一船只，则为集装箱到达提单所述港口或买方可以接受的其他港口的日期。

For other means of transport it will be the date each delivery is made to the place stated in the contract.

如系其他运输方式，则为交付至合同所述地点的日期。

41 'Dispute' or 'difference' relating to a contract will include any argument, disagreement or question about how to interpret the contract, or the rights or responsibilities of anyone bound by the contract.

41 涉及一项合同的"纠纷"或"异议"包括关于如何解释合同或受合同约束的任何一方的权利或义务的任何争议、不同意或质疑。

42 'False packed bale' is a bale containing:

42 "掺次包装货包"是指货包包含以下成分：

substances which are not cotton;

非棉物质；

damaged cotton;

受损棉花；

good cotton on the outside and inferior cotton on the inside; or
外层为优质棉，而里层为劣质棉；或

pickings or linters instead of cotton.
下脚或棉短绒，而非棉花。

43 'Far East cotton' means cotton grown in Bangladesh, Burma, China, India or Pakistan.

43 "远东棉"是指生长于孟加拉国、缅甸、中国、印度或巴基斯坦的棉花。

44 'Foreign matter' means anything that is not part of the cotton plant.

44 "异物"指任何不属于棉花植株的东西。

45 'Full container load' and 'FCL' mean an arrangement which uses all the space in a container. 'Less than container load' and 'LCL' mean a parcel of cotton which is too small to fill a container and which is grouped by the carrier at the container freight station with similar cargo going to the same destination.

45 "整箱"和"FCL"指一种利用集装箱全部空间的安排。"拼箱"和"LCL"指棉花货量过小不足以装满一个集装箱，由承运人在集装箱货运站与其他去往同一目的地的类似货物拼装一箱。

46 'House to', 'container yard to' and 'door to' mean loading controlled by the shipper at the place (house, CY or door) of his choice. Whoever books the freight must pay all costs beyond the point of loading and the cost of providing the containers at the house, CY or door.

46 "仓库至"、"集装箱堆场至"和"户至"指托运人在其指定地点（仓库、集装箱堆场或户）控制货物装载。预定一方须支付装载后的全部费用，以及在仓库、集装箱堆场或户提供集装箱的费用。

47 'Immediately' means within three days.

47 "立即"指在三天内。

48 'Institute Cargo Clauses' and 'Institute Commodity Trades Clauses' mean the clauses of the Institute of London Underwriters.

48 "协会货运保险条款"和"协会商品贸易保险条款"指伦敦保险商协会条款。

49 'Internal moisture' or 'Moisture regain' mean the weight of moisture in the cotton expressed as a percentage of the weight of the fibre when totally dry.

49 "内含水分"或"回潮"指棉花中的水分重量，以其占纤维完全干燥时重量的百分比为表示。

50 'Lot' is a number of bales placed under one mark.

50 "批"指同一唛头下存放的多个货包。

51 'Mixed packed bale' is a bale containing many different grades, colour or staple.

51 "混杂包装货包"指货包中包含多种不同等级的颜色及纤维长度。

52 'Marine cargo insurance' and 'transit insurance' mean insurance against the risks covered by the Marine Policy Form (MAR form) used in conjunction with the Institute Cargo Clauses, or covered by similar first-class policies in other insurance markets.

52 "海上运输保险"和"运送保险"指水险保单表（MAR 表）涵盖的各项风险保险，与协会货运保险条款联合使用，或指其它保险市场的同类一级保单涵盖的各项风险保险。

53 'Micronaire' means a measurement of the combination of fineness and maturity of raw cotton fibre.

53 "马克隆值"指原棉纤维细度及成熟度组成的测试。

54 'No control limit' and 'NCL' mean that no control limit is allowed.

54 "无公差"和"NCL"指不允许发生公差。

55 'On-board bill of lading' means a bill which is signed by the captain or his agent when the cotton has been loaded on the ship.

55 "已装运提单"指棉花装船后由船长或其代理签字的提单。

56 'Percentage allowance' means a percentage of the invoice price.

56 "百分比允许值"指发票价格的百分比。

57 'Pier to', 'container freight station to' and 'container base to' mean that the carrier controls the loading. The cotton must be delivered to the carrier at the pier, container freight station or container base.

57 "码头至"、"集装箱货运站至"和"集装箱基地至"指由承运人负责装载。棉花须发送至码头，集装箱货运站或集装箱基地交给运货人。

58 'Plated bale' is a bale in which a layer of very different quality cotton appears on the outside of at least one side.

58 "夹次棉包"指货包至少一边的外侧露出一层品质完全不同的棉花。

59 'Point of destination' means the exact place where the cotton is delivered to the person who has ordered it, or is delivered to his agent, and where the carrier's

responsibility ends.

59 "目的地"指将棉花送交订货人或其代理的具体地点，此时运货人的责任终止。

60 'Point of origin' means the exact place where the carrier or his agent receives the cotton and where the carrier's responsibility begins.

60 "发货地"指承运人或其代理接收棉花的确切地点，承运人的责任在此处开始。

61 'Prompt' means within 14 days (two weeks).

61 "及时"指在 14 天（两周）内。

62 'Shipment' means the loading of cotton onto any means of transport for delivery from the seller or his agent to the buyer, or to a carrier who can provide a bill of lading or a combined transport document.

62 "装运"指将棉花装载上任何运输工具，以供卖方或其代理交付予买方，或交付予能够提供提单或联合运输单据的承运人。

63 'Shipper's load and count' means the shipper is responsible for the contents of the container.

63 "托运人自行装货点件"指托运人对集装箱装载的内容承担责任。

64 'Shipping' or 'shipped' means loading or loaded for shipment.

64 "装运中"或"已装运"指装运的装载过程或已装载。

65 'Shipping documents' means the document of title showing how the cotton is to be shipped under the contract.

65 "装运单据"指表明依合同棉花应如何装运的物权文件。

66 'Strikes, riots and civil commotions insurance' means insurance against the risks set out in the Institute Strikes Clauses (Cargo) or Institute Strikes Clauses (Commodity Trades), or similar clauses of other first-class insurance markets.

66 "罢工、暴动及民变保险"指针对协会罢工保险条款（货物）或协会罢工保险条款（商品贸易）或其它一级保险市场的同类条款规定的险种进行投保。

67 'Tare' means the weight of wrapping, bands, ropes or wires used to cover cotton bales.

67 "皮重"指用以包装货包的包装物、包装带、绳索或金属线的重量。

68 'To house', 'to container yard' and 'to door' mean delivery to the warehouse or mill selected by the person who booked the freight.

68 "至仓库"，"至集装箱堆场"和"至户"指发送到由预定方选择的仓库或工厂。

69 'To pier', 'to container freight station' and 'to container base' mean that the carrier will unload (de-van) at his warehouse in the port of destination, in a container freight station or container base.

69 "至码头"、"至集装箱货运站"和"至集装箱基地"指运货人在其位于目的港、集装箱货运站或集装箱基地的仓库卸载（开箱）。

70 'Usual control limit' and 'UCL' mean the variation allowed in readings to account for the normal variation expected from different instruments, even if the same cotton is used.

70 "一般控制界限"和"UCL"指对同一棉纤维，允许不同仪器测量的读数有正常波动。

71 'War risks insurance' means insurance against the risks set out in the Institute War Clauses (Cargo) or Institute War Clauses (Commodity Trades), or similar clauses of other first-class insurance markets.

71 "兵险保险"指针对协会兵险条款（货物）或协会兵险条款（商品贸易）或其它一级保险市场的同类条款规定的险种进行投保。

General Bylaws
一般规章

Bylaw 101

规章 101

These Bylaws and Rules apply to all parties contracting under our Bylaws and Rules.

本规章和规则适用于依我们的规章和规则缔约的所有当事方。

Bylaw 102

规章 102

1　If a contract is made under our Bylaws and Rules:

1　如依据我们的规章和规则制定了一项合同，则：

All of the Bylaws in this book will apply to the contract and no amendment by the buyer and seller is allowed; but the buyer and seller can agree terms in their contract which are different to any of the Rules.

本手册中的所有规章均适用于合同，买卖双方不得加以修正；但买卖双方可以在合同中议定与任何规则不一致的条款。

2　If we change any of the Bylaws or Rules after the date of the contract, the change will not apply to the contract unless the buyer and seller agree otherwise. This is with the exception of those Bylaws in Section 3 covering arbitration timescales, notices, fees and other procedures. In such cases, the procedures to be used for arbitration or appeal will be those in force at the time of making the application.

2 如在合同签订后我们对任何规章或规则作了变更，则该等变更不适用于合同，除非买卖双方同意。第 3 部分有关仲裁时间安排、通知、费用及其他程序的规章除外。在此情况下，用于仲裁或申诉的程序将为提出申请时有效的程序。

3 All other changes will apply when we say so.

3 所有其他变更根据我们的规定适用。

Bylaw 103

规章 103

1 If there is a doubt or difference in the meaning between any translation and the English, the Bylaws and Rules in English will apply.

1 若译文版本与英文版本间涵义上出现疑问或不符，则以英文版规章和规则为准。

2 We are not responsible for any mistakes in any version of the Rule Book.

2 我们不对本规则手册任何译文版本中的任何错误负责。

Bylaw 104

规章 104

The powers which the Bylaws and Rules give to the President are also given to the First Vice-President, Second Vice-President and any acting President.

规章和规则赋予会长的权力同样赋予第一副会长、第二副会长和任何代理会长。

Bylaw 105

规章 105

In these Bylaws and Rules:

在该等规章和规则中：

If something must be done within a fixed number of days of an event, the number of days will not include the day of the event itself. Days allowed will run continuously.

如在某事件发生后某一固定数量的日期内必须完成某事，则日期数不包含事件发生之日本身。所允许的日期连续计算。

Unless the buyer and seller agree otherwise, a kilogram will equal 2.2046 pound weight (lb).

除非买卖双方另行协商一致，否则一千克相当于 2.2046 磅（lb）。

'He', 'him' and 'his' mean 'she', 'her' and 'hers' if necessary.

在必要时"他"和"他的"意指"她"和"她的"。

Words referring to people can also refer to firms if necessary.

指称人的词汇在必要时亦得指称公司。

Words in the singular also cover the plural. Words in the plural also cover the singular.

单数词汇亦可指复数。复数词汇亦可指单数。

Time is expressed in terms of the 24 hour clock. All times are given in Universal Time (Greenwich Mean Time).

时间按 24 小时制表示。所有时间以世界时（格林威治平均时）表示。

The Contract
合同

The application of Bylaws and Rules
规章与规则的适用

Bylaw 200
规章 200

Every contract made under our Bylaws and Rules will be deemed to be a contract made in England and governed by English law.

依据我们的规章和规则订立的所有合同均视为在英格兰订立并受英国法调整。

Bylaw 201
规章 201

1　Subject to Bylaws 302 and 330 the following clauses will apply to every contract made under our Bylaws and Rules, or containing words to similar effect:

1　依据规章 302 和 330，以下条款适用于所有依据我们的规章和规则订立、或包含类似效力措辞的合同：

The contract will incorporate the Bylaws and Rules of the International Cotton Association Limited as they were when the contract was agreed. If any contract has not been, or will not be performed, it will not be treated as cancelled. It will be closed by being invoiced back to the seller under our Rules in force at the date of the contract.

双方议定合同之时，合同即归并国际棉花协会有限公司规章和规则于其中。如有合同未履行，或者将不会履行的，不得按撤销处理。该合同应依据合同签署之时有效的规章和规则，以向卖方回开发票的方式终止。

All disputes relating to the contract will be resolved through arbitration in accordance with the Bylaws of the International Cotton Association Limited. This agreement incorporates the Bylaws which set out the Association's arbitration procedure.

有关合同的一切争议按国际棉花协会有限公司规章仲裁解决。仲裁协议包括规定协会仲裁程序的各项规章。

Neither party will take legal action over a dispute suitable for arbitration, other than to obtain security for any claim, unless they have first obtained an arbitration award from the International Cotton Association Limited and exhausted all means of appeal allowed by the Association's Bylaws. The words 'all disputes' can be changed to read 'quality disputes' or 'technical disputes'. But if nothing else is agreed, the words 'all disputes' will apply.

任何一方在取得索赔保全之前不得就适合仲裁的争议采取法律措施，除非其已事先从国际棉花协会有限公司处取得仲裁裁决书并已穷尽所有协会规章所允许的上诉途径。"所有争议"一词可以变更为"品质事项争议"或"技术事项争议"。但若未就任何其他事项达成一致，则将适用"所有争议"一词。

2 Attention is drawn to Bylaws 302 and 330 which allow the Directors to deny arbitration, if, on the day before the date of the contract giving rise to the dispute, either party has its name circulated on the ICA List of Unfulfilled Awards in accordance with Bylaw 366.

2 如在导致争议的合同签署之日前一天，有任何一方的姓名依规章 366 被列入 ICA 未执行裁决书名单，则董事会可依据规章 302 和 330 拒绝仲裁。

3 This Bylaw will apply even if the contract is held to be invalid or ineffective, or was not concluded.

3 在合同被认定为无效或不生效、或者未成立的情况下，本条规章仍然适用。

Bylaw 202
规章 202

Unless the buyer and seller agree otherwise, the provisions of the following will not apply to contracts made under our Bylaws and Rules:

除非买卖双方另行达成一致，否则以下规定将适用于依据我们的规章和规则订立的合同：

The Uniform Law on International Sales Act (1967); and the 1980 Vienna Convention on Contracts for the International Sale of Goods.

《国际销售法案统一法》(Uniform Law on International Sales Act)(1967)；和《1980 年国际货物销售合同维也纳公约》(1980 Vienna Convention on Contracts for the International Sale of Goods)。

Closing contracts in special cases
特殊情况中结清合同

Bylaw 203
规章 203

1　If a buyer or seller (in circumstances not covered by other regulations):
1　如买方或卖方有如下情形（在不属于其他规定覆盖范围的情况下）：

suspends payment;
拖欠付款；

enters into an arrangement with his creditors;
与其债权人达成某种安排；

has a receiver or administrator appointed to run his business;
指定接收人或管理人代营其业务；

is asked to wind up the company through a petition; or
被要求申请停业清算；或

is judged by the Directors to be unable to continue to manage his affairs (or dies);
被董事会裁定为无力继续管理其事务（或去世）；

either party may give the President full written details and ask for the contract to be closed. The President may then appoint a tribunal to decide whether it should be closed. The President will fix a fee for the arbitrators which must be paid by the party who asked the President to take action. If the party paying the fee is not a Principal Firm they must pay us an extra fee set by the Directors.

则任何一方均可向会长提供完整的详细书面说明，并请求终止合同。会长可以指定一个仲裁庭裁决合同是否应当终止。会长将确定要求会长采取行动一方必须支付的仲裁员费用。如支付费用的当事方不是主公司，则其必须向我们支付由董事会确定的额外收费。

2 If the arbitrators decide the contract should be closed, they will fix the prices and terms for closing. Either party can appeal to the Directors against the arbitrators' decisions. The appeal must be made in writing to the Secretary, accompanied by the reasons for appeal, within seven days (one week).

2 如仲裁员裁定合同应当终止,则将确定结算的价格和条款。任何一方均可就仲裁员的裁决向董事会提起上诉。上述必须在七天(一周)内以书面形式提交秘书,并附上诉理由。

Section 2 RULES
第2部分 规则

Rules are the non-mandatory provisions of the Association and can be varied by the mutual agreement of the parties.

规则为本协会非强制性规定,可由当事各方协商一致后变更。

Shipment and the Bill of Lading
装运与提单

Rule 200

规则 200

A signed bill of lading will be evidence of the date of shipment.

经签字的提单可作为装运日期的证据。

Rule 201

规则 201

1 The seller must provide an invoice or full and correct details of marks, ships' names and other facts contained in the bill of lading within the time set out in the contract. If the seller does not do so, the buyer can close all or part of the contract covered in the bill of lading and invoice it back to the seller as laid down in our Rules. The buyer must do this within 14 days (two weeks) of the deadline set out in the contract. If the seller provides the invoice or details after the deadline, and the buyer intends to close the contract or any part of it, he must let the seller know within three days.

1 卖方必须在合同规定的时限内提供发票或提单所载的完整和正确的唛头、船名和其他情况细节。如卖方未提供,则买方可按我们规则之规定结清提单所覆盖之全部或部分合同,并向卖方回开发票。买方必须在合同规定的 14

天（两周）期限内按此操作。如卖方在时限过后方提供发票或提单详细情况，则买方如欲终止合同或其任何一部分，必须在三天内通知卖方。

2　If there is no time limit set in the contract and the seller does not provide the invoice or details within 21 days (three weeks) of the date of the bill of lading, the above will apply.

2　如合同中未规定时限，卖方在提单出具之日起 21 天（三周）内未提供发票或详情，则可适用上述规定。

3　Shipping Instructions and Letters of Credit must be issued for the full value of the quantity of the shipment, notwithstanding the allowed variation in weight of the shipment. (Please see Rule 220).

3　装运指示和信用证必须按装运量全额开具，无论是否有装运量允许差值。（请参见规则 220）。

4　In the event that Letters of Credit are opened late, or Shipments have not been made as stipulated in the contract, then both parties may agree to extend the shipment period. If the parties cannot agree to extend the shipment period, then Rule 237 and Rule 238 apply.

4　如信用证开具时间延后，或未按合同规定装运，则双方可以协商同意延长装运期。如双方不能就装运期延展达成一致，则适用规则 237 和规则 238。

5　Slight differences in marks will not be relevant.

5　唛头有轻微差异不构成影响。

Rule 202

规则 202

If the buyer can prove that the details set out in the bill of lading are incorrect or do not meet the terms of the contract, he can take the matter to arbitration. The arbitrators will decide whether the buyer should accept the cotton with an allowance or have a chance to close out the contract. For shipments over land, the buyer must apply for arbitration within 42 days (six weeks) of receiving the details. For shipments by sea, he must apply within 28 days (four weeks) of receiving the details.

如买方能证明提单规定的详情出错或不符合合同条款，可将相关事项提交仲裁。仲裁员将裁定买方是否应当在接受补贴的情况下接受棉花，还是可以终止合同。如系陆地运输，则买方必须在收到提单详情后 42 天（六周）内提出仲裁申请。如系海运，则买方必须在收到提单详情后 28 天（四周）内提出仲裁申请。

Rule 203

规则 203

The contract will not be closed if the cotton, or part of it, is shut out from the named ship, as long as the bill of lading is correct and fits the definition given in Bylaw 100. This only applies to contracts for shipment, not to contracts for sailing or clearance.

如棉花或其部分滞留未装上指定货船，但只要提单正确并符合规章 100 所规定之定义，则合同不得终止。以上仅适用于货运合同，不适用于航行或通关合同。

Rule 204

规则 204

If there is a dispute over a contract for the shipment of American cotton in containers from US ports it will be settled under the 'Container Trade Rules' set out in Appendix B of our Rule Book.

如就从美国港口起运的集装箱装载美国棉货运合同发生争议，则可按我们的规则手册附录 B 规定的"集装箱贸易规则协议"处理。

Insurance
保险

Rule 205

规则 205

When a buyer or seller takes out insurance on a shipment of cotton under a contract made under our Bylaws and Rules, the insurance must include:

如买方或卖方为依据我们的规章和规则而订立的合同项下之棉花货运投保，则该等保险须包括：

'Marine cargo insurance' and 'transit insurance' in line with the Institute Cargo Clauses (A) or Institute Commodity Trades Clauses (A);

符合协会货运保险条款（A）或协会商品贸易保险条款（A）的"海上运输保险"和"运送保险"；

'War Risks Insurance' in line with the Institute War Clauses (Cargo) or the Institute War Clauses (Commodity Trades);

符合协会战争险条款（货物）或协会战争险条款（商品贸易）的"兵险保险"；

'Strikes, riots and civil commotions insurance' in line with the Institute Strikes, Clauses (Cargo) or Institute Strikes Clauses (Commodity Trades), and cover the invoice value of the shipment plus 10%.

符合协会罢工险条款（货物）或协会罢工险条款（商品贸易）的"罢工、暴动及民变保险"；并且保额覆盖货品全价另加10%。

Rule 206

规则 206

Unless otherwise agreed between the parties, the seller shall be responsible for country damage, subject to the limitations detailed in Rule 208 (b).

除非双方另有约定，否则卖方须对产地污损负责，但以规则 208(b)所述之限制为前提。

Rule 207

规则 207

The following conditions apply to contracts where the seller is responsible for providing marine cargo insurance, transit insurance and country damage insurance:

以下条件适用于卖方负责提供海上运输保险、运送保险和产地污损保险的合同：

a　There must be a policy document or certificate of insurance. This document or certificate must be produced as one of the shipping documents.

a　必须有保单文件或保险证明。该文件或证明必须作为货运文件之一制作。

b　If the cotton is country-damaged when it arrives, the buyer must separate the damaged bales and must make a claim against the seller within seven days (one week) of weighing or devanning, whichever is later, notwithstanding that the claim must be made within 42 days (six weeks) of arrival of the conveyance at the location or point of delivery stated on the bill of lading. The parties must try to agree on an allowance. If they cannot do so, a Lloyd's Agent, or a qualified surveyor recognised by the insurance company shall be appointed to inspect the damaged cotton. The cost of the survey shall be for buyer's account in the first instance. If the survey confirms country damage, the seller's insurance shall be called upon to pay:

b　如棉花到达时发生产地污损，则买方必须将已受损的棉包分离，并且必须在七天（一周）内就重量或损毁对卖方提出索赔，以后发生者为准，惟索赔必须在货物运抵提单所述地点或交付地后42天（六周）内提出。双方必须尝试就补贴事宜进行协商。如双方无法达成一致，则须指定保险公司认可的劳合社代理人或适格调查员对受损棉花进行检验。调查费用先由买方承担。如调查确认发生产地污损，则卖方之保险人须当：

The buyer, for the market value of country damaged cotton removed from the bales as set out in the surveyor's report, plus any reasonable charges incurred in the separation of the country damaged cotton; and the cost of the survey.

按调查员报告中所述从棉包分离的受产地污损的棉花的市场价值向买方赔付，外加在分离受损棉花过程中所发生的合理费用；以及支付调查费用。

c If a charge is made for collecting the insurance claim and the buyer pays it, the seller must refund the buyer. If the loss is not covered by seller's insurance the seller must pay.

c 如因提起保险索赔而发生费用且买方已先行支付，则卖方须补偿买方。如卖方保险不足以覆盖全部损失，则卖方必须支付剩余部分。

Rule 208
规则 208

The following conditions apply to contracts where the buyer is responsible for providing marine cargo insurance or transit insurance, and the seller is responsible for providing country damage insurance:

以下条件适用于买方负责提供海上货物险或运送保险，卖方提供产地污损险的合同：

a So that the buyer can arrange insurance, the seller must give the buyer the necessary details of each shipment.

a 为让买方安排投保起见，卖方必须向买方提供每批货品的必要详情。

b If the cotton is country-damaged, the buyer must separate the damaged bales and must make a claim against the seller within seven days (one week) of weighing or devanning, whichever is later notwithstanding that the claim must be made within 42 days (six weeks) of arrival of the conveyance at the location or point of delivery stated on the bill of lading.

b 如棉花到达时发生产地污损，则买方必须将已受损的棉包分离，并且必须在七天（一周）内就重量或损毁对卖方提出索赔，以后发生者为准，惟索赔必须在货物运抵提单所述地点或交付地后 42 天（六周）内提出。

The parties must try to agree on an allowance. If they cannot do so, a Lloyd's Agent, or a qualified surveyor recognised by the insurance company shall be appointed to inspect the damaged cotton. The cost of the survey shall be for buyer's account in the first instance. If the survey confirms country damage and that the damage is greater than 1.0% (one percent) of the total weight of the shipment, subject to a minimum claim of US$ 500.00, the seller's insurance shall be called upon to pay:

双方必须尝试就补贴事宜进行协商。如双方无法达成一致，则须指定保险公司认可的劳合社代理人或适格调查员对受损棉花进行检验。调查费用先由买方承担。如调查确认发生产地污损，并且损害超过货品总重的 1.0%（百分之一），在最小索赔额为 500.00 美元之前提下，卖方之保险人须：

The buyer, for the market value of any country damaged cotton removed from the bales as set out in the surveyor's report, plus any reasonable charges incurred in the separation of the country damaged cotton; and the cost of the survey.

按调查员报告中所述从棉包分离的受产地污损的棉花的市场价值向买方赔付，外加在分离受损棉花过程中所发生的合理费用；以及支付调查费用。

c If a charge is made for collecting the insurance claim and the buyer pays it, the seller must refund the buyer. If the loss is not covered by the seller's insurance the seller must pay.

c 如因提起保险索赔而发生费用且买方已先行支付，则卖方须补偿买方。如卖方保险不足以覆盖全部损失，则卖方必须支付剩余部分。

Rule 209
规则 209

1 The seller must refund the buyer any extra charge or premium which the buyer has to pay if:

1 如发生下列情况，则卖方必须向买方退还其必须不得不支付的任何额外费用和保险金：

The buyer is responsible for marine insurance;
买方负责购买海事险；

The seller is responsible for booking the freight;
卖方负责预定货运；

The seller books the freight on a different ship from the one the buyer has asked for; and

卖方预定之货轮与买方所要求的不同；以及

The ship is subject to an additional premium under the terms of the Institute

Classification clause of the Institute of London Underwriters or another similar clause in force when the buyer learns the name of the ship.

根据伦敦保险商协会"协会船级条款"或买方获悉该船名时有效的其他类似条款之规定须额外多付保险费。

2 The buyer must pay the seller any extra charge or premium if:

2 在下列情况下，买方必须向卖方支付一切额外的费用或保险费：

the seller is responsible for marine insurance;

卖方负责购买海事险；

the buyer is responsible for booking the freight;

买方负责预定货运；

the buyer books the freight on a different ship from the one the seller has asked for; and

买方预定之货轮与卖方所要求的不同；以及

the ship is subject to an additional premium under the terms of the Institute Classification clause of the Institute of London Underwriters or another similar clause in force when the seller learns the name of the ship.

根据伦敦保险商协会"协会船级条款"或卖方获悉该船名时有效的其他类似条款之规定须额外多付保险费。

Invoicing and payment
开票及付款

Rule 210
规则 210

When the shipment arrives, the payment must be made on arrival or within 49 days (seven weeks) of the date on the bill of lading or shipping documents, whichever is earlier. Upon first presentation of the contracted shipping documents, the payment must be made within three working days unless otherwise agreed by the parties.

货船抵达后，必须在到达之日或者提单或货运文件日期后49天（七周）内支付货款，以先到者为准。合同所约定的货运文件首次展示后，须在三个工作日内付款，除非双方另有约定。

Rule 211
规则 211

Claims that are made in accordance with the terms of the contract must be paid within 21 days (three weeks) of the claim date. If the party responsible for the payment does not do so, they will also have to pay interest on the final amount of the claim at a rate agreed by both parties. If the parties cannot agree, the claim amount and interest rate will be fixed by arbitration under our Bylaws.

依据合同条款提出的索赔必须在索赔之日起 21 天（三周）内支付。如负有付款义务的一方未按此履行，则还须按双方议定的利率根据最终索赔金额支付利息。如双方不能达成一致，则索赔金额和利息将依据我们的规章通过仲裁确定。

Rule 212

规则 212

Claims for clerical errors in invoices will be accepted if there is evidence to support.

如有证据支持，可以接受发票笔误的主张。

Rule 213

规则 213

The price of cotton set out in the contract will not include any Value Added Tax due, unless the contract says that it does.

合同约定的棉花价格不包括任何应付的增值税，除非合同明确规定包含在内。

Sales 'on call'
"叫价"销售

Rule 214

规则 214

1　On buyer's call:

1　应买方叫价：

　i　For sales on call Intercontinental Exchange ('ICE') Cotton Contract No. 2 Futures:

　i　对于美国洲际交易所（ICE）二号棉花期货合约叫价销售：

The final price of cotton sold on call will be fixed based on the ICE Cotton Contract No. 2 Futures contract month specified in the sales contract.

叫价销售的棉花最终价格将以销售合同所定 ICE 二号棉花期货合约当月定价为基础确定。

The buyer should communicate to the seller an executable fixation instruction.The seller should communicate to the buyers in writing any filled fixation and the resultant fixed price.

买方应将可执行的定价指示通知卖方。卖方应以书面方式通知买方已填报的定价细节和定价结果。

Unless agreed otherwise by the parties:
除非双方另行协商一致，否则：

Cotton must be fixed no later than the ICE Cotton Contract No. 2 Futures on the day prior to first notice day for the futures contract month specified in the sales contract.

棉花定价必须在销售合同所定 ICE 二号棉花期货合约的期货合约月份的第一个通知日之前确定。

If cotton has not been fixed by this time, the final price shall be based on the ICE Cotton Contract No. 2 Futures closing price on the day prior to first notice day of the futures contract month specified in the sales contract.

如届时棉花仍未定价，则最终价格须以销售合同所定的期货合约月份的第一个通知日前一日的 ICE 二号棉花期货合约收盘价格为基础。

ii For sales on call with reference to products other than the ICE Cotton Contract No. 2 Futures Market:

ii 对于 ICE 二号棉花期货合约市场以外产品叫价销售：

The final price of cotton sold on call will be fixed based on the quotation of the product specified in the sales contract.

叫价销售的棉花的最终价格将以销售合同所定产品报价为基础确定。

The buyer should communicate to the seller an executable fixation instruction. The seller should communicate to the buyers in writing any filled fixation and the resultant fixed price.

买方应将可执行的定价指示通知卖方。卖方应以书面方式通知买方已填报的定价细节和定价结果。

Unless agreed otherwise by the parties:
除非双方另行协商一致，否则：

Cotton must be fixed prior to the expiration of the product.
棉花必须在产品到期日之前确定。

If cotton has not been fixed prior to the expiration of the product then the fixation shall be based on the last published quotation of the product, or if no expiration date then on the date of shipment/delivery.

如在产品到期日之前棉花仍未确定,则须在产品最后一次公开报价的基础上确定价格,或者,如无到期日,则按装运/交付日确定。

2 On seller's call, the roles of the buyer and seller are reversed.

2 应卖方叫价,买卖双方的角色互换。

Bale Tare and Weight
皮重和重量

Rule 215

规则 215

1 Unless the seller declares and guarantees otherwise, all cotton must be sold on actual tare.

1 除非卖方另有宣告和保证,否则所有棉花必须按实际皮重出售。

2 The buyer can insist that the actual tare be established at the time of delivery. The actual tare must be measured within 28 days (four weeks) of the date of arrival of the cotton and must be carried out by the buyer under the supervision of the seller's representatives. This will then be the measurement of tare applied to the weight adjustment.

2 买方可以坚持要求实际皮重按交付之时为准。实际皮重必须在棉花到货之日起 28 天(四周)内称量,并且必须在卖方代表监督下由买方执行。此即为将来进行重量调整的皮重称量。

3 If the buyer insists that the tare be established after arrival and it proves to be not more than the allowance in the contract or invoice, the buyer will have to pay the costs of taring, otherwise, the seller must pay these costs.

3 如买方坚持在到货后确定皮重,并且证明未超过合同或发票规定的补贴额,则买方必须支付皮重成本,否则,由卖方支付成本。

Rule 216

规则 216

1 To calculate actual tare, a minimum of 5% of the bales, subject to a minimum of five bales of each type of tare composed in any one lot or mark must be checked.

1 为计算实际皮重,必须检验至少 5%的棉包,并且对同一批次或唛头的每一种棉包至少抽检 5 个。

2 Actual tare is established by ascertaining the average weight of the wrapping, bands, ropes or wires from each type of the different tares comprising the

lot or mark and multiplying the average weight of each type of tare by the total number of bales in the shipment.

2 在确定实际皮重时，先确定同一批次或唛头所包含每一种不同棉包的包装物、包装带、绳索或金属线的平均重量，再以每一种棉包皮重的平均重量乘以货品总棉包数。

3 Repaired bales must be tared separately.

3 经修复的棉包必须单独称量皮重。

Rule 217

规则 217

All cotton must be weighed 'gross weight' on a bale by bale basis unless otherwise agreed. The tare is to be deducted from the gross weight.

所有棉花都必须逐包称量"毛重"，除非另有约定。皮重从毛重中扣除。

Rule 218

规则 218

1 Gross Shipping Weights-must be established by an independent weighing organisation or other organisation as determined in writing between the buyer and seller within 28 days (four weeks), or any other time period as agreed between buyer and seller, after sampling and before shipment.

1 运货毛重-必须由买卖双方于货物装运前、样品提供后28天（四周），或双方议定的其他期间内以书面方式确定的独立称量组织或其他组织确定。

2 Gross Landing Weights-all cotton must be weighed by the buyer (for buyer's cost), under the supervision of the seller's representatives (for seller's cost) at the agreed point of delivery or other location as determined by the buyer and seller, in any event within 28 days (four weeks) of the date of arrival of the cotton. If the cotton has already been sampled, a weight allowance must be made for the samples taken.

2 卸货毛重-所有棉花均必须于到达之日起28天（四周）内，在议定的交付地点或买卖双方商定的其他地点，由买方（买方承担费用）在卖方代表（卖方承担费用）监督下进行称重。如棉花已取样品，则必须对样品部分进行重量补正。

3 Both the buyer and the seller can appoint representatives at their own cost to supervise any weighing. The party arranging the weighing must advise the other party where and when it will take place, allowing a reasonable time to enable the representative to attend.

3 买卖双方均可自担费用任命代表监督称重工作。安排称重的一方必须将称重的时间地点通知另一方，以给予其合理的时间派代表参加。

Rule 219

规则 219

1 The weight of bales which are condemned, short-landed, burst, wrongly marked or not marked will be calculated according to the average gross weight of the landed bales, as long as at least 25% of the lot has been landed in good condition. If less than 25% is in good condition, the weight of these bales will be calculated according to the average invoice weight.

1 报废、短缺、破裂、唛头错误或未打唛头的棉包重量按到岸棉包的平均毛重计算，只要到岸的同一批次货品中至少有 25% 处于良好状态。若处于良好状态者不足 25%，则该等棉包按平均发票重量计算。

2 If the buyer accepts bales which are wrongly marked or not marked, those bales will be weighed, and the weights shown separately.

2 若买方接受唛头错误或未打唛头的棉包，则该等棉包将进行称重，其重量独立标示。

3 If the buyer does not weigh the total shipment within 28 days (four weeks) of the date of the arrival of the cotton, the unweighed bales will be calculated according to the average gross weight of the weighed bales, as long as at least 90% of the lot has been weighed. If less than 90% of the lot has been weighed, the weight of the unweighed bales will be calculated according to the average invoice weight.

3 若在棉花到达后 28 天（四周）内买方仍未对全部货品称重，则未称重部分的棉包按已称重部分棉包平均毛重计算，只要该批次已有至少 90% 进行了称重。若已称重不足 90%，则未称重部分棉包按平均发票重量计算。

4 If the shipment is by container and all the containers are loaded onto one ship, the 25% referred to in paragraph (1) of this Rule will apply to the total number of bales delivered.

4 若以集装箱装运，并且所有集装箱装载于一艘货船，则本条规则第(1)节所述 25% 比例适用于已交付棉包总数。

5 If the shipment is by container and the containers are loaded onto more than one ship, the 25% referred to in paragraph (1) of this Rule will apply to the number of bales delivered in each ship.

5 若以集装箱装运，而所有集装箱分装于多艘货船，则本条规则第(1)节所述25%比例分别适用于各艘货船已交付棉包总数。

Rule 220

规则 220

When contracts are made for shipments or deliveries of specified quantities during various shipment/delivery periods, each shipment or delivery should fall within the allowed variation. Each month's shipment or delivery shall form one weight settlement, even if shipped or arriving by more than one conveyance. Proof of any variation in weight, must be sent to the other party within 49 days (seven weeks) of the date of arrival of the cotton. Compensation for variation in weight will normally be based on the invoice price. But, if the variation is more than the amount allowed for in the contract, the buyer may then demand compensation for the market difference over that amount of variation, based on the market value of the cotton on the date of arrival of the cotton. If the contract does not specify an allowable variation, the variation allowed will be 3%.

如合同约定在多次货运/交货期间内交运特定数量的货品,则每一批货运或交货均须符合差值标准。每个月的货运或交货构成一个称量结算单位,即便货运或到货涉及多次运输。重量差值的证明必须在棉花达到之日起49天(七周)内送交对方。重量差值的补偿一般以发票价格为基础。但若差值量超过合同允许范围,则买方可以要求按市场差价对差值量进行补偿,以棉花到达之日棉花的市场价值为基础计算。如合同未规定允许的差值量,则允许的量为3%。

Quality of the cotton delivered

棉花交货质量

Rule 221

规则 221

Unless 'average' has been stated in the contract, when cotton is sold on the description of grade, the cotton must be equal to or better than contracted quality.

除非合同中注明"平均",否则在按等级描述出售棉花时,棉花必须相当于或优于合同约定的质量。

Rule 222

规则 222

1 The buyer and seller can say in the contract what the grade, length, micronaire, strength and other fibre characteristics of the delivered cotton must be. The contract can also lay down what allowances, differences, limits and so on apply,

and, where applicable, what type of instruments must be used to establish the characteristics in the event of a dispute.

 1 买卖双方可以在合同中约定所交付的棉花必须具备一定的等级、长度、马克隆、强度和其他纤维属性。合同也可以规定补贴、差额、限制及如何适用，以及，如可适用，在发生争议的情况下使用何种手段确定问题的性质。

 2 If the buyer and seller disagree about a claim, the dispute will be settled by arbitration under our Bylaws.

 2 如买卖双方就某项索赔发生争议，则将依据我们的规章通过仲裁解决。

Sampling
采样

Rule 223
规则 223

 1 Sampling must take place at the point of delivery or other location as determined between buyer and seller. The buyer's and seller's representatives must supervise the sampling. The seller must give the name of his representative to the buyer: before sending the buyer an invoice; or with the invoice.

 1 采样必须在买卖双方确定的交付地点或其他地点进行。取样必须有由买卖双方的代表监督。卖方必须在向买方寄送发票前告知代表姓名；或者发票与代表姓名同时送达。

 2 Samples for arbitration must be drawn, sealed and marked in the presence of both the buyer and seller and/or their respective representatives.

 2 仲裁采样必须在买卖双方和/或各自代表在场的情况下采集、封存和标注。

Rule 224
规则 224

 1 A sample from a bale of cotton should weigh about 150 grams.

 1 从一个棉包取出的样品重量应为约 150 克。

 2 For manual classification claims and/or arbitration, American and Australian cotton must be sampled 100%. Unless otherwise agreed, other cottons need only be sampled on the basis of 10% representative samples from each lot or mark as defined on the seller's commercial invoice.

 2 若发生手工评级索赔和/或仲裁，则美国棉和澳大利亚棉均须 100%采样。除非另有约定，其他棉品只须从卖方商业发票所注明的各批次或唛头中抽取 10%的代表样品即可。

3 Samples may be drawn from part lots and/or shipments; however, a claim may only be made on the number of bales available at the time of sampling.

3 样品可以从部分批次和/或货物中抽取；但索赔只能针对采样时适用的棉包数提起。

4 For instrument testing, claims and/or arbitrations may only be made on individual bales specified by the party applying for instrument testing.

4 就仪器检测而言，索赔和/或仲裁只能针对申请仪器检测方所指明的个别棉包提起。

For arbitration 100% of the bales claimed must be sampled.

在仲裁中，提出索赔的棉包 100%都必须取样。

5 In the event that a quality arbitration award is made, the cost of drawing, supervision of drawing and dispatch of samples shall be:

5 在作出一项质量仲裁裁决的情况下，样品抽取、抽取监督和样品传送的费用按如下方式处理：

for the party whose final written offer for amicable settlement is furthest from the quality arbitration award; or

由在友好协商过程中所提出最终书面报价距离质量仲裁裁决内容最远的一方承担；或

for the buyer if the quality award is less than the seller's final offer for amicable settlement; or

如质量裁决的内容低于卖方在友好协商中提出的最终报价的条件，则由买方承担；或

shared in equal proportions if neither party has made a written offer for amicable settlement.

若在友好协商中各方均未提出书面报价，则按相同比例承担。

6 If the buyer or seller believes that the cotton or cotton waste is false packed, mixed packed or in plated bales, every bale must be sampled, and samples must be drawn from each side of the bale.

6 若买方或卖方认为棉花或废棉系掺次包装、混杂包装或夹次棉包，则每个棉包都必须取样，并且样品必须从棉包两侧采集。

Rule 225

规则 225

The buyer must not sample the bales before weighing without the seller's permission.

未经卖方许可,买方不得在称重前对棉包取样。

Rule 226

规则 226

If the seller takes a set of samples, he must pay for them at the contract price of the cotton.

如卖方采集了一系列样品,则他必须按棉花合同价支付价款。

Claims
索赔

False packed, mixed plated bales and bales containing foreign matter
掺次包装、混杂夹次包装和含异物的货包

Rule 227

规则 227

1   The buyer must claim for false packed, mixed packed or plated bales within six months (26 weeks) of the date of arrival of the cotton. The bales must be set aside for inspection for 28 days (four weeks) after the claim is made and the inspection must be done by an agreed expert. If the seller tells the buyer within 14 days (two weeks) of the claim being proved that he intends to take back this cotton, he has the right to do so. If the buyer has already paid for the cotton, the seller must buy it back at the market value of good cotton on the date the claim is proved and repay the buyer his substantiated expenses.

1   买方必须在棉花到货之日起六个月(26周)内就掺次包装、混杂包装或夹次棉包提出索赔。索赔提出后,棉包必须留存28天(四周)供检验,并且检验必须由经认可的专家进行。若卖方在索赔证实后14天(两周)内告知买方其欲收回棉花,则他有权如此行事。若买方已经支付了棉花价款,则卖方必须按索赔证实当日优良棉花市场价买回,并向买方赔偿其经证明的支出。

2   If the seller does not take back the cotton, the claim must be settled based on the market value of good cotton on the date the claim is proved to the seller. The seller must also repay the buyer his substantiated expenses.

2   若卖方不收回棉花,则索赔必须按向卖方证实当日优良棉花的市场价进行结算。卖方还必须向买方赔偿其经证明的支出。

3   The buyer must claim for unmerchantable cotton within six months (26 weeks) of the date of arrival of the cotton. The bales must be set aside for

inspection for a further 28 days (four weeks) after the claim is made and the inspection must be done by an agreed expert. The buyer will be able to claim reasonable and substantiated expenses from the seller for opening the bales and separating the merchantable from the unmerchantable cotton. The buyer can also claim the value of any unmerchantable cotton removed from the bales.The value must be based on the market value of the merchantable cotton on the date the claim is proved to the seller.

3 买方必须在棉花到货之日起六个月（26周）内就不适销的棉花提出索赔。索赔提出后，棉包必须留存28天（四周）供检验，并且检验必须由经认可的专家进行。买方可以就打开棉包以及分离适销棉花与不适销棉花所发生的合理和经证实费用向卖方索赔。买方还可以就从棉包中分离的不适销棉花的价值提出索赔。价值必须按索赔向卖方证实当日适销棉花的市场价值确定。

4 Foreign matter-the buyer must claim for foreign matter in the cotton within six months (26 weeks) of the date of arrival of the cotton. The bales must be set aside for inspection for 28 days (four weeks) after the claim is made and the inspection must be done by an agreed expert. The buyer will be able to claim reasonable substantiated expenses from the seller for removal of the foreign matter.

4 异物-买方必须在棉花到货之日起六个月（26 周）内就棉花中的异物提出索赔。索赔提出后，棉包必须留存28天（四周）供检验，并且检验必须由经认可的专家进行。买方可以就去除异物所发生的合理经证实的支出向卖方提出索赔。

Rule 228

规则 228

The buyer must give notice of any claim for country damage as detailed in Rule 207 or Rule 208 and the survey shall be completed within 14 days (two weeks) of the notice of the claim, or within 56 days (eight weeks) of the date of arrival of the cotton, whichever is earlier.

买方必须就规则207或规则208详述的产地污损索赔发出通知，并且调查工作须在索赔通知发出后14天（两周）内，或在棉花到货之日起56天（八周）内完成，以先到者为准。

Rule 229

规则 229

The following will apply when sampling bales to test for internal moisture:

对棉包采样进行内含水分检验时适用如下规定：

Samples of at least 250 grams must be taken from each bale to be sampled. These samples must be taken by the representative of the party who has asked for the test, and in the presence of a representative of the other party (if it appoints one). The samples must be taken at the time of weighing.

必须从每个待采样棉包中采集至少 250 克样品。样品必须由要求检验的一方代表采集，另一方代表在场（若其指定代表）。样品必须在称重时采集。

Representative samples must be taken from 5% of the bales in each lot (at least three bales). These bales must be selected at random. Samples must be taken from at least two different parts of each bale from a depth of about 40 centimetres inside the bale. The samples must be placed at once in dry, hermetically-sealed containers and labelled to show the identity of the bale the samples have come from.

代表性样品必须从每一批次的 5%棉包（至少三个棉包）中采集。棉包必须随机选择。样品必须从每个棉包至少两处不同的部分采集，深度为深入棉包约 40 厘米。样品必须立即置于干燥、完全密封的容器中并贴上标签显示样品采自棉包的识别号。

The samples must be sent immediately to a testing laboratory mutually acceptable to both parties.

样品必须立即送往双方共同接受的检验实验室。

Rule 230
规则 230

1　The buyer must:

1　买方必须：

give notice of any claim for internal moisture within 42 days (six weeks); and

在 42 天（六周）内就内含水分索赔发出通知；并且

produce a report from a mutually agreed laboratory and final claim within 63 days (nine weeks) of the date of arrival of the cotton.

在 63 天（九周）内提供双方同意的实验室出具的报告以及最终的索赔；以上时限按棉花到货日起计算。

2　The allowance given to the buyer will be based on the laboratory's report. The allowance will be the difference between:

2　给予买方的补贴将以实验室报告为依据确定。补贴应为以下二者的差价：

the weight of the absolutely dry fibre in the lot plus the percentage of moisture regain set out in the contract; and the total weight of the lot. This allowance will also be based on the invoice price.

该批次中完全干燥纤维加上合同中规定回潮水分的百分比；与该批次的总重量。此项补贴也以发票价格为基础确定。

Rule 231

规则 231

The party claiming and asking for the moisture test will have to pay the cost of sampling and all related charges. If the claim is proved, sampling, courier and laboratory charges will be reimbursed by the other party.

提出索赔并要求进行水分检验的一方必须支付取样成本和所有相关费用。若索赔获得证实，则对方将退还取样、运输和实验室费用。

Extending time limits
延长期限

Rule 232

规则 232

The Directors can extend any time limit stated in Rule 218, 220, 227, 228 or 230 but only if the firm concerned can show that substantial injustice would otherwise be done:

董事会可以延长规则 218、220、227、228 或 230 中所述的时限，但仅限于所涉及的公司能够表明不如此就会导致重大不公正的情形：

because it could not reasonably have anticipated the delay; or

因为其无法合理预期该等迟延；或

because of the conduct of the other firm.

因为对方公司的行为。

Applications must be made to us in writing. The Directors will take the other firm's comments into account before they make a decision.

申请必须以书面方式向我们提出。董事会在作出决定前将考虑对方公司的意见。

Instrument testing
仪器检测

Rule 233

规则 233

This Rule applies to all quality disputes regarding testing of cotton samples of any origin by instruments.

本规则适用于所有涉及使用仪器对任何原产地棉花样品进行检测的质量争议。

1   High Volume Instrument testing or classification shall be carried out in accordance with the approved practices and procedures listed in the latest version of the Universal Cotton Standards Agreement between the United States Department of Agriculture and the international signatories.

1   高容量仪器检测或评级须按美国农业部和国际签署方签订的最新版本《全球棉花标准》所规定的经核准操作和程序进行。

2   At least two tests shall be made on each sample. The average result of the tests shall be the test result.

2   对每份样品须至少检验两次。各次检测的结果平均值为检验结果。

3   If sealed samples have already been taken for manual arbitration in accordance with Rule 223, the same samples can be used for the tests, provided they have been resealed.

3   若封存样品已经取出供进行规则 223 规定的手工检验仲裁，则同一份样品可以用于检测，只要其有再封存。

4   A first set of tests will be done in a laboratory agreed between the buyer and seller. If there is no agreement, the tests will be undertaken in a certified laboratory selected by the party applying for the test.

4   首个系列的检测将在一家买卖双方同意的实验室中进行。如双方未达成协议，则检测将在申请检测方选择的一家经认证实验室进行。

5   In the event the first test was undertaken in a certified laboratory it will be final, and no request for a second test will be allowed.

5   若首次检测系在一家经认证实验室进行，则即为终局结果，不得再请求作二次检测。

6   The laboratory which does the first test will issue a test report signed and/or stamped by its authorised personnel. The test report will show the results of the test. If testing is performed by a non-certified laboratory the samples will be resealed by the laboratory and retained for up to 35 days (five weeks) in case a second test by a certified laboratory is called for.

6   进行首次检测的实验室将签发一份由其经授权人员签署和/或盖章的

检验报告。检验报告将表明检测结果。如检测系由一家非认证实验室进行，则实验室将重新封存样品，保留至多 35 天（五周），以备有认证实验室作二次检测。

7　Subject to paragraph (5) above, either firm can request a second test within 21 days (three weeks) of the first results being dispatched. If no request is lodged, the information on the test report will be final.

7　在上文第（5）节前提下，任何一方公司均可在首次检验结果发布后 21 天（三周）内要求作二次检验。如无人提起请求，则检测报告所载内容即为最终结果。

8　Any request for a second test must apply to the total number of bales in the first test. A second test may only be undertaken in a certified laboratory agreed between the parties. In the event of no agreement, the claimant will indicate the certified laboratory to be used. The tests will be made on samples of cotton drawn from the original resealed samples. The party applying for the second test shall pay for the resealed samples to be dispatched to the certified laboratory designated for the second test.

8　二次检验请求必须适用于首次检验中的全部棉包。二次检验仅可在双方同意的一家经认证实验室进行。若无法达成协议，则由申请人指定所使用的认证实验室。检测将使用从原来再封存的样品中采集的样品。申请进行二次检验的一方须支付再封存样品送至指定承担二次检验的认证实验室的费用。

9　Test reports will be issued and signed and/or stamped by the laboratory's authorised personnel.

9　检验报告由实验室经授权人员签署和/或盖章后签发。

10　In the event the parties cannot reach agreement on the allowances to be applied, or the interpretation of the results, arbitrator(s) may be appointed by, or on behalf of, both parties.

10　若双方对所适用的补贴或检验结果的解释无法达成一致意见，则可由双方任命仲裁员或仲裁庭进行裁决。

11　A contract may say how much variation is acceptable in the fibre characteristics determined by the laboratory tests. Control limits should be stipulated in the contract.

11　合同中可以说明由实验室检测确定的纤维特性可接受的数值变动范

围。合同中应规定控制界限。

12 For micronaire, unless the parties agree otherwise, the usual control limit of 0.3 will apply.

12 对于马克隆,除非双方另有约定,否则适用一般控制界限 0.3。

13 For strength, unless the parties agree otherwise, the usual control limit of 2.0grams/tex or 3000 psi will apply.

13 对于强度,除非双方另有约定,否则适用一般控制界限 2.0 克/特或 3000 磅/平方英寸。

14 Whichever party asks for the tests must pay the laboratory the whole cost. But, if the buyer pays, the seller must repay the cost of testing every bale which does not come within the control limit set out in the contract or, where the control limit is not stated in the contract, within the UCL specified in paragraph (12) and paragraph (13) above.

14 要求进行检测的一方必须支付全部实验室费用。但是,若买方支付费用,卖方必须就不符合合同约定控制界限的棉包,或者在合同未规定控制界限的情况下、上文第(12)和(13)节所述 UCL 的每个棉包向买方补偿成本。

Micronaire and allowances
马克隆及允许值

Rule 234

规则 234

1 The Rules apply to all disputes relating to micronaire, including disputes relating to American cotton. Its terms are intended to be consistent with a micronaire agreement between us and the American Cotton Shippers Association, but if there is any conflict between the two, the terms of this Rule will take priority after the terms of the contract.

1 本规则适用于所有涉及马克隆的争议,包括涉及美国棉的争议。其中条款旨在与我们和美国棉商协会之间达成的马克隆协议相一致,但若二者间有任何冲突,则本规则条款之优先性低于合同条款。

2 If the contract states 'micronaire' but does not say whether it should be the 'minimum' or 'maximum', it will be taken to mean 'minimum micronaire'. However, both parties can agree otherwise in writing before they send the samples

for testing.

2　若合同提及"马克隆",但并未说明系"最小值"还是"最大值",则将理解为"最小马克隆"。但是,双方可以在送交样品供检测之前以书面方式达成另一一致意见。

3　A contract may say how much variation is acceptable in the other fibre characteristics that can be determined by recognised laboratories.

3　合同中可以说明由经认可实验室检测确定的其他纤维特性可接受的数值变动范围。

Rule 235

规则 235

1　In any dispute about micronaire, the procedure in Rule 233 will apply unless the parties agree otherwise.

1　在涉及马克隆的任何争议中,均适用规则 233 所述之程序,除非双方另有约定。

2　Unless the buyer and seller agree otherwise:

2　除非双方另有约定,否则:

For American cotton:

对于美国棉:

For contracts which set out a minimum micronaire value, the allowances for bales which do not reach this minimum will be as follows:

对于规定最小马克隆值的合同,未达到最低值的棉包允许值如下:

Micronaire value below the control limit by 马克隆值低于控制界限	Percentage allowance 百分比允许值
0.1	1.0
0.2	2.0
0.3	3.0
0.4	4.0
0.5	5.0
0.6	6.0
and so on by 1% for each 0.1 micronaire. 以此类推,每 1%提高 0.1 个马克隆。	

But if the contract sets out a minimum of 3.5 (3.5 NCL or 3.8 UCL) or higher:
但若合同规定最小值为 3.5（3.5 NCL 或 3.8 UCL）或更高，则：

on cotton reading 2.9 to 2.6 inclusive, the percentage allowance will be increased to 3% for each 0.1 micronaire below 3.0; and on cotton reading 2.5 or below, the percentage allowance will be increased to 4% for each 0.1 micronaire below 2.6.

对于读数为 2.9 至 2.6（含）的棉花，在 3.0 以下每 0.1 个马克隆提高百分比允许值至 3%；并且对于读数为 2.5 或以下的棉花，每 0.1 个马克隆提高百分比允许值至 4%。

For contracts which set out a maximum micronaire value, the allowances for bales which go over this maximum will be as follows:
对于规定最大马克隆值的合同，超过最大值的棉包允许值如下：

Micronaire value above the control limit by 马克隆值高于控制界限	Percentage allowance 百分比允许值
0.1	0.5
0.2	1.0
0.3	2.0
0.4	3.0
0.5	4.0
0.6	5.0
and so on by 1% for each 0.1 micronaire. 以此类推，每 1% 提高 0.1 个马克隆。	

But if the contract specifies a maximum micronaire reading of 4.9 or lower:
但若合同规定了最大马克隆读数为 4.9 或以下，则：

on cotton reading 5.6 or higher, the percentage allowance will be increased to 3% for each 0.1 micronaire above 5.6.

对于读数为 5.6 及以上的棉花，在 5.6 以上者每 0.1 个马克隆提高百分比允许值至 3%。

For non-American cotton:
对于非美国棉：

For contracts which set out a minimum micronaire value, the allowances for bales which do not reach this minimum will be as follows:
对于规定最小马克隆值的合同，未达到最低值的棉包允许值如下：

Micronaire value below the control limit by 马克隆值低于控制界限	Percentage allowance 百分比允许值
0.1	0.5
0.2	1.0
0.3	2.0
0.4	3.0
0.5	4.0
0.6	5.0
and so on by 1% for each 0.1 micronaire. 以此类推,每 1%提高 0.1 个马克隆。	

But if the contract sets out a minimum of 3.5 (3.5 NCL or 3.8 UCL) or higher:

但若合同规定最小值为 3.5（3.5 NCL 或 3.8 UCL）或更高，则：

on cotton reading 2.9 to 2.6 inclusive, the percentage allowance will be increased to 3% for each 0.1 micronaire below 3.0; and on cotton reading 2.5 or below, the percentage allowance will be increased to 4% for each 0.1 micronaire below 2.6.

对于读数为 2.9 至 2.6（含）的棉花，在 3.0 以下每 0.1 个马克隆提高百分比允许值至 3%；并且对于读数为 2.5 或以下的棉花，每 0.1 个马克隆提高百分比允许值至 4%。

For contracts which set out a maximum micronaire value, the allowances for bales which go over this maximum will be as follows:

对于规定最大马克隆值的合同，超过最大值的棉包允许值如下：

Micronaire value above the control limit by 马克隆值高于控制界限	Percentage allowance 百分比允许值
0.1	0.5
0.2	1.0
0.3	2.0
0.4	3.0
0.5	4.0
0.6	5.0
and so on by 1% for each 0.1 micronaire. 以此类推,每 1%提高 0.1 个马克隆。	

But if the contract specifies a maximum micronaire reading of 4.9 or lower:

但若合同规定了最大马克隆读数为 4.9 或以下，则：

on cotton reading 5.6 or higher, the percentage allowance will be increased to 3% for each 0.1 micronaire above 5.6.

对于读数为 5.6 及以上的棉花，在 5.6 以上者每 0.1 个马克隆提高百分比允许值至 3%。

Strength and allowances
强度及允许值

Rule 236

规则 236

1　In any dispute about strength, the procedure in Rule 233 will apply unless the parties agree otherwise.

1　在涉及强度的任何争议中，均适用规则 233 所述之程序，除非双方另有约定。

2　Unless the buyer and seller agree otherwise, for contracts which set out a minimum strength value, the allowances for bales which do not reach this minimum will be as follows (HVI):

2　除非买卖双方另有约定，否则对于规定了最低强度值的合同，未达到最低值的棉包允许值按如下规定 (HVI)：

grams/tex below the control limit by 克/特值低于控制界限	minimum 最低	maximum 最高	Percentageallowance 百分比允许值
	1.1	2.0	1.0
	2.1	3.0	1.5
	3.1	4.0	3.0
	4.1	5.0	5.0
	5.1	6.0	8.0
Plus 4% for each gram/tex below 6 低于 6 则每克/特值另加 4%			

Closing contracts
结清合同

Rule 237

规则 237

1 If for any reason a contract or part of a contract has not been, or will not be, performed (whether due to a breach of the contract by either party or due to any other reason whatsoever) it will not be cancelled.

1 如因任何原因，某一合同的全部或部分未履行，或将不会履行（无论由于任何一方违约或其他任何原因），该合同不得撤销。

2 The contract or part of a contract shall in all instances be closed by being invoiced back to the seller in accordance with our Rules in force at the date of the contract.

2 该合同之全部或部分须依据合同签署当日有效的规则以向卖方回开发票的方式结清。

Rule 238

规则 238

Where a contract or part of a contract is to be closed by being invoiced back to the seller, then the following provisions will apply:

若合同之全部或部分以向卖方回开发票的方式结价，则使用以下规定：

1 If the parties cannot agree upon the price at which the contract is to be invoiced back to the seller, then that price will be determined by arbitration, and if necessary, appeal.

1 若双方无法就向卖方回开发票的价格达成一致，则通过仲裁确定价格，如有必要可以申诉。

2 The date of closure is the date when both parties knew, or should have known, that the contract would not be performed. In determining that date the arbitrators or appeal committee will take into account:

2 结算日期为双方知晓、或应当知晓合同无法继续履行之日。 在判定该日期的规程中，仲裁员或申诉委员会将考虑如下因素：

a the terms of the contract;

a 合同的条款；

b the conduct of the parties;

b 双方的行为；

c　any written notice of closure; and
c　任何书面终止通知；以及
d　any other matter which the arbitrators or appeal committee consider to be relevant.
d　仲裁员或申诉委员会认为相关的任何其他因素。

3　In determining the invoicing back price, the arbitrators or Technical Appeal Committee shall have regard to the following:
3　在判定回开发票价格时，仲裁员或技术申诉委员会须考虑如下因素：
a　the date of closure of the contract as determined in paragraph (2) above;
a　按上文第(2)节所述判定的合同结算日期；
b　the terms of the contract; and
b　合同的条款；以及
c　the available market price of the cotton which is the subject of the contract, or such like quality, on the date of closure.
c　结算当日合同系争棉花或类似品质棉花的可适用市场价格。

4　The settlement payable on an invoicing back will be limited to the difference (if any) between the contract price and the available market price at the date of closure.
4　就回开发票达成的应付和解金额限于合同价格与终止当日可适用市场价格之间的差额（如有）。

5　Any settlement due and payable on an invoicing back of a contract closed in accordance with Rules 237 and 238 will be calculated and shall be paid regardless of whether the party receiving or making the payment is considered to be responsible for the non-performance and/or breach of the contract.
5　如依据规则237和238结清合同，所开还发票的应付及到期的结算金额，在计算和支付过程中不考虑收款方或付款方是否被认为应对合同的未履行和/或违约负责。

Other claims and losses
其他索赔和损失

6　Any other losses or claims expressly agreed between the parties as recoverable will not be included in an invoicing back price. Such losses or claims should be settled by amicable settlement, or claimed at arbitration or appeal.
6　双方明确同意可以补偿的任何其他损失或索赔不计入回开发票价格。该等损失或索赔应通过友好协商或仲裁及申诉程序确定。

Rule 239
规则 239
Claims for consequential damages will not be allowed.
不允许对间接损害提出索赔。

Rule 240
规则 240

1 The arbitrators will set the invoicing back weight if:
1 若发生如下情况，则仲裁员将确定回开发票所开重量：

the seller has not provided an invoice; or
卖方未提供发票；或

no actual weights are available; or
没有可适用的实际重量；或

the parties cannot agree the weight.
双方无法就重量达成一致。

2 For the purpose of determining the invoicing back weight, when part of the contract has already been fulfilled, weight tolerances will not apply to the balance.
2 就判定回开发票所开重量之目的而言，如有部分合同已经履行，则重量偏差允许值不适用于余额。

Appendix A
附录 A
Contract Conditions
合约条件

1 Growth and quality
All cotton provided must be of even running quality (ICA Rule 221)
1 品种及质量
全部棉花供货必须保证连续、平均的质量（ICA 规则 221）。

2 Micronaire and Strength
2 马克隆值及强度

Unless we agree otherwise, any dispute about micronaire will be settled under ICA Rules 234 and 235, and any dispute about strength will be settled under ICA Rule 234. If we have not agreed percentage allowances or the use of market differences, or a control limit, the percentage allowances or control limit in the Bylaws will apply.

除非我们另行同意，所有关于马克隆值的争议将根据 ICA 规则 234 及 235

解决，所有关于强度的争议将根据 ICA 规则 234 解决。若我们未同意折让百分比或使用市场差价，或控制界限，则适用规章规则的折让百分比和控制界限。

3　Quantity　Unless we agree otherwise, cotton is to be supplied in high density compressed bales.
　　3　数量　除非我们另行同意，棉花应以高密度压缩货包的形式供货。

4　Shipment　The seller must get any export licence necessary.
　　4　装运　卖方必须获得所需出口许可。

The buyer must get any import licence necessary and must tell the seller that he has this licence before the first permitted shipment date.
买方必须获得所需进口许可，并告知卖方他已于允许装运的最早日期前获此许可。

5　Insurance (ICA Rules 205 - 209)
　　5　保险　（ICA 规则 205 - 209）

According to whichever box is ticked in Section 11 of this form:
根据本表格第 11 部分中所选项目：
　　a　The seller must take out marine cargo insurance covering risk to the mill or warehouse, war risks insurance, and strikes, riots and civil commotions insurance for the invoice value plus 10%. The seller must take out this insurance through Lloyd's or another first class insurance company; or
　　a　卖方须承担海上运输保险，包括工厂或仓库风险、兵险，以及罢工、暴乱及民变保险，数额为发票价加 10%。卖方须通过劳埃德或其它一流保险公司承担此项保险；或
　　b　The buyer must take out marine cargo insurance, war risks insurance, and strikes, riots and civil commotions insurance for the invoice value plus 10%. The buyer must take out this insurance through Lloyd's or another first class insurance company; or
　　b　买方须承担海上运输保险、兵险，以及罢工、暴乱及民变保险，数额为发票价加 10%。买方须通过劳埃德或其它一流保险公司承担此项保险；或
　　c　The seller will be responsible for insuring the cotton until it is delivered to the shipping company or its agent; or
　　c　棉花发送至船运公司或其代理商之前，由卖方负责保证其安全；或
　　d　The seller will be responsible for insuring the cotton for non-containerised

shipments only.

 d 棉花装运上船之前由卖方负责保证其安全。

In the case of (b) and (d), the seller must tell the buyer the ship's name as soon he knows it.

在（b）及（d）两种情况下，卖方获知船名后须立即通知买方。

In the case of (c), the seller must tell the buyer the date of delivery as soon as he knows it.

在（c）情况下，卖方获知发货日期后须立即通知买方。

The buyer is responsible for marine insurance on any amount over the invoice value plus 10%.

超过发票价加 10%数额部分的全部海运保险责任由买方承担。

6 Quality differences and quality arbitration (ICA arbitration Bylaws)

6 质量差别及质量仲裁（ICA 仲裁条款）

International Cotton Association official differences will apply unless we agree otherwise. If the quality of the cotton is not as it should be, the seller must pay the buyer an allowance. We will try to agree the amount with you. But if there is no agreement, the dispute must be resolved through quality arbitration under the Bylaws of the International Cotton Association Limited.

除非我们另行同意，将适用国际棉花协会制定的官方差价。若棉花质量与所期不符，卖方须向买方支付折价。我们将尽量与您达成赔偿数额协议。但若未能达成一致，争议须根据国际棉花协会有限公司规章经质量仲裁加以解决。

If quality arbitration is required, samples for arbitration must be taken within 42 days (six weeks) of the date of arrival of the cotton. Arbitration must be commenced in line with ICA Bylaw 329 within 49 days (seven weeks) of the date of arrival of the cotton. Samples must be sent off to the place of arbitration within 70 days (ten weeks) of the date of arrival of the cotton (ICA Bylaw 337).

若需质量仲裁，须在棉花到货 42 日（六周）之内进行仲裁采样。根据 ICA 规章 329，仲裁过程须在棉花到货 49 日（七周）之内开始。采样须在棉花到货 70 日（十周）之内送至仲裁地点。（ICA 规章 337）

These deadlines can be extended if we agree, or an application can be made to the International Cotton Association for an extension under Bylaw 337. Each lot will be treated separately for arbitration.

经我方同意，这些期限可以延长，或可根据国际棉花协会规章 337 申请延长期限。仲裁时各批货物分别处理。

7 Shipping documents
7 装运单据

The seller must give the buyer a detailed invoice within 14 days (two weeks) of the date of the clean onboard bill of lading or other negotiable document of title.

卖方须在清洁装运提单或其它可转让物权文件签署 14 日（两周）之内向买方提供详细的发票清单。

The required shipping documents are:
所需装运单据为：

a full set of clean on-board bills of lading or other document of title. The document must show the buyer's name and address as the consignee. Otherwise, the consignee must be shown as 'To order' and blank endorsed;

全套清洁装运提单或其它可转让物权文件。文件须在收货人处写明买方的姓名及地址。否则，收货人须注明"待定"并空白背书；

a minimum of three copies of the invoice signed by the seller which sets out the total weight, the amount of tare and the total weight less tare; and

至少三份卖方签字的发票复印件，规定总重，皮重，及去皮总重；以及

under CIF terms only, a marine cargo, war, and strikes, riots and civil commotions' insurance risk insurance policy or certificate.

海上运输，兵险，及罢工，暴乱和民变保险保单或证明（仅对到岸价格条款）。

8 Weight
8 重量

Provisionally, the cotton will be invoiced on shipping weights. If net landed weights are stipulated, tare must be allowed for. If net landed weights are stipulated and the net landed weight of the cotton is different, the seller must compensate the buyer or the buyer must compensate the seller, as appropriate.

棉花暂时按船运重量开票。若规定了卸货净重，必须考虑皮重。若规定了卸货净重，且与棉花的卸货净重不符，须根据情况，或由卖方偿付买方，或由买方偿付卖方。

9 Tare
9 皮重

If the buyer thinks that the seller has not allowed enough for tare in the invoice, the actual tare can be established under Rules 215 and 216. The seller must not use sisal bagging.

若买方认为卖方未能在发票中适量考虑皮重,可根据规则215和216制定具体皮重。卖方不得使用剑麻包装。

10 Claims
10 索赔

Claims under Rule 227 for false packed, mixed packed or plated bales, for unmerchantable cotton and for foreign matter must be made within six months of the date of arrival of the cotton. Notice of any claim under Rule 228 for country damage must be given in accordance with Rules 206, 207 and 228. Unless we agree otherwise, all claims (including insurance claims) must be settled in the country the cotton is delivered to. Claims must also be settled in the currency of the contract.

根据规则227,对掺次包装,混杂包装或夹次棉包,不适销的棉花及异物提出索赔,须在棉花到货之日六个月之内进行。按规则228作出的对产地污损的索赔须同时按照规则206、207和228。除非我们另行同意,所有索赔(包括保险索赔)须在棉花到货国之内解决。赔款还须用合约货币支付。

11 Damage
11 损坏

If the cotton arrives country damaged or having damage which appears to have been caused before shipment, we must try to agree on a settlement in accordance with Rule 206 or 207, as appropriate.

若棉花到货时有产地污损,或有迹象表明在装运之前即损坏,我们须尽力依据规则206和207达成适当的解决协议。

Appendix B
附录 B

Container Trade Rules Agreement
集装箱贸易规则协议

Section A: Definitions
A 部分:定义

In this agreement, unless there be something in the context inconsistent therewith, the following expressions shall have the following meanings:

本协议中，除非上下文有出入，下列术语应作如下解释：

1 ' Container yard' or 'CY' mean a location where containers may be parked, picked-up or delivered full or empty. A container yard may further be a place of loading/stuffing by a shipper or unloading/de-vanning by a receiver of cargo, and/or where water carrier accepts custody and control of cargo at origin.

1 "集装箱堆场"或"CY"指满载或空载集装箱停放、提货或交货的地点。集装箱堆场也可作为托运公司装载/填货，或货物接收方卸载/开箱的地点，及/或水运公司最初接受货物监管及控制的地点。

2 'Container freight station' or 'CFS' mean a location where the water carrier and/or its agent is loading or unloading containers under their control.

2 "集装箱货运站"或"CFS"指水运公司及/或其代理装载或卸载其控制下的集装箱的地点。

3 'House to', 'container yard to' or 'door to' mean shipper-controlled loading at a location determined by the shipper. All costs beyond point of loading, as well as the cost of providing containers, at House/CY/Door are for the account of the party responsible for freight booking.

3 "仓库至"、"集装箱堆场至"或"户至"指在托运公司指定地点由托运公司控制的货物装载。在仓库/CY/户装载后的一切费用，以及提供集装箱的费用，均由运费责任方支付。

4 'Pier to' or 'container freight station to' mean carrier-controlled loading where the cargo is delivered to the carrier at a pier or container freight station.

4 "码头至"或"集装箱货运站至"指由运货方控制的货物装载，货物从码头或集装箱货运站发送至运货方。

5 'To house' or 'to container yard' or 'to door' mean deliver to consignee's location (warehouse or mill) upon arrival at port of destination.

5 "至仓库"和"至集装箱堆场"或"至户"指货物达到目的港后发送至收货地点（仓库或厂房）。

6 'To pier' or 'to container freight station' mean carrier will de-van container at pier at port of destination or at a container freight station.

6 "至码头"或"至集装箱货运站"指运货方将在目的港码头或集装箱货运站开箱。

7 'Mini-bridge' means cargo carried by rail or substitute transportation from US port area to another US port area for onward transportation in containers on water. Intermodal bill of lading is issued by the water carrier at originating port covering transport to the overseas destination.

7 "迷你陆桥"指货物经由铁路或其它替代运输方式从美国某港口地区运至美国另一港口地区，再从此地由集装箱水运。协调联运提单由水运公司在货物始发港签发，覆盖至海外目的地的运输全程。

8 'Micro-bridge' means cargo moving directly from interior point by rail or substitute transportation (either in containers or other equipment) to port for onward transportation in containers on water. Intermodal bill of lading is issued by the water carrier at interior loading point covering transport to the overseas destination.

8 "微型桥"指货物从内陆地点经由铁路或其它替代运输方式（集装箱或其它设备）直接运抵港口，再由集装箱水运。协调联运提单由水运公司在内陆装货处签发，覆盖至海外目的地的运输全程。

9 'Land-bridge' means cargo arriving by water carrier, and moving from one coast to another via rail for onward transportation on water.

9 "大陆桥"指货物经水运抵达，经铁路从某海岸运抵另一海岸，再经水运。

10 'Free carrier - named point', 'interior point intermodal' or 'IPI' mean the seller fulfils his responsibility when he delivers the cargo into the custody of the water carrier at the named point. If no precise point can be mentioned at the time of contract of sale, the parties should refer to the place or range where the water carrier should take the cargo into his charge.

10 "运货方自由指定地点"、"内陆地点协调联运"或"IPI"指卖方在指定地点将货物交付给水运公司，从而完成所承担责任。若签约时尚未确定具体地点，双方应参考该地点，或另行安排运货方接货地点。

11 'Shippers load and count' means the shipper assumes responsibility for the contents of the container (CY loading).

11 "托运人自行装货点件"指由托运方对集装箱内货物（CY装载）承担责任。

12 'Inter-modal bill of lading' or 'combined transport document' mean a negotiable document issued by a water carrier after receipt of container or cotton on board a rail car or other transport equipment.

12 "协调联运提单"或"联合运输单据"指水运公司在接收到由铁路或其它运输设备运抵的集装箱或棉花后签发的可转让文件。

13 'Bunker adjustment factor', 'BAF', 'fuel adjustment factor' or 'FAF' mean a charge added to the base freight rate to cover extraordinary increases in fuel costs which are beyond the control of the carrier.

13 "燃油调整费"('BAF')、"燃料调整费"('FAF')指在基本运费上加收一笔费用,支付水运公司无法控制的额外燃料费用。

14 'Currency adjustment factor' or 'CAF' mean a charge, generally expressed as a percentage of base freight, that attempts to compensate for extraordinary fluctuations in currency relationships to the US Dollar which is the 'tariff currency'.

14 "货币调整费"或"CAF"用于补偿货币相对于美元(即"关税货币")的异常波动,通常用基本运费的百分比表示。

15 'Terminal receiving charge', 'TRC', 'terminal handling charge', 'THC', 'Container yard charge' or 'CYC' mean a charge, added to the base freight rate by the carrier, which reflects the costs of handling cotton from place of receipt at the terminal to on board vessel.

15 "终点接收费"('TRC'),"终点操作费"('THC'),"集装箱堆场费"('CYC')指运货方在基本运费上加收一笔费用,反映从港口接收到装船的棉花处置费。

16 'Origin receiving charge or 'ORC' mean a charge, added to the base freight rate, which reflects the cost of handling cotton from place of receipt at origin to on board intermodal conveyance.

16 "初始接收费"或'ORC'指在基本运费上加收一笔费用,反映从初始接收地点到装船协调联运的棉花处置费。

Section B: Trade Rules
B 部分:贸易规则

Every contract for the shipment of US cotton in containers from US ports shall, unless there be anything inconsistent therewith explicitly or impliedly stated in the contract or subsequently agreed thereto by the parties to the contract, be deemed to provide that should there be a dispute concerning such contract, it shall be settled between the parties or by arbitration in accordance with the following rules:

有关美国产棉花从美国港口经集装箱船运的任何合约,除非与合约明文或隐含内容不符并经合约双方同意,都须规定:若对合约有任何争议,须根据以下规则由双方协商解决或仲裁解决：

1 Shipment: Cotton may be shipped by water and/or intermodal transportation at the option of the party responsible for freight booking. All charges imposed by the carrier, whether included in the freight rate, shown as separate item(s) in the bill of lading, or billed separately, are for the account of the party responsible for the freight booking. However if the seller elects to use a CFS facility, then the difference between CFS and CY charges at such location shall be for seller's account.

1 装运：棉花可水运及/或协调联运，由负责预定的一方决定。运货方的所有收费，无论是包含在运费提单中的各单项，还是分别计费，均由负责预定的一方支付。但若卖方选择使用 CFS 设施，则 CFS 与 CY 间的差价须由卖方支付。

2 Providing containers and transport: The party responsible for freight booking is obliged to provide containers in time for transport and loading within contracted shipping month at the port(s) or point of origin stated in the contract.

2 提供集装箱及运力：负责预定的一方有责任在合约规定的港口或始发地于合约托运月之内为运输及装载及时提供集装箱设备。

3 Date of shipment: In case of intermodal transportation, the date of the intermodal bill of lading shall constitute the date of shipment.

3 装运日期：在协调联运的情况下，协调联运提单日期即为装运日期。

4 Insurance: In case of FOB/FAS/C&F or "Free Carrier - (Named Point)" sales, buyer's insurance to cover all risks from the time the cotton is shipped or on board or is accepted into the custody and control of the water carrier, whether advised or not.

4 保险：若为 FOB/FAS/C&F 和"运货方自由指定地点"的销售，自棉花装运或装船或交付给水运公司监管之时起的全部风险保险，无论是否知悉，应由买方承担。

5 Full container load (FCL):

5 整箱（FCL）：

a Unless otherwise stated, sales should be based on freight rates for full forty-foot container loads. Any extra charges for overflow bales or minimum charges shall be paid by the party responsible for freight booking.

a 除非另有说明，运费应根据四十英尺满载集装箱的费率计算。过载货包加收费用及最小收费均应由负责预定的一方承担。

b　If quantity is expressed in containers it shall mean:

b　若货物量是以集装箱数表示，它应表明：

i　origin Gulf Area: about 78 bales per forty-foot container;

i　原产海湾地区：每一四十英尺集装箱约 78 包；

ii　origin West Coast: about 83 bales per forty-foot container; Containers other than forty-footers may be substituted for 'house to pier' or 'pier to pier' shipments only.

ii　原产西海岸地区：每一四十英尺集装箱约 83 包；不同于四十英尺的集装箱仅可在"仓库至码头"或"码头至码头"运输中代用。

6　Loading and unloading: It shall be seller's choice to load at 'house/CY' or 'pier/CFS', and buyer's choice to unload at 'house/CY' or 'pier/CFS'. However, seller shall 'ship to pier', unless specifically instructed by buyer to 'ship to house'.

6　装载及卸载：由卖方决定是在"仓库/CY"还是在"码头/CFS"装载，由买方决定是在"仓库/CY"还是在"码头/CFS"卸载。但是，除非买方特别指定"运至仓库"，卖方应"运至码头"。

7　Weighing: Unless otherwise agreed, 'pier to house' and 'house to house' shipment shall be understood to mean 'net certified shipping weights final'.

7　重量：除非另有协议，"码头至仓库"及"仓库至仓库"的运输应被理解为"最终核准的运货净重"。

8　Sampling:

8　采样：

a　Buyer may ask seller to by-load samples, subject to seller's agreement. Any extra charges shall be for the buyer's account.

a　经卖方同意，买方可要求卖方对货物进行采样。由买方支付全部附加费用。

b　In case of 'pier to house' or 'house to house' shipments, normal arbitration rules shall apply, except that sampling may take place on buyer's premises under supervision. Sampling expenses are for the buyer's account.

b　若为"码头至仓库"或"仓库至仓库"运输，除采样是在买方地点监督进行外，适用一般的仲裁规则。由买方支付采样费用。

9　Missing bales: In case of shipper's load and count, seller is liable for the contents of the container. Unless otherwise agreed between buyer and seller, any claim must be supported by certificates issued by seller's controller stating the container serial and seal number and certifying that the seal was intact. However, in shipments involving 'pier to house' or 'house to house' movements and when seals are broken by customs or other authorities at port of entry container must be re-sealed and both the original seal and new seal numbers provided to shipper's controller.

9 遗失货包：在由托运人自行装货点件的情况下，卖方应对集装箱内货物承担责任。除非买卖双方之间另有协议，任何索赔要求均应出示卖方检验员签发的证明，说明集装箱序号及封条号，证明封条完好无损。但是，若货运过程涉及"码头至仓库"或"仓库至仓库"的运输，并且海关或其它机关在港口或入境处开启了封条，必须重新密封集装箱，并向托运方检验员提供新老封条号。

10 Payment:

10 付款：

a Letter of credit payment: Letter of credit must allow inter-modal bill of lading.

a 信用状付款：信用状必须允许使用协调联运提单。

b Cash against documents on first presentation: Buyer must pay against inter-modal bill of lading.

b 初次交单时凭单据付款：买方须支付协调联运提单。

c Cash on Arrival: Buyer shall pay against the bill of lading upon arrival of the vessel at the destination named in the bill of lading. However, if the containers are on-carried by feeder vessels or other means, payment shall be made upon arrival of the feeder vessels or on-carrying conveyance at the final destination named in the contract. In case of seller's freight booking, if any containers are not on board the vessel named in the bill of lading, buyer shall have the right to claim against the seller for refund of interest until actual arrival of the container(s). This is not applicable if shipment by container vessel is required by buyer subsequent to entering into the contract.

c 货到付现：货船到达提单指定的目的地时买方须支付提单。但是，若集装箱经由支线船舶或其它运输方式转运，须在支线船舶或其它转运工具达到合约指定的最终目的地时付款。若由卖方预定货运，若集装箱不在提单指明的船舶上，买方有权要求卖方赔偿利息，直至集装箱真正运抵。若买方在合约签署后才要求集装箱船运，则该规则不适用。

Section 3 Arbitration Bylaws
第3部分 仲裁规章
ARBITRATION BYLAWS
行政管理规章

Bylaws are the mandatory provisions of this Association which cannot be changed or varied by the parties.

规章为本会强制性规定,当事方不得更改或变动。

Any dispute arising out of, or in connection with, a contract which incorporates and provides for arbitration under these Bylaws shall be referred to arbitration. Arbitrators, an umpire, a technical appeal committee or a quality appeal committee (as the case may be) will determine all matters placed before them in accordance with the following Bylaws.

凡合同中规定依本规章进行仲裁,因该等合同发生的任何争议,须提交仲裁。仲裁员、裁定人、技术申诉委员会或质量申诉委员会(依情况而定)将依据以下规章对交由其处理的所有事项进行裁决。

Introduction
简介

Bylaw 300
规章 300

1   We will conduct arbitration in one of two ways:

1   我们将按以下两种方式之一进行仲裁:

Quality arbitrations will deal with disputes arising from the manual examination of the quality of cotton and/or the quality characteristics which can only be determined by instrument testing. Bylaws especially applicable to quality arbitrations and appeals are set out herein.

质量仲裁将处理因棉花品质的手工检验和/或仅可通过仪器检测判断的品质特征而产生的争议。本文规定特别适用于质量仲裁和申诉的规章。

Technical arbitrations will deal with all other disputes. Bylaws especially applicable to technical arbitrations and appeals are set out herein.

技术仲裁将处理所有其他争议。本文规定特别适用于技术仲裁和申诉的规章。

2   The law of England and Wales and the mandatory provisions of the Arbitration Act 1996 (the Act) shall apply to every arbitration and/or appeal under

these Bylaws. The non-mandatory provisions of the Act shall apply save insofar as such provisions are modified by, or are inconsistent with, these Bylaws.

2 依本规章提起的仲裁和/或申诉适用英格兰和威尔士法律以及《1996年仲裁法》(以下称"仲裁法")的强制性规定。仲裁法的非强制性规定须予以适用,唯经本规章修改或与本规章冲突者除外。

3 The seat of our arbitrations is in England. No one can decide or agree otherwise.

3 我们的仲裁地为英格兰。任何人不得决定或同意在别处审理。

4 Disputes shall be settled according to the law of England and Wales wherever the domicile, residence or place of business of the parties to the contract may be.

4 无论合同当事方的住所、居住地或营业地在何处,争议须依英格兰和威尔士法律裁决。

5 If parties have agreed to arbitration under our Bylaws, then, subject to paragraph (6) below, they must not use any court at all unless we have no further power to do what is required, or the Act allows, in which case they must apply to the courts in England or Wales.

5 如当事方已经同意依本规章仲裁,则在遵守以下第(6)节规定的前提下,当事方不得再诉诸任何法院,除非仲裁庭没有进一步的权力实施需要进行的或仲裁法允许的工作,在这种情况下必须向英格兰或威尔士的法院提出申请。

6 A party can apply to a court anywhere to obtain security for its claim while arbitration or an appeal is taking place.

6 仲裁或申诉开始后,一方当事人可以向任一法院申请诉请保全。

7 If a party is prevented from proceeding with an arbitration as a result of the application of the provisions of Bylaw 302 (3) or Bylaw 330 (1), it is free to apply to any court which is willing to accept jurisdiction.

7 如一方当事人因规章第302(3)款或第330(1)款之规定适用而不得进入仲裁程序,则可诉诸同意接受司法管辖的任何法院。

Notices
通知

Bylaw 301
规章 301

1 Any notice or other communication that may be or is required to be given by a party under these Bylaws shall be in writing and shall be delivered by registered

postal or recognised international courier service or transmitted by fax, e-mail or any other means of telecommunication that provide a record of its transmission.

  1 根据本规章一方可能或需要发出的任何通知及其他通讯须以书面形式，且须以挂号或知名国际快递服务寄送或以传真、电子邮件或任何其他提供传送记录的电信方式传送。

  Where service of notices or other documents on parties by a tribunal or appeal committee via the Secretariat using e-mail or fax is concerned, the day after the date of despatch of an e-mail or fax shall be deemed to be the date of service on the party. Service on agents, brokers or representatives shall be deemed proper service under these Bylaws. So far as concerns such notices, this Bylaw over-rides any other provisions concerning notices in the parties' contract.

  如仲裁庭或申诉委员会通过秘书使用电子邮件或传真向仲裁当事人送达通知或其他文件，电子邮件或传真发出之日次日须视为送达日期。根据本规章，送至代理、中间人或代表须视为恰当送达。有关此等通知，本规章凌驾于双方合同有关通知的任何其他条款。

  2 A party's last-known residence or place of business or last known e-mail or fax address during the arbitration shall be a valid address for the purpose of any notice or other communication in the absence of any notification of a change to such address by that party to the other parties, the Tribunal, Appeal Committee or Secretariat.

  2 若未通知其他当事人、仲裁庭、申诉委员会或秘书其最近所知的居住地或营业地或最近所知的电子邮件或传真地址变更，则仲裁期间一方的最近所知的居住地或营业地或最近所知的电子邮件或传真地址将视为送达任何通知或其他通讯的有效地址。

  3 For the purpose of determining the date of commencement of a time limit, a notice or other communication shall be treated as having been received on the day after it is delivered or deemed to have been delivered. If we give notice that something must be done within a set period, the period begins the day when it is deemed that the relevant notice has been delivered.

  3 为确定时限开始之日期，通知或其他通讯须视为已于送达或视同送达之日收到。如我们发出通知要求在某一规定期限内必须做某事，则该期间从相关通知视为送达之日起开始计算。

  4 For the purpose of calculating a period of time under these Bylaws, such period shall begin to run on the day following the day when a notice or other

communication is delivered or deemed to have been delivered. If the last day of such period is an English bank (official) holiday, the period is extended until the first business day which follows. English bank (official) holidays or non-business days occurring during the running of the period of time are included in calculating that period.

4 为根据本规章计算时期,此等时期须于通知或其他通讯送达或视同送达之日次日开始。如此等时期最后一日为英格兰银行(官方)假期或收信方居住地或营业地的非营业日,该时期延至接下来的第一个营业日。计算时期时将计入时期内出现的英格兰银行(官方)假期或非营业日。

5 The Tribunal or Appeal Committee may at any time extend (even where the period of time has expired) time prescribed under these Bylaws for the conduct of the arbitration, including any notice or communication to be served by one party on any other party.

5 仲裁庭或申诉委员会可随时延长(甚至在时期已过的情况下)本规章规定之仲裁进行时间,包括一方向另一方送达任何通知或通讯。

6 If something is to be given or paid to us by a set date or within a period, it must arrive on or before 23; 59 hours on the last day it is due. If it is something that is delivered to us by hand, this must be done during our office hours. If money is paid by cheque or something similar and the bank refuse to pay us the amount due, we will consider that it was not paid on the date it was received by us.

6 如在指定日期前或时期内需要向我们提供任何物品或支付任何款项,必须于最终到期日期23时59分或之前送达。如果亲手递送任何物品,必须于我们的办公时间内送达。如果使用支票或类似票据付款,而银行拒绝支付应付金额,我们将此视为收讫支票或类似票据当日并未付款。

Technical Arbitration
技术仲裁

Commencement of Arbitration
仲裁的启动

Bylaw 302
规章 302

1 Any party wishing to commence arbitration under these Bylaws ("the claimant") shall send us a written request for arbitration ("the request"), and we shall copy the Request to the other party ("the respondent").

1 任一方（下称"申诉方"）如希望起始本规章所述仲裁程序，须向我们提交书面的仲裁申请，我们将复制申请书送给对方（下称"被诉方"）。

2 When sending the request, the claimant shall also send:

2 申诉方提交申请书时，须同时提交：

the name, address including email address, telephone and facsimile number of the respondent,

被诉方姓名/名称、地址（包括电子邮件地址）、电话和传真号码，

a) a copy of the contract signed by both parties; or

a) 双方签署的合同复印件；或

b) a copy of the arbitration agreement signed by both parties if not included in the contract; or

b) 双方签署的仲裁协议复印件（如合同中没有仲裁条款）；或

c) a copy of the contract with any additional supporting evidence,

c) 附其他支持性证据的合同复印件，

the name of their nominated arbitrator, or, if appropriate, the name of the sole arbitrator agreed by the parties, and such application fee as may be due under Appendix C of the Rule Book.

提名指定的仲裁员姓名，或者，如系独任仲裁，则为双方一致同意的独任仲裁员姓名，以及依规则手册附录 C 可能应缴纳的仲裁申请费。

3 We may refuse arbitration facilities where one of the parties to the dispute has been suspended from the Association or expelled. Arbitration will also be refused where either the name of one of the parties appeared on the Association's List of Unfulfilled Awards at the time that the contract under dispute was entered into, or the penalty of denial of arbitration services has been imposed on one of the parties pursuant to Article 27 or Bylaw 418.

3 如争议的一方当事人已被暂停或开除协会资格，则我们可以拒绝仲裁。如在系争合同签署之时有任何一方当事人被列入协会未执行裁决书名单，或者一方当事人依章程第 27 条或规章 418 之规定受到过拒绝可得仲裁的处罚的，我们亦可拒绝仲裁。

The Tribunal
仲裁庭
Bylaw 303
规章 303

Disputes which fall to be determined under these Bylaws shall be heard by a tribunal of three arbitrators or, if both parties agree, by a sole arbitrator who, for the purposes of these Bylaws, shall be deemed to be a qualified Chairman. Each party shall appoint one arbitrator and we shall appoint the third arbitrator who shall serve as Chairman of the tribunal. The tribunal shall ensure that the parties are treated with impartiality and equality and that each party has the right to be heard and is given a fair opportunity to present its case as directed within the Chairman's directions. The tribunal shall conduct the proceedings with a view to expediting the resolution of the dispute

依据规章进行裁决的争议须由三名仲裁员组成的仲裁庭审理,或者如双方同意,由一名独任仲裁员审理,该独任仲裁员视为规章所述的合资格的首席仲裁员。双方各自指定一名仲裁员。我们指定第三名仲裁员,由其担任仲裁庭首席仲裁员。仲裁庭须保证公平公正的原则,各方均获得公平的机会在首席仲裁员主持下陈述本方理由。仲裁庭须以方便快速解决争议为宗旨主持仲裁程序。

Appointment of Arbitrators
仲裁员的任命
Bylaw 304
规章 304

1 Upon receipt of a Request made in accordance with Bylaw 302, we shall ask the respondent to appoint their arbitrator or to agree to the appointment of a sole arbitrator within 14 days (two weeks) and to notify us and the claimant of the name of their arbitrator. If the respondent fails to appoint an arbitrator within this timescale, we will appoint an arbitrator and give notice of the name of the arbitrator so appointed to the parties.

1 收到依规章302之规定提交的仲裁申请书后,我们须要求被诉方在14天(二周)内指定本方仲裁员,或同意任命独任仲裁员,被诉方应将其选定的仲裁员姓名通知给我们和申诉方。如被诉方未在该时限内指定仲裁员,则我们将指定一名仲裁员并将指定仲裁员的姓名告知双方。

2 We shall appoint the third arbitrator who shall serve as Chairman of the tribunal within seven days (one week) of the appointment of the second arbitrator, whether appointed by us or the respondent. The Chairman will be selected from those arbitrators nominated as such by the Directors.

2 第二名仲裁员指定后的七天(一周)内(无论由我们还是被诉方指定),我们将指定第三名仲裁员担任仲裁庭的首席仲裁员。该名首席仲裁员将从董事提名的仲裁员中选出。

3 We may appoint an observer for training purposes who will not form part of the tribunal.

3 我们可出于培训目的委任观察员，该观察员不构成仲裁庭的一部分。

4 Arbitrators must be Individual Members of our Association when they are appointed. Arbitrators must additionally be qualified to the standards set by the Directors from time to time before they may accept such appointments.

4 仲裁员接受任命时必须为协会个人会员。此外，仲裁员在接受任命前还必须满足董事会不时规定的资质要求。

5 If a vacancy arises because any arbitrator dies, resigns, refuses to act, ceases to hold the requisite qualifications or becomes incapable of performing his functions, the vacancy shall be filled by the method set out in paragraph (1) above.

5 如因仲裁员死亡、辞职、拒绝担任仲裁员、丧失必须的资质或无法继续履行职责而出现空缺，则须按上文第(1)节规定的方法填补空缺。

6 By accepting appointment (whether by a party or by us) an arbitrator binds himself to the Association to act in accordance with the Bylaws and Articles.

6 仲裁员一旦接受任命（无论由一方当事人还是我们任命），均须遵守规章和章程、按协会规定履行职责。

7 If either firm:

7 如有任何一方公司：

fails to nominate an arbitrator within 14 days (two weeks) of being requested to do so; or

未能在被要求时于 14 日（两周）内提名仲裁员；或

fails to agree on a replacement arbitrator within 14 days (two weeks) of a substantiated and valid objection to a nomination, the other firm can ask the President of the Association to make an appointment on behalf of the firm that has failed to nominate an arbitrator, or failed to agree on a replacement arbitrator within the time allowed.

在提名遭到实质性和有效异议时，未能在 14 日（两周）内同意仲裁员的更换人选，另一方公司可以请求协会会长代表未在时限内提名仲裁员或同意仲裁员的更换人选的一方公司指定仲裁员。

8 The Association will give notice of the President's intention. If the firm in default does not nominate an arbitrator acceptable to the other firm within 14 days

(two weeks) of that notice being given, the President may act.

8 协会将发出关于会长意向的通知。如在通知发出后 14 日（两周）内，违约一方公司未提名对方公司可接受的仲裁员人选，则主席可以任命仲裁员。

9 If either firm objects to an arbitrator or any member of a tribunal or an observer, it must do so within seven days (one week) of notice being given of the relevant appointment. Any objection must be made in writing, accompanied by the reasons for objection. An objection to an appointment will only be valid if the President decides that substantial injustice could result.

9 如任何一方公司对某一仲裁员或仲裁庭任何成员有异议，则必须在相关的任命通知发出后 7 天（一周）内提出。异议必须为书面形式，并附有异议的理由。对任命的异议仅在会长判定如此可能导致重大不公正的情况下有效。

10 If an objection is not acted on and not withdrawn, the President must be asked to decide whether it is to be valid.

10 如异议既未通过也没有撤回，则必须要求会长裁定其是否有效。

11 If new evidence comes to light after the normal time limits for raising an objection have expired, an objection may still be raised. The President will decide whether it will be heard and whether it is valid.

11 如在提出异议的正常时限届满后出现新证据，则仍可提出异议。由会长决定是否听取理由以及异议是否有效。

12 If a firm disagrees with the President's intention or decision it can appeal to the Directors but it must do so within seven days (one week) of notice having been given of the Presidents' decision. The Directors can use any of the powers given to the President at paragraph (6) and paragraph (7) above.

12 如一方公司不同意会长的意向或决定，可以向董事会提起申诉，但须在会长的决定下达后 7 天（一周）内提出。董事会可以运用上文第(6)和(7)节赋予会长的权力。

13 If the President should have a possible conflict of interest, he will not appoint arbitrators under these Bylaws. In that situation, the Vice-President or an acting President, will have the same powers of appointment as the President.

13 如会长有可能的利害关系，则不得依据规章任命仲裁员。在这种情况下，副会长或代理会长拥有与会长相同的任命权。

Revoking the authority of an arbitrator or appeal committee member
撤销仲裁员或申诉委员会成员的权力
Bylaw 305
规章 305

1　Once an arbitrator or appeal committee member has been appointed, his authority cannot be revoked by either firm unless both firms agree.

1　仲裁员或申诉委员会成员一旦任命，任何一方不得撤销其权力，除非双方一致同意撤销其权力。

2　If an arbitrator or appeal committee member ceases to be a Member of the International Cotton Association, he cannot continue to act in whatever capacity he was appointed unless the Directors agree.

2　如仲裁员或申诉委员会成员不再担任国际棉花协会会员，均不得继续以任何身份履行任何曾被委任的职责，但董事会同意除外。

3　The President may revoke an appointment and appoint an alternative:

3　在下列情况下，主席可以撤销任命并转任他人：

If substantial injustice will be caused by him not doing so; or
如会长不如此履行将会导致重大不公正；或

if requested to do so by either firm in the following circumstances:
由任一方公司在下列情况下提出请求：

if he upholds an objection under Bylaw 304;
如其根据规章304 所规定赞同异议；

if an appointed arbitrator dies, refuses or becomes unable to act;
如有任命的仲裁员死亡、拒绝或不能履行职务；

if a sole arbitrator does not make an award within 56 days (eight weeks) of having received the final written submissions from the parties; or if the tribunal do not make an award within 56 days (eight weeks) of having received the final written submissions from the parties.

如独任仲裁员在收到双方最后提交书面材料后56日（八周）内仍不作出裁决；或如仲裁庭在收到双方最后提交书面材料后56日（八周）内仍不作出裁决。

4　The Association will give notice of the President's intention. If a firm disagrees with the President, it can appeal to the Directors but it must give its reasons in writing within seven days (one week) of notice having been given. The

Directors can use any of the powers given to the President.

4　协会将发出关于会长意向的通知。如一方公司不同意会长的决定，则可向董事会申诉，但必须在通知发出后7天（一周）内提出书面理由。董事会可以行使赋予会长的任何权力。

5　The timeframes indicated in paragraph (3) above shall not be construed so as to undermine or overrule the arbitrators' duty under the Act to allow each party reasonable opportunity to reply to any query or order from the tribunal subsequent to the closure of final written submissions.

5　上文第(3)节所述的时间限制不得被解释为损害或否定仲裁员依仲裁法所具有的、允许各方当事人在最终书面材料提交后享有合理的机会就仲裁庭提出的任何询问或命令作出答复的职责。

Jurisdiction
司法管辖权
Bylaw 306
规章 306

Without prejudice to the provisions of the Act relating to jurisdiction, the tribunal may rule on its own jurisdiction, that is, as to whether there is a valid arbitration agreement, whether the tribunal is properly constituted and what matters have been submitted to arbitration in accordance with the arbitration agreement.

在不违反仲裁法关于司法管辖权规定的前提下，仲裁庭可以就其司法管辖权问题作出裁定，即关于是否存在有效的仲裁协议、仲裁庭的构成是否适当、以及依据仲裁协议将何事项提交仲裁。

Conduct of the arbitration
仲裁的进行
Bylaw 307
规章 307

1　It shall be for the Chairman, having consulted his fellow arbitrators:

1　首席仲裁员须与其他仲裁员协商：

to determine whether the Tribunal has jurisdiction; and
确定仲裁庭是否具有司法管辖权；以及

to decide all procedural and evidential matters, subject to the right of the parties to agree any matter.
决定所有程序和证据事项，以双方当事人有权同意为前提。

2 The Chairman shall ensure the prompt progress of the arbitration, where appropriate by the making of Orders.

2 首席仲裁员须确保仲裁程序的迅速进行，有必要时可以发出命令。

3 As soon as the Chairman has issued directions and determined a timetable for proceedings, we shall notify the parties.

3 一旦首席仲裁员发出指令并确定程序时间表，我们将立即通知双方当事人。

4 The parties have a duty to do all things necessary for the proper and expeditious conduct of the proceedings, including complying without delay with any order or direction of the tribunal as to procedural and evidential matters.

4 当事人有义务执行所必需的事项促进仲裁程序的适当和快速进展，包括不迟延地遵守仲裁庭的关于程序和证据事项的任何命令或指示。

All communications between either party and the tribunal shall be simultaneously copied to the other party.

任一方与仲裁庭之间的一切沟通信息均须同时抄送给对方。

5 If either party fails to comply with any procedural order of the tribunal, the tribunal shall have power to proceed with the arbitration and make an Award.

5 如有任何一方不遵守仲裁庭的任何程序性命令，则仲裁庭有权继续仲裁并作出裁决。

6 Decisions, Orders and Awards shall be made by all or a majority of the arbitrators, including the Chairman. The view of the Chairman shall prevail in relation to a decision, order or Award in respect of which there is neither unanimity nor a majority.

6 决定、命令和裁决须由全体或多数仲裁员确定，包括首席仲裁员在内。如在作出决定、命令和裁决过程中无法达成一致同意或无法使多数同意，则以首席仲裁员的意见为准。

7 All statements, contracts and documentary evidence must be submitted in the English language. Whenever documentary evidence is submitted in a foreign language, unless otherwise directed by the tribunal, this must be accompanied by an officially certified English translation.

7 所有陈述、合同和文件证据均须以英文提交。如文件证据以外语提交，必须附有经官方认证的英文翻译件，除非仲裁庭有其他指示。

8 We will not accept submissions directly from legal firms or independent

lawyers.

8 我们不接受律师事务所或独立律师直接提交的文件。

Oral hearings

口头庭审

Bylaw 308

规章 308

1 Where either party or both parties request an oral hearing, they shall apply in writing to the tribunal. The tribunal may grant or decline the request without giving reasons. Their decision shall be final. If a request is granted, the Chairman, having consulted his fellow arbitrators, shall decide the date, time and place of the hearing and the procedure to be adopted at the hearing.

1 一方或双方当事人如要求口头庭审，须以书面方式向仲裁庭提出。仲裁庭可以同意或拒绝该请求，无须说明理由。其决定为终局决定。如请求获得同意，则首席仲裁员须在询问其他仲裁员后决定庭审的日期、时间和地点，以及所采用的庭审程序。

2 The Chairman, having consulted his fellow arbitrators, may, in advance of the hearing, give detailed directions with any appropriate timetable for all further procedural steps in the arbitration, including (but not limited to) the following:

2 首席仲裁员在询问其他仲裁员后，可以于庭审前发布详细的指示，以及适用于未来仲裁所有程序性步骤的相应时间表，包括（但不限于）下列：

written submissions to be advanced by or on behalf of any party,

需要审理的任何一方提交的书面材料，

examination of witnesses,

证人盘问，

disclosure of documents.

文件公布。

3 The Chairman may impose time limits on the length of oral submissions and the examination or cross-examination of witnesses.

3 首席仲裁员可以规定口头陈述和证人盘问及交叉盘问的时间限制。

4 Parties may be represented by one of their employees, or by an Individual Member of the Association, but they may not be represented by a solicitor or barrister, or other legally qualified advocate. Parties may be accompanied by a legal representative at any oral hearing. Such legal representative may advise the party

but may not address the tribunal.

4　当事方可以由自己的雇员或协会个人会员代理出庭，但不得由律师或其他具有法定资质的代理人代理。口头庭审时当事方可以由法律代表陪同出席。该法律代表可以向当事方提供建议，但不得向仲裁庭发言。

Technical Arbitration Awards
技术仲裁裁决书

Bylaw 309
规章　309

1　An Award shall be in writing on our official form dated and signed by all members of the tribunal or the sole arbitrator as applicable and shall contain sufficient reasons to show why the tribunal has reached the decisions contained in it, unless the parties agree otherwise or the Award is by consent. The Chairman will be responsible for drafting the Award but can delegate this responsibility to a qualified member of the tribunal. The members of the tribunal need not meet together for the purpose of signing their award or for effecting any corrections thereto.

1　裁决书须符合我们的官方格式，署明日期并视情况由仲裁庭全体成员或独任仲裁员签署，并须充分说明仲裁庭作出裁决的理由，除非当事方同意不说明理由或裁决书系由双方同意作出。首席仲裁员将负责草拟仲裁书，但可将该责任委托于一名合资格的仲裁庭成员。各仲裁庭成员无需为签署前述裁决书或确认任何修订生效而会面。

2　Any Award shall state that the seat of the arbitration is in England and the date by which we must receive notice of appeal.

2　裁决书须说明仲裁地点在英格兰，以及我们必须收到申诉通知的日期。

3　All Awards made under our Bylaws will be treated as having been made in England, regardless of where matters were decided, or where the Award was signed, despatched or delivered to the firms in dispute.

3　依据规章作出的所有裁决将视为在英格兰作出，无论系争事项在何处裁定，或者裁决书在何处签署及交寄或交付争议方公司。

4　We will stamp every Award in our offices on the date of the Award, and apply the scale of fees laid down in Appendix C of the Rule Book.

4　我们将在裁决作出之日于我们的办公室对每一份裁决书盖章，并应用规则手册附录 C 规定的收费标准。

5　An Award will only become effective and binding when we stamp it.

5　盖章后裁决书方生效并具有约束力。

6　After we stamp an Award, we will notify all of the parties concerned.

6　裁决书盖章完毕后，我们将通知所有相关当事方。

7　The Award will only be released upon payment of the stamping fee and any outstanding fees, costs and expenses.

7　盖章费及任何应付收费、成本和费用支付完毕后，方可发出裁决书。

8　The Award must be honoured within 28 days (four weeks) from notification to all of the parties under paragraph (6) above.

8　裁决书必须于依上文第(6)节通知各当事方后28天（四周）内执行。

9　The Association will keep a copy of every Award.

9　每份裁决书协会将保留一份副本。

Interest on Awards
裁决书裁定的利息

Bylaw 310
规章 310

The tribunal and technical appeal committee can award simple or compound interest from such dates and at such rates as they consider meets the justice of the case.

仲裁庭和技术申诉委员会可以裁定按照其认为公平的利率从某日起支付单利或复利。

Corrections to Awards
裁决书修订

Bylaw 311
规章 311

1　The tribunal, sole arbitrator or appeal committee may on its own initiative or on the application of a party or secretariat:

1　仲裁庭、独任仲裁员或申诉委员会可自行或在一方或秘书申请下：

correct an award so as to remove any clerical mistake or error arising from an accidental slip or omission or clarify or remove any ambiguity in the award, or

修订裁决书，以移除任何因意外失误或疏忽而引致的笔误，或澄清或消除裁决书中的任何模糊不清，或者

make an additional award in respect of any claim (including a claim for interest or costs) which was presented to the tribunal but was not dealt with in the award.

就任何呈交仲裁庭但未在仲裁书中予以处理的申索（包括利息或费用申索）追加裁决书。

2　These powers shall not be exercised without first affording the parties a reasonable opportunity to make representations to the tribunal.

2　必须首先给予当事人合理机会向仲裁庭进行陈述,方可行使上述权力。

3　Any application for the exercise of those powers must be made within 28 days of the date of the award or such longer period as the parties may agree.

3　行使上述权力的任何申请必须于裁决书作出日期 28 天内或双方可能一致同意的较长期限内提交。

4　Any correction of an award shall be made within 28 days of the date the application was received by the tribunal or, where the correction is made by the tribunal on its own initiative, within 28 days of the date of the award or, in either case, such longer period as the parties may agree.

4　裁决书任何修订须于仲裁庭收到申请之日 28 天内作出,如仲裁庭自行作出修订,则为裁决书作出日期 28 天内作出,或者不论何种情况,于双方可能一致同意的较长期限内作出。

5　Any additional award shall be made within 56 days of the date of the original award or such longer period as the parties may agree.

5　任何追加裁决书须于原裁决书作出日期 56 天内或双方可能一致同意的较长期限内作出。

6　Any correction of an award shall form part of the award.

6　裁决的任何修正应构成裁决的一部分。

Technical Appeals

技术申诉

Bylaw 312

规章 312

1　If either party disagrees with the tribunal's Award, it can appeal to us within the period specified in the Award. It must send Notice of Appeal to us.

1　如任何一方对仲裁庭的裁决不满,可以在裁决书规定的期限内向我们提出申诉。当事方必须向我们提交申诉通知。

2　Upon receipt of the Notice of Appeal we may demand that sums of money be deposited with us by the appellant, by way of deposit against any fees, costs or expenses in connection with or arising out of the Appeal. The appellant must also deposit any costs or stamping fees that the tribunal's Award ordered them to pay. Failure to pay within the specified period will result in the Appeal being dismissed.

2 收到申诉通知后,我们可能要求申诉人存入一笔款项作为与申诉有关的收费、成本或费用的押金。申诉人还必须存入仲裁庭的裁决书要求其支付的任何费用及盖章费。未在规定期限内支付的,申诉会被驳回。

3 The Directors, or appeal committee if appointed, can extend the time limits in paragraph (2) above, but only if the firm concerned can show that substantial injustice would otherwise be done and the request for an extension is reasonable in all the circumstances. Any request for an extension should be made in writing and should outline the reasons why substantial injustice may occur if an application is refused.

3 董事会或申诉委员会(如有任命)可以延长上述第(2)节所述的时限,但前提是所涉及的当事方能够证明不如此将导致重大不公正,并且在所有情势下提出延长时限申请都是合理的。延长时限申请应以书面方式提出,并应简要说明为何如果申请被拒绝就会导致重大不公正的原因。

Oral hearings

口头庭审

Bylaw 313

规章 313

1 Where either party or both parties request an oral hearing, they shall apply in writing to the tribunal. The tribunal may grant or decline the request without giving reasons. Their decision shall be final. If a request is granted, the Chairman, having consulted his fellow arbitrators, shall decide the date, time and place of the hearing and the procedure to be adopted at the hearing.

1 如一方当事人或双方要求口头庭审,须以书面形式致函申诉委员会申请。申诉委员会可批准或驳回请求,无需提供理由。其决定为最终决定。如请求获批准,主席咨询其他仲裁员后,须决定庭审日期、时间及地点,以及庭审中采取的程序。

2 The Chairman, having consulted his fellow arbitrators, may, in advance of the hearing, give detailed directions with any appropriate timetable for all further procedural steps in the arbitration, including (but not limited to) the following:

2 主席咨询其他仲裁员后,可在庭审之前作出所有仲裁程序步骤的详细指示及任何恰当时间安排,包括(但不限于)以下:

written submissions to be advanced by or on behalf of any party,
任一方或其代表需要提交的书面材料，

examination of witnesses,
证人询问，

disclosure of documents.
披露文件。

3   The Chairman may impose time limits on the length of oral submissions and the examination or cross-examination of witnesses.

3   主席可规定口头提交及询问或交叉询问证人的时限。

4   Parties may be represented by one of their employees, or by an Individual Member of the Association, but they may not be represented by a solicitor or barrister, or other legally qualified advocate. Parties may be accompanied by a legal representative at any oral hearing. Such legal representative may advise the party but may not address the tribunal.

4   当事人可由其雇员或协会个人成员代表，惟该个人成员不得担当过争议的仲裁员，但其不可由事务律师或出庭律师或其他具备法律资质的辩护人代表。当事人可在任何口头庭审中由一名法律代表陪同。此等法律代表可为当事人提供建议，但不可向申诉委员会发言。

Technical Appeal Committee
技术申诉委员会

Bylaw 314
规章 314

1   As soon as the appellant has paid all fees due under Bylaw 312 (2) and served its case for appeal the Directors shall appoint a Technical Appeal Committee ('appeal committee').

1   一旦申诉人支付规章 312 (2) 所规定的所有费用并提交申诉材料，则董事会须任命一个技术申诉委员会（下称"申诉委员会"）。

2   A Director cannot be involved in any decision about an appeal or be on an Appeal Committee if he has acted as an arbitrator in the dispute or if substantial injustice could result.

2   如有董事已作为仲裁员参与争议解决，或者可能导致重大不公正，则该董事不得涉及申诉裁决事务或加入申诉委员会。

3   An Individual Member cannot be on an appeal committee if he has acted as

an arbitrator in the dispute, or substantial injustice could result.

3　如有个人会员已作为仲裁员参与争议解决,或者可能导致重大不公正,则该会员不得加入申诉委员会。

4　An appeal committee will consist of a Chairman and four other people, who must be Individual Members when they are appointed. All appeal committee members must additionally be qualified to the standards as set by the Directors from time to time.

4　申诉委员会由一名主席和四名其他人员(任职时必须为个人会员)组成。此外,所有申诉委员会成员必须符合董事会不时规定的标准。

5　We may appoint an observer for training purposes who will not form part of the technical appeal committee.

5　我们可出于培训目的委任观察员,该观察员不构成技术申诉委员会的一部分。

6　A member of an appeal committee may only attend and vote at committee meetings if he has been present at all previous meetings.

6　申诉委员会成员只有在已经参加所有此前会议的前提下方可出席委员会会议并投票。

7　At any meeting of an appeal committee, a quorum must comprise the Chairman and three, or at the Chairman's discretion, two members. In the event that there is no quorum, the Directors will appoint a new appeal committee. However, the provisions of this paragraph may be varied by the Directors if both parties agree in writing.

7　任何申诉委员会会议的法定人数构成必须包括主席和三名或(根据主席的决定)两名成员。如不能构成法定人数,则董事会应任命新的申诉委员会。不过,如双方以书面方式表示同意,则董事会可以变更本节规定。

8　If the Directors appoint an appeal committee, either party can object to the Chairman or any member of the committee but must do so within seven days (one week) of notice being given of the relevant appointment. Any objection must be made in writing, accompanied by the reasons for objection. An objection to an appointment will only be valid if the President decides that substantial injustice could result.

8　如董事会任命申诉委员会,任一当事方可以对主席或委员会的任何成员提出异议,但必须在相关任命下达后七天(一周)内提出异议。异议必须为书面形式,并附有异议的理由。对任命的异议仅在主席判定如此可能导致重大不公正的情况下有效。

9 If the Directors uphold an objection, they shall immediately nominate a substitute.

9 如董事会同意某项异议,则须立即提名替代者。

10 An appeal involves a new hearing of the dispute and the appeal committee can allow new evidence to be put forward. It may confirm, vary, amend or set aside the award of the first tribunal and make a new award covering all of the matters in dispute.

10 申诉程序中包括对争议的重新审理,并且申诉委员会允许提出新证据。申诉委员会可以确认、变更、修改或推翻第一仲裁庭作出的裁决书,并作出一份覆盖所有系争事项的新裁决书。

11 The appeal committee will decide the issues by a simple majority vote. Every member, including the Chairman will have one vote. If both sides have the same number of votes, the Chairman will vote again to decide the issue.

11 申诉委员会以简单多数投票作出决定。所有成员,包括主席在内,均享有一票权。如双方所获票数相等,则由主席重新投票决定。

12 All of the appeal committee arbitrators will sign the award.

12 所有上诉委员会仲裁员将签署该裁决。

Appeal timetable
申诉时间表

Bylaw 315
规章 315

1 The appellant must submit its Notice of Appeal to us within the time specified in the Award. The appellant must then submit all fees due under Bylaw 312 (2) and its case for appeal within 14 days (two weeks) of the Association receiving its Notice of Appeal.

1 申诉人必须在裁决书说明规定的时间内向我们提交申诉通知。申诉人必须于协会收到申诉通知后14日(两周)内提交规章312(2)规定的所有费用和申诉材料。

2 If the respondent intends to comment, it should do so within 14 days (two weeks) of receiving a copy of the appellant's case.

2 如被申诉人需要作出回应,则应在收到申诉人材料副本后 14 天(两周）内提出。

3   If the respondent replies, the appellant is allowed to make further comment, but must do so within seven days (one week) of receiving a copy of the respondent's reply.

3   如被申诉人作出答复，则申诉人可以作出进一步的回应，但必须在收到被申诉人答复副本后 7 天（一周）内作出。

4   The respondent is allowed to make final comment, but must do so within seven days (one week) of receiving a copy of the appellant's further comment.

4   被申诉人可以作出最后一次回应，但必须在收到申诉人进一步回应材料副本后 7 天（一周）内作出。

5   The Directors, or appeal committee if appointed, can extend these time limits, but only if the firm concerned can show that substantial injustice would otherwise be done and the request for an extension is reasonable in all the circumstances. Any request for an extension should be made in writing and should outline the reasons why substantial injustice may occur if an application is refused.

5   董事会或申诉委员会（如有任命）可以延长上述时限，但前提是所涉及的当事方能够证明不如此将导致重大不公正，并且在所有情势下提出延长时限申请都是合理的。延长时限申请应以书面方式提出，并应简要说明为何如果申请被拒绝就会导致重大不公正的原因。

6   Applications for extensions must be made before time limits expire.

6   延长时限申请必须至少在时限届满前提出。

7   Further submissions may only be allowed if both parties agree, or the appeal committee decides that substantial injustice will be caused by rejecting them; then the appellant is allowed to make further comment, but must do so within seven days (one week) of receiving a copy of the respondent's further comments; and the respondent is allowed to make final comment, but must do so within seven days (one week) of receiving a copy of the appellant's further comments.

7   仅在双方同意或者申诉委员会认为如拒绝将会导致重大不公正的情况下，允许进一步提供材料；随后申诉人可以作出进一步的回应，但必须在收到被申诉人进一步答复副本后 7 天（一周）内作出；并且被申诉人可以作出最后一次回应，但必须在收到申诉人进一步回应材料副本后 7 天（一周）内作出。

8   Unless circumstances otherwise dictate, the Association shall arrange for the appeal to be heard no later than 14 days (two weeks) after final submissions

have been received by the appeal committee.

8　除非情况不允许，否则协会须在申诉委员会收到最终提交材料后 14 天（两周）内安排申诉开庭。

9　Either party may nominate, in writing, a representative, who must be an Individual Member, to act on their behalf in any matter concerned with an appeal, provided the Individual Member has not acted as arbitrator in the dispute. We will then communicate with them and no-one else.

9　各方可以书面形式指定一名代表，此人须为个人会员，在申诉过程中代表当事方处理一切事务。此后我们将与此人而非其他任何人沟通信息。

10　All appeal material must be submitted to us by:

10　所有申诉材料必须由下列人员提交给我们：

the firms in dispute; or our Individual Members acting as appointed representatives.

系争议公司；或任命为代表的个人会员。

11　We will not accept submissions directly from legal firms or independent lawyers.

11　我们不接受律师事务所或独立律师直接提交的文件。

Small Claims Technical Arbitration
(for disputes with a value at, or less than, US$ 25,000)
小额索赔技术仲裁
（争议金额等于或小于 25000 美元）

Bylaw 316

规章 316

1　Disputes which fall to be determined under these Bylaws shall be restricted to all disputes related to a total value not exceeding US$ 25,000 (twenty five thousand United States Dollars) but excluding those disputes for any contract that has not been, or will not be performed, and is to be closed by being invoiced back to the seller under our Rules in force at the date of the contract.

1　依据本条规章裁决的争议限于争议总金额等于或小于 US$25000（贰万伍仟美元）的事项，但不包括因尚未执行完毕、不会被执行、或依合同当时有效规章通过向出卖人回开发票而即将终结的合同而引发的争议。

2　A sole arbitrator appointed by us will hear such disputes. The sole arbitrator shall ensure that the parties are treated with equality and that each party is given a fair opportunity to present its case. The sole arbitrator shall conduct the proceedings with a view to expediting the resolution of the dispute. All communications between either party and the sole arbitrator shall be simultaneously copied to the other party.

2 争议由我们任命的一名独任仲裁员审理。独任仲裁员须确保各方得到公平对待，并且各方获得公平的机会阐明自己的理由。独任仲裁员须以方便快速解决争议为宗旨主持仲裁程序。任何一方当事人与独任仲裁员之间的沟通信息均须同时抄送给对方。

3 If upon receipt of the submissions from both parties the sole arbitrator considers that the matter is not within the remit of the small claims procedure or the matter is too complex for a sole arbitrator to consider, he will advise the parties of this and they will have the right to proceed to a full tribunal hearing to resolve the dispute.

3 如收到双方提交的材料后，独任仲裁员认为案件不属于小额索赔程序，或者案情复杂不能由一名独任仲裁员裁决，独任仲裁员可以向双方提出建议，双方当事人有权请求由完整的仲裁庭审理解决争议。

4 The previously appointed sole arbitrator will act as the tribunal Chairman unless either party objects. Any objection must be made in writing within seven days (one week) of notice being given of the relevant appointment and accompanied by the reasons for objection. An objection to an appointment will only be valid if the President decides that substantial injustice could result. Each party will appoint their own arbitrator within 14 days (two weeks) of being requested to do so by us. If either party fails to appoint an arbitrator within the stated period, the President will appoint an arbitrator and give notice of the appointment to the parties.

4 此前受任命的独任仲裁员将担任仲裁庭主席，除非有一方当事人提出异议。异议必须在相关任命通知发出后七天（一周）内以书面方式提出，并应说明异议理由。对任命的异议仅在及会长判定如此可能导致重大不公正的情况下有效。各方应在收到要求后14天（两周）内指定各自的仲裁员。如任何一方未在上述期间内任命仲裁员，则会长将任命一名仲裁员，并向各方发出任命通知。

Commencement of Arbitration
仲裁的启动
Bylaw 317
规章 317

1 Any party wishing to commence arbitration under these Bylaws ("the claimant"), who must be a Member Firm of the Association at the time of commencement of arbitration, shall send us a written request for arbitration ("the request"), and we shall copy the Request to the other party ("the respondent").

1 任一方（下称"申诉方"）如希望启动本规章所述仲裁程序，其在仲裁启动之时必须为协会成员公司，须向我们提交书面的仲裁申请（下称"申请"），我们将复制申请书并送给对方（下称"被诉方"）。

2 When sending the request, the claimant shall also send:

2 申诉方提交申请书时，须同时提交：

the name, address including email address, telephone and facsimile number of the respondent,

被诉方姓名/名称、地址（包括电子邮件地址）、电话和传真号码，

a) a copy of the contract signed by both parties; or

a) 双方签署的合同复印件；或

b) a copy of the arbitration agreement signed by both parties if not included in the contract; or

b) 双方签署的仲裁协议复印件（如合同中没有仲裁条款）；或

c) a copy of the contract and any supporting evidence,

c) 合同复印件及支持证据，

details of the claim value which must be no more than US$ 25,000, and

索赔金额不超过 25000 美元的详细情况，

such application fee as may be due under Appendix C of the Rule Book.

以及依规则手册附录 C 可能应缴纳的仲裁申请费。

3 We may refuse arbitration facilities where one of the parties to the dispute has been suspended from the Association or expelled. Arbitration will be refused where either the name of one of the parties appeared on the Association's List of Unfulfilled Awards at the time that the contract under dispute was entered into, or the penalty of denial of arbitration services has been imposed on one of the parties pursuant to Article 27 or Bylaw 418.

3 如争议的一方当事人已被暂停或开除协会资格，则我们可以拒绝仲裁。如在系争合同签署之时有任何一方当事人被列入协会未执行裁决书名单，或者一方当事人依章程第 27 条或规章 418 之规定受到过拒绝可得仲裁处罚的，我们亦可拒绝仲裁。

Appointment of a sole arbitrator
独任仲裁员的任命
Bylaw 318
规章 318

1 Upon receipt of a Request made in accordance with Bylaw 317, we will nominate the sole arbitrator within seven days (one week).

1 收到依规章 317 提出的请求后，我们将于七天（一周）内询问被诉方是否同意对独任仲裁员的提名。

2 The sole arbitrator must be an Individual Member of our Association when appointed. The arbitrator must additionally be qualified to the standards set by the Directors from time to time before he may accept such an appointment.

2 独任仲裁员接受任命之时必须为协会个人会员。此外，仲裁员在接受任命前还必须满足董事会不时规定的资质要求。

3 If the sole arbitrator dies, resigns, refuses to act, ceases to hold the requisite qualifications or becomes incapable of performing his functions, a replacement sole arbitrator will be appointed by the President.

3 如独任仲裁员死亡、辞职、拒绝履行职务、不再符合必要的资质、或无法履行自己的职责，则由会长指定更换人选。

4 By accepting appointment (whether by the parties or by us) a sole arbitrator binds himself to the Association to act in accordance with the Bylaws.

4 独任仲裁员一旦接受任命（无论由一方当事人还是我们指定），均须遵守规章规定履行职责。

5 If either party raises an objection to a nominated sole arbitrator, it must do so within seven days (one week) of notice being given of the relevant appointment. Any objection must be made in writing, accompanied by the reasons for objection. An objection to an appointment will only be valid if the President decides that substantial injustice could result. If the objection is upheld, the President shall appoint a replacement sole arbitrator.

5 如任何一方对独任仲裁员提名人选有异议，则必须在相关的任命通知发出后七天（一周）内提出。异议必须为书面形式，并附有异议的理由。对任命的异议仅在会长判定如此可能导致重大不公正的情况下有效。如异议成立，则会长须任命一名新的独任仲裁员。

6 If new evidence comes to light after the normal time limits for raising an objection have expired, an objection may still be raised. The President will decide whether it will be heard and whether it is valid.

6 如在提出异议的正常时限届满后出现新证据，则仍可提出异议。由会长决定是否听取理由以及异议是否有效。

7 If a party disagrees with the President's intention or decision it can appeal to the Directors but it must do so within seven days (one week) of notice having been given of the Presidents' decision. The Directors can use any of the powers given to the President at paragraph (5) and paragraph (6) above.

7 如一方不同意会长的意向或决定，可以向董事会提起申诉，但须在会长的决定下达后7天（一周）内提出。董事会可以行使上文第(5)和(6)节赋予会长的权力。

8 If the President should have a possible conflict of interest, he will not appoint the sole arbitrator under these Bylaws. In that situation, the Vice-President or an acting President will have the same powers of appointment as the President.

8 如会长有可能的利害关系，则不得依据规章任命独任仲裁员。在这种情况下，副会长或代理会长拥有与会长相同的任命权。

Revoking the authority of a sole arbitrator
撤销独任仲裁员的权力

Bylaw 319
规章319

1 Once a sole arbitrator has been appointed, his authority cannot be revoked by either party unless both parties agree.

1 一旦独任仲裁员获得任命，其权力不得由一方当事人撤销，双方同意除外。

2 If a sole arbitrator ceases to be a Member of the International Cotton Association, he cannot continue to act in whatever capacity he was appointed unless the Directors agree.

2 如独任仲裁员不再担任国际棉花协会会员，均不得继续以任何身份履行任何曾被委任的职责，董事会同意除外。

3   The President may revoke an appointment and appoint an alternative:
3   在下列情况下，会长可以撤销任命并转任他人：

if substantial injustice will be caused by him not doing so; or
如会长不如此履行将会导致重大不公正；或

if requested to do so by either party in the following circumstances:
由任一方在下列情况下提出请求：

if he upholds an objection under Bylaw 318;
如其赞同规章 318 所规定的异议；

if an appointed arbitrator dies, refuses or becomes unable to act;
如有任命的仲裁员死亡、拒绝或不能履行职务；

if a sole arbitrator does not make an award within 56 days (eight weeks) of having received the final written submissions from the parties.
如独任仲裁员在收到双方最后提交书面材料后 56 日（八周）内仍不作出裁决。

4   If, upon appointment as the tribunal Chairman, the sole arbitrator declines to act, he must give notice in writing and the President will appoint a replacement within seven days (one week) of notice having been given.

4   如独任仲裁员被任命为首席仲裁员，但本人不愿任职的，他必须发出书面通知。会长收到通知后七天（一周）内另行指定人选。

5   The Association will give notice of the President's intention. If a party disagrees with the President, it can appeal to the Directors but it must give its reasons in writing within seven days (one week) of notice having been given. The Directors can use any of the powers given to the President.

5   协会将发出关于会长意向的通知。如一方不同意会长的决定，则可向董事会申诉，但必须在通知发出后 7 天（一周）内提出书面理由。董事会可以行使赋予会长的任何权力。

6   The timeframes indicated in paragraph (3) above shall not be construed so as to undermine or overrule the arbitrators' duty under the Act but to allow each party reasonable opportunity to reply to any query or order from the sole arbitrator subsequent to the closure of final written submissions.

6   上文第(3)节所述的时间限制不得被解释为损害或否定仲裁员依仲裁法所具有的、允许各方当事人在最终书面材料提交后享有合理的机会就独任仲裁员提出的任何询问或命令作出答复的职责。

Association's fees and deposits on account of Small Claims Arbitration fees
因小额索赔而产生的协会收费以及仲裁费的押金
Bylaw 320
规章 320

1 Sole arbitrators shall be entitled to charge fees which shall be fixed by reference to the total amount of time reasonably devoted to the arbitration and shall be in accordance with the fees laid down in Appendix C of the Rule Book.

1 独任仲裁员有权按其为处理仲裁案件花费的合理时间计算并收取费用，并须符合规则手册附录 C 所规定的收费标准。

2 Where the sole arbitrator finds it necessary to obtain legal advice on any matter arising from an arbitration, reasonable legal fees thereby incurred will be payable by the parties, as specified in the Award.

2 如独任仲裁员认为有必要就仲裁中发生的任何事项寻求法律意见的，由此发生的合理法律费用应由双方承担，并在裁决书中列明。

3 When an Award is presented for stamping in accordance with Bylaw 321, the sole arbitrator shall invoice us for all fees, clearly stating their applicable hourly rate. The sole arbitrator is required to submit a time sheet in a format approved by the Directors.

3 裁决书依规章 321 提交盖章时，独任仲裁员须向我们出具发票：包含所有费用，列明相应的小时费率。独任仲裁员需要提交一份格式经董事批准的时间记录表。

4 The only expenses a sole arbitrator shall be entitled to claim are courier fees, up to a maximum of £50.

4 独任仲裁员唯一有权报销的开支为快递费，最多不可超过 50 英镑。

5 The time sheet shall be forwarded to both parties by the Secretariat within 14 days (two weeks) of the award being released.

5 时间记录表须在裁决书发出后 14 天（两周）内由秘书发给双方当事人。

6 The payment of fees and expenses to the sole arbitrator is conditional upon the Association's receipt of the time sheet.

6 协会收到时间记录表后，方可向独任仲裁员支付费用和开支。

7 Subject to the foregoing, the sole arbitrator shall be entitled to prompt

payment of fees and expenses following release of the Award. If, following a review under Bylaw 359 the Directors determine that any fees or expenses are unreasonable, the sole arbitrator shall act in accordance with the decision of the Directors.

7 在上述前提下,裁决书发出后独任仲裁员有权要求支付费用和开支。在董事会依据规章359进行审核后如果发现费用或支出有不合理情形,则独任仲裁员须按董事会的决定采取行动。

Jurisdiction
司法管辖权
Bylaw 321
规章321

Without prejudice to the provisions of the Act relating to jurisdiction, the sole arbitrator may rule on his jurisdiction, that is, as to whether there is a valid arbitration agreement and what matters have been submitted to arbitration in accordance with the arbitration agreement.

在不违反仲裁法关于司法管辖规定的前提下,独任仲裁员可以就其司法管辖问题作出裁定,即关于是否存在有效的仲裁协议以及依据仲裁协议将何事项提交仲裁。

Conduct of the Small Claims Technical Arbitration
小额索赔技术仲裁的进行
Bylaw 322
规章 322

1　The conduct of the small claims arbitration will be based on documentary evidence only.

1　小额索赔仲裁仅在书面证据的基础上进行。

2　It shall be for the sole arbitrator;

2　适用独任仲裁员;

to determine whether he has jurisdiction; and
裁定独任仲裁员是否具有司法管辖权;以及

to decide all procedural and evidential matters,
决定所有程序和证据事项,

subject to the right of the parties to agree any matter.

以双方当事人有权同意为前提。

3 The sole arbitrator shall ensure the prompt progress of the arbitration, where appropriate by the making of orders.

3 独任仲裁员应确保仲裁程序的迅速进行,有必要时可以发出命令。

4 As soon as the sole arbitrator has determined a timetable for proceedings, we shall notify the parties.

4 一旦独任仲裁员确定程序时间表,我们将立即通知双方当事人。

5 The parties have a duty to do all things necessary for the proper and expeditious conduct of the proceedings, including complying without delay with any order or direction of the sole arbitrator as to procedural and evidential matters.

5 当事人有义务执行所必需的事项促进仲裁程序的适当和快速进展,包括迅速遵守独任仲裁员关于程序和证据事项的任何命令或指示,不予以延误。

6 If either party fails to comply with any procedural order of the sole arbitrator, the arbitrator shall have power to proceed with the arbitration and make an Award.

6 如有任何一方不遵守独任仲裁员的任何程序性命令,则仲裁员有权继续仲裁并作出裁决。

7 All statements, contracts and documentary evidence must be submitted in the English language. Whenever documentary evidence is submitted in a foreign language, unless otherwise directed by the sole arbitrator, this must be accompanied by an officially certified English translation.

7 所有陈述、合同和文件证据均须以英文提交。如文件证据以外语提交,必须附有经官方认证的英文翻译件,除非独任仲裁员有其他指示。

8 We will not accept submissions directly from legal firms or independent lawyers.

8 我们不接受律师事务所或独立律师直接提交的文件。

Small Claims Technical Arbitration Awards
小额索赔技术仲裁裁决书

Bylaw 323
规章 323

1 An Award shall be in writing, dated and signed by the sole arbitrator and shall contain sufficient reasons to show why he has reached the decisions contained in it, unless the parties agree otherwise or the Award is by consent.

1 裁决书须符合我们的官方格式，署明日期并视情况由独任仲裁员签署，并应充分说明仲裁庭作出裁决的理由，除非当事方同意不说明理由或裁决书系由双方同意作出。

2 Any Award shall state that the seat of the arbitration is in England and the date by which we must receive notice of appeal.

2 裁决书须说明仲裁地点在英格兰，以及我们必须收到申诉通知的日期。

3 All Awards made under our Bylaws will be treated as having been made in England, regardless of where matters were decided, or where the Award was signed, despatched or delivered to the firms in dispute.

3 依据规章作出的所有裁决将视为在英格兰作出，无论系争事项在何处裁定，或者裁决书在何处签署及交寄或交付争议方公司。

4 We will stamp every Award in our offices on the date of the Award, and apply the scale of fees laid down in Appendix C of the Rule Book.

4 我们将在裁决作出之日于我们的办公室对每一份裁决书盖章，并应用规则手册附录 C 规定的收费标准。

5 An Award will only become effective and binding when we stamp it.

5 盖章后裁决书方生效并具有约束力。

6 After we stamp an Award, we will notify all of the parties concerned.

6 裁决书盖章完毕后，我们将通知所有相关当事方。

7 The Award will only be released upon payment of the stamping fee and any outstanding fees, costs and expenses.

7 盖章费及任何应付收费、成本和费用支付完毕后，方可发出裁决书。

8 The Award must be honoured within 28 days (four weeks) from notification to all of the parties under paragraph (6) above.

8 裁决书必须于依上文第(6)节通知各当事方后 28 天（四周）内执行。

9 The Association will keep a copy of every Award.

9 每份裁决书协会将保留一份副本。

Interest on Awards
裁决书裁定的利息

Bylaw 324
规章 324

A sole arbitrator or small claims appeal committee can award simple or compound interest from such dates and at such rates as he or they consider appropriate.

独任仲裁员或小额索赔申诉委员会可以裁定按其认为适当的日期和利率支付单利或复利。

Costs
费用
Bylaw 325
规章 325

The general principle is that costs follow the event, but subject to the overriding discretion of the sole arbitrator and small claims appeal committee as to which party will bear what proportion of the costs of the arbitration or appeal. In the exercise of that discretion the sole arbitrator or small claims appeal committee shall have regard to all the material circumstances.

一般原则是根据发生情况支付费用,但独任仲裁员和小额索赔申诉委员会有充分权利裁定双方按何种比例承担仲裁或申诉费用,该决定具凌驾性。独任仲裁员或小额索赔申诉委员会在行使该等裁量权的过程中须考虑所有重要的情况。

Small Claims Technical Appeals
小额索赔技术申诉
Bylaw 326
规章 326

1 If either party disagrees with the sole arbitrator's Award, it can appeal to us within the period specified in the Award. It must send Notice of Appeal to us.

1 如任何一方对独任仲裁员的裁决不满,可以在裁决书规定的期限内向我们提出申诉。当事方必须向我们提交申诉通知。

2 Upon receipt of the Notice of Appeal we may demand that sums of money be deposited with us by the appellant, by way of deposit against any fees, costs or expenses in connection with or arising out of the Appeal. The appellant must also deposit any costs or stamping fees that the Tribunal's Award ordered them to pay. Failure to pay within the specified period will result in the Appeal being dismissed.

2 收到申诉通知后,我们可能要求申诉人存入一笔款项作为与申诉有关的收费、成本或费用的押金。申诉人还必须存入仲裁庭的裁决书要求其支付的任何费用及盖章费。未在规定期限内支付的,申诉会被驳回。

3 The Directors, or appeal committee if appointed, can extend the time limits

in paragraph (2) above, but only if the firm concerned can show that substantial injustice would otherwise be done and the request for an extension is reasonable in all the circumstances. Any request for an extension should be made in writing and should outline the reasons why substantial injustice may occur if an application is refused.

3 董事会或申诉委员会（如有任命）可以延长上述第(2)节所述的时限，但前提是所涉及的当事方能够证明不如此将导致重大不公正，并且在所有情势下提出延长时限申请都是合理的。延长时限申请应以书面方式提出，并应简要说明为何如果申请被拒绝就会导致重大不公正的原因。

Small Claims Technical Appeal Committee
小额索赔技术申诉委员会
Bylaw 327
规章 327

1 The conduct of the small claims technical appeal will be based on documentary evidence only.

1 小额索赔技术申诉将仅依据书面证据进行。

2 As soon as the appellant has paid all fees due under Bylaw 326 (2) and served its case for appeal, the Directors shall appoint a Small Claims Technical Appeal Committee ('appeal committee').

2 一旦申诉人支付规章 326 (2)规定的所有费用并提交申诉材料，则董事会须任命一个小额索赔技术申诉委员会（下称"申诉委员会"）。

3 A Director cannot be involved in any decision about an appeal or be on an appeal committee if he has acted as the arbitrator in the dispute or if substantial injustice could result.

3 如有董事已作为仲裁员参与争议解决或可能导致重大不公正，则该董事不得涉及申诉裁决事务或加入申诉委员会。

4 An Individual Member cannot be on an appeal committee if he has acted as the arbitrator in the dispute, or substantial injustice could result.

4 如有个人会员已作为仲裁员参与争议解决，或者可能导致重大不公正，则该会员不得加入申诉委员会。

5 An appeal committee will consist of a Chairman and two other people, who must be Individual Members when they are appointed. All appeal committee members must additionally be qualified to the standards as set by the Directors from time to time.

5 申诉委员会由一名主席和两名其他人员（任职时必须为个人会员）组成。此外，所有申诉委员会成员必须符合董事会不时规定的标准。

6 At any meeting of an appeal committee, the Chairman and both members must be present. In the event a member of the committee cannot continue to act, the Directors will appoint a new appeal committee member. However, the provisions of this paragraph and paragraph (5) above may be varied by the Directors if both parties agree in writing.

6 在申诉委员会的任何会议中，主席和两位成员都必须出席。如有委员会成员不能继续履行职务，则董事会将任命一位新的申诉委员会成员。不过，如双方以书面方式表示同意，则董事会可以变更本节及上文第(5)节的规定。

7 If the Directors appoint an appeal committee, either party can object to the Chairman or any member of the committee but must do so within seven days (one week) of notice being given of the relevant appointment. Any objection must be made in writing, accompanied by the reasons for objection. An objection to an appointment will only be valid if the President decides that substantial injustice could result.

7 如董事会任命申诉委员会，任一当事方可以对主席或委员会的任何成员提出异议，但必须在相关任命下达后七天（一周）内提出。异议必须为书面形式，并附有异议的理由。对任命的异议仅在主席判定如此可能导致重大不公正的情况下有效。

8 If the Directors uphold an objection, they shall immediately nominate a substitute.

8 如董事会赞同某项异议，则须立即提名替代者。

9 An appeal involves a new hearing of the dispute and the appeal committee can allow new evidence to be put forward. It may confirm, vary, amend or set aside the award of the sole arbitrator and make a new award covering all of the matters in dispute.

9 申诉程序中包括对争议的重新审理，并且申诉委员可以允许提出新证据。申诉委员会可以确认、变更、修改或推翻独任仲裁员作出的裁决书，并作出一份覆盖所有系争事项的新裁决书。

10 The appeal committee will decide the issues by a simple majority vote. Every member, including the Chairman will have one vote.

10 申诉委员会以简单多数投票作出决定。所有成员，包括主席在内，均享有一票权。

Appeal timetable
申诉时间表
Bylaw 328
规章 328

1 The appellant must submit its Notice of Appeal to us within the time specified in the Award. The appellant must then submit all fees due under Bylaw 326 (2) and its case for appeal within 14 days (two weeks) of the Association receiving its Notice of Appeal.

1 申诉人必须在裁决书说明规定的时间内向我们提交申诉通知。申诉人必须于协会收到申诉通知后 14 日（两周）内提交规章 326 (2)规定的所有费用以及申诉材料。

2 If the respondent intends to comment, it should do so within 14 days (two weeks) of receiving a copy of the appellant's case.

2 如被申诉人需要作出回应，则应在收到申诉人材料副本后 14 天（两周）内提出。

3 If the respondent replies, the appellant is allowed to make further comment, but must do so within seven days (one week) of receiving a copy of the respondent's reply.

3 如被申诉人作出答复，则申诉人可以作出进一步的回应，但必须在收到被申诉人答复副本后 7 天（一周）内作出。

4 The respondent is allowed to make final comment, but must do so within seven days (one week) of receiving a copy of the appellant's further comment.

4 被申诉人可以作出最后一次回应，但必须在收到申诉人进一步回应材料副本后 7 天（一周）内作出。

5 The Directors, or appeal committee if appointed, can extend these time limits, but only if the firm concerned can show that substantial injustice would otherwise be done and the request for an extension is reasonable in all the circumstances. Any request for an extension should be made in writing and should outline the reasons why substantial injustice may occur if an application is refused.

5 董事会或申诉委员会（如有任命）可以延长上述时限，但前提是所涉及的当事方能够证明不如此将导致重大不公正，并且在所有情势下提出延长时限申请都是合理的。延长时限申请应以书面方式提出，并应简要说明为何如果申请被拒绝就会导致重大不公正的原因。

6　Applications for extensions must be made before time limits expire.

6　延长时限申请必须至少在时限届满前提出。

7　Further submissions may only be allowed if both parties agree, or the appeal committee decides that substantial injustice will be caused by rejecting them; then the appellant is allowed to make further comment, but must do so within seven days (one week) of receiving a copy of the respondent's further comments; and the respondent is allowed to make final comment, but must do so within seven days (one week) of receiving a copy of the appellant's further comments.

7　仅在双方同意或者申诉委员会认为如拒绝将会导致重大不公正的情况下，允许进一步提供材料；随后申诉人可以作出进一步的回应，但必须在收到被申诉人进一步答复副本后7天（一周）内作出；并且被申诉人可以作出最后一次回应，但必须在收到申诉人进一步回应材料副本后7天（一周）内作出。

8　Unless circumstances otherwise dictate, the Association shall arrange for the appeal to be heard no later than 14 days (two weeks) after final submissions have been received by the appeal committee.

8　除非情况不允许，否则协会须在申诉委员会收到最终提交材料后14天（两周）内安排申诉开庭。

9　Either party may nominate, in writing, a representative, who must be an Individual Member, to act on their behalf in any matter concerned with an appeal, provided the Individual Member has not acted as arbitrator in the dispute. We will then communicate with them and no-one else.

9　各方可以书面形式指定一名代表，此人须为个人会员，在申诉过程中代表当事方处理一切事务，该个人会员此前不得在该争议中担任过仲裁员。此后我们将与此人而非其他任何人沟通信息。

10　All appeal material must be submitted to us by:

10　所有申诉材料必须由下列人员提交给我们：

the firms in dispute; or our Individual Members acting as appointed representatives.

系争公司；或任命为代表的个人会员。

11　We will not accept submissions directly from legal firms or independent lawyers.

11　我们不接受律师事务所或独立律师直接提交的文件。

Quality Arbitration
质量仲裁

Commencing arbitration
仲裁的启动

Bylaw 329
规章 329

If an application is required, it must be accepted by us before arbitration can commence. If that is done or if an application is not required, arbitration will commence when one firm tells the other in writing that it intends to go to arbitration and:

如需申请,则必须在仲裁程序起始之前经我们批准。如申请已被批准或者不需要申请,则当一方公司以书面方式告知另一方其有意提交仲裁,并采用下列措施时,仲裁程序即为开始:

asks the other firm to agree to use a sole arbitrator and suggests the name of an arbitrator; or

询问对方公司是否同意采用独任仲裁员并提出了仲裁员人选;或

names his arbitrator and asks the other firm to do the same.

提出本方的仲裁员人选并要求对方公司也提出自己的人选。

Bylaw 330
规章 330

1 If firms agree to quality arbitration under our Bylaws, our Individual Members can arbitrate and hear appeals. We will assist with the arbitral process. This applies to both registered and non-registered firms subject to the following:

1 如双方同意依本规章所接受仲裁,则我们的个人会员可以承担仲裁工作并审理申诉。我们将协助完成仲裁流程。注册和非注册公司均可提起仲裁,规则如下:

Non-registered firms must apply for arbitration. We can refuse to accept such applications. The applicant has a right of appeal to the Directors. Their decision is final.

非注册公司必须提出仲裁申请。我们可以拒绝批准此类申请。申请方有权向董事会申诉。董事会的决定是最终决定。

If a firm was not registered on the date of the contract giving rise to the dispute, an application fee may be due. Details are set out in Appendix C.

如在系争合同签署之日该公司并非注册公司,则可以收取申请费。详细内容见附录 C。

If, on the day before the date of the contract giving rise to the dispute, either party has its name circulated on the ICA List of Unfulfilled Awards in accordance with Bylaw 366, application for arbitration must be made to the Association. If the applicant is a non-registered firm, we will refuse to accept such applications. The applicant has a right of appeal to the Directors. Their decision is final.

在系争合同签署日之前，如果任一方依规章 366 被列入 ICA 未执行裁决书名单，则必须向协会提出仲裁申请。如果申请方系非注册公司，则我们可以拒绝批准此类申请。申请方有权向董事会申诉。董事会的决定是最终决定。

A registered firm of the Association which has entered into a contract with a party whose name, on the day before the date of the contract, appeared on the ICA List of Unfulfilled Awards will be subject to the provisions of Bylaw 418, or, if applicable, the provisions and procedures laid down in the Association's Articles.

如与其签约的对方在合同签署日之前被列入 ICA 未执行裁决书名单，协会注册公司将按规章 418 处理，或者，如果适用，按协会章程的规定和程序处理。

If a firm has been suspended or expelled, or has been refused re-registration, we will not accept an application for arbitration from it.

如某公司被暂停或开除会员资格，或被拒绝再次注册，则我们不会接受其提出的仲裁申请。

2 If an application for arbitration is required under this Bylaw, no Individual Member can act as an arbitrator until informed that the application has been accepted and any fee due has been paid.

2 如依据本规章需要提出仲裁申请，则在得到申请已被接受且应付费用已付清的通知之前，任何个人会员不得作为仲裁员采取行动。

Appointment of arbitrators
仲裁员的任命
Bylaw 331
规章 331

1 Quality arbitration will be conducted by two arbitrators unless the firms in dispute agree that one arbitrator is sufficient.

1 将由两名仲裁员负责质量仲裁，除非当事公司同意一名仲裁员已经足够。

2 If two arbitrators are appointed and they cannot agree, an umpire will make a decision.

2 如果任命了两名仲裁员，二人出现意见分歧，则由裁定人作出决定。

3 Arbitrators and umpires must be Individual Members of our Association when they are appointed.

3 仲裁员和裁定人接受任命时必须为协会个人会员。

4 Either firm can ask the President of the Association to appoint an arbitrator on its behalf.

4 任一方当事公司均可请求协会主席代表自己任命一名仲裁员。

Bylaw 332

规章 332

1 If one firm commences arbitration in line with Bylaw 329 and asks the other firm to agree to a sole arbitrator, then within 14 days (two weeks) the other firm must:

1 如果一方公司依规章 329 起始了仲裁程序，并要求对方公司同意由独任仲裁员仲裁，则该对方公司在 14 天（两周内）内必须：

either accept the name of the suggested arbitrator; or agree the name of another sole arbitrator; or

要么接受对方建议的仲裁员人选；或同意选任另一名独任仲裁员；

either say that it does not agree to using a sole arbitrator; name its own arbitrator; and may object to the arbitrator named by the first firm.

要么表示本方不同意使用独任仲裁员；提名自己的仲裁员；并可以反对第一家公司提出的仲裁员人选。

2 If the second firm names its own arbitrator, the first firm must object to the nomination within seven days (one week) or it will be considered to have been accepted.

2 如果第二家公司提名了自己的仲裁员，则第一家公司必须在七天（一周）内提出反对，否则视为接受提名。

3 If the second firm does not respond, the arbitration cannot proceed with a sole arbitrator. Arbitrators must be appointed by or on behalf of both firms.

3 如第二家公司没有回应，则不能采用独任仲裁员形式继续仲裁流程。仲裁员必须由双方任命或代表双方行事。

Bylaw 333

规章 333

If one firm commences arbitration in line with Bylaw 329 but does not ask the other firm to agree to a sole arbitrator, the other firm must nominate its arbitrator in writing within 14 days (two weeks). Unless a reasoned objection is made in writing within seven days (one week), any arbitrator nominated by either firm will be considered to have been accepted by the other.

如一家公司依规章 329 之规定起始了仲裁程序，但并未询问对方公司是否同意独任仲裁员,则对方公司必须在14日(两周)内以书面方式提名其仲裁员。如果在七日(一周)内没有以书面方式提出合理的异议,则各方提名的仲裁员视为已被对方接受。

Bylaw 334

规章 334

Once the arbitrator or arbitrators have been nominated and the periods allowed for objections have expired, and any objections dealt with, the arbitrator or arbitrators will be considered to have been appointed. Firms must then allow arbitrators to act independently in accordance with the law.

仲裁员被提名且异议期届满,而提出的任何异议已经处理后,则视为仲裁员已获得任命。此时当事方必须允许仲裁员按仲裁法的规定独立采取行动。

Bylaw 335

规章 335

1 If one firm raises an objection to an arbitrator nominated by the other it must do so within seven days (one week) of notice being given of the relevant appointment. Any objection must be made in writing, accompanied by the reasons for objection. An objection to an appointment will only be valid if the President decides that substantial injustice could result.

1 如任何一方对对方提名的仲裁员有异议,则必须在相关的任命通知发出后七天(一周)内提出异议。异议必须为书面形式,并附有异议的理由。对任命的异议仅在主席判定如此可能导致重大不公正的情况下有效。

2 If either firm:

2 如有任何一方公司:

fails to nominate an arbitrator within 14 days (two weeks) of being requested to do so, or

未能在被要求时于14日(两周)内提名仲裁员;或

fails to agree on a replacement arbitrator within 14 days (two weeks) of a substantiated and valid objection to a nomination, the other firm can ask the President to make an appointment on behalf of the firm that has failed to nominate an arbitrator, or failed to agree on a replacement arbitrator within the time allowed.

在提名遭到实质性和有效异议时,未能在14日(两周)内同意更换仲裁员,另一方公司可以请求协会会长代表未在时限内提名仲裁员或同意仲裁员更换人选的一方公司任命仲裁员。

3 The Association will give notice of the President's intention. If the firm in default does not nominate an arbitrator acceptable to the other firm within 14 days (two weeks) of that notice being given, the President may act.

3 协会将发出关于会长意向的通知。如在通知发出后 14 日（两周）内，违约一方公司未提名对方公司可接受的仲裁员人选，则会长可以任命仲裁员。

4 Either firm can object to the Chairman or any member of a Quality Appeal Committee, but must do so within seven days (one week) of notice being given of the relevant appointment. Any objection must be made in writing, accompanied by the reasons for objection. An objection to an appointment will only be valid if the President decides that substantial injustice could result.

4 如任何一方公司对质量申诉委员会主席或任何成员有异议，则必须在相关的任命通知发出后七天（一周）内提出异议。异议必须为书面形式，并附有异议的理由。对任命的异议仅在主席判定如此可能导致重大不公正的情况下有效。

5 If an objection is not acted on and not withdrawn, the President must be asked to decide whether it is valid.

5 如异议既未通过也没有撤回，则必须要求会长裁定其是否有效。

6 If new evidence comes to light after the normal time limits for raising an objection have expired, an objection may still be raised. The President will decide whether it will be heard and whether it is valid.

6 如在提出异议的正常时限届满后出现新证据，则仍可提出异议。由会长决定是否听取理由以及异议是否有效。

7 If a firm disagrees with the President's intention or decision it can appeal to the Directors but it must do so within seven days (one week) of notice having been given. The Directors can use any of the powers given to the President at paragraph (3) and paragraph (4) above.

7 如一方公司不同意会长的意向或决定，可以向董事会提起申诉，但须在会长的决定下达后 7 天（一周）内提出。董事会可以行使上文第(3)和(4)节赋予会长的权力。

8 If the President should have a possible conflict of interest, he will not appoint arbitrators under these Bylaws. In that situation, the Vice-President or an acting President will have the same powers of appointment as the President.

8 如会长有可能的利害关系，则不得依据规章任命仲裁员。在这种情况下，副会长或代理会长拥有与会长相同的任命权。

Revoking the authority of an arbitrator, umpire or appeal committee member
撤销仲裁员、裁定人或申诉委员会成员的权力
Bylaw 336
规章 336

1 Once an arbitrator, umpire or appeal committee member has been appointed, his authority cannot be revoked by either firm unless both firms agree.

1 仲裁员、裁定人或申诉委员会成员一旦任命,任何一方公司不得撤销其权力,除非双方一致同意撤销其权力。

2 If an arbitrator, umpire or appeal committee member ceases to be a Member of the International Cotton Association, he cannot continue to act in whatever capacity he was appointed unless the Directors agree.

2 如仲裁员、裁定人或申诉委员会成员不再担任国际棉花协会会员,均不得继续以任何身份履行任何曾被委任的职责,董事会同意除外。

3 The President may revoke an appointment and appoint an alternative:

3 在下列情况下,会长可以撤销任命并转任他人:

if substantial injustice will be caused by him not doing so; or
如会长不如此履行将会导致重大不公正;或

if requested to do so by either firm in the following circumstances:
由任一方公司在下列情况下提出请求:

if he upholds an objection under Bylaw 335;
如其赞同规章 335 所规定的异议;

if an appointed arbitrator dies, refuses or becomes unable to act;
如有任命的仲裁员死亡、拒绝或不能履行职务;

if a sole arbitrator does not make an award within 21 days (three weeks) of having been appointed or the arrival of the samples at the place of arbitration, whichever is the later;
如独任仲裁员在被任命或样品到达仲裁地后(以后到者为准)21 日(三周)内仍未作出裁决;

if the two arbitrators do not make an award or appoint an umpire within 21 days (three weeks) of both having been appointed or the arrival of the samples at the place of arbitration, whichever is the later; or
如两名仲裁员在被任命或样品到达仲裁地后(以后到者为准)21 日(三周)内仍未作出裁决或任命裁定人;或

if the umpire does not make an award within seven days (one week) of the date of his appointment.

如裁定人在被任命后七天（一周）内仍未作出裁决。

4 The Association will give notice of the President's intention. If a firm disagrees with the President, it can appeal to the Directors but it must give its reasons in writing within seven days (one week) of notice having been given. The Directors can use any of the powers given to the President.

4 协会将发出关于会长意向的通知。如一方公司不同意会长的决定，则可向董事会申诉，但必须在通知发出后 7 天（一周）内提出书面理由。董事会可以行使赋予会长的任何权力。

Timetable
时间表

Bylaw 337
规章 337

1 In manual quality arbitrations:

1 手工检验质量仲裁：

samples to be used must be taken within 42 days (six weeks) of the date of arrival of the cotton;

棉花到达后，必须在 42 天（六周）内采集样品；

arbitration must be commenced in line with Bylaw 329 within 49 days (seven weeks) of the date of arrival of the cotton; and

仲裁必须依规章 329 之规定在棉花到达后 49 天（七周）内起始；并且

samples must be sent to the place of arbitration within 70 days (10 weeks) of the date of arrival of the cotton.

样本必须在棉花到达后 70 日（十周）内送往仲裁地。

2 In instrument test based arbitrations:

2 仪器检测质量仲裁

samples to be used must be taken within 42 days (six weeks) of the date of arrival of the cotton;

棉花到达后，必须在 42 天（六周）内采集样品；

samples must be sent to the place of testing within 70 days (10 weeks) of the date of arrival of the cotton; and

样本必须在棉花到达后 70 日（十周）内送往检验地；并且

arbitration must be commenced within 21 days (three weeks) of the date the test results are published.

仲裁必须在检测结果公布后 21 天（三周）内起始。

3　The Directors can extend these limits, but only if the firm concerned can show that substantial injustice would otherwise be done and that the request for an extension is reasonable in all the circumstances. Applications must be made to us in writing. The Directors will take the other firm's comments into account before it makes a decision.

3　董事会可以延长上述时限，但前提是所涉及的当事方能够证明不如此将导致重大不公正，并且在所有情势下提出延长时限申请都是合理的。申请必须以书面方式向我们提出。董事会在作出决定前须考虑对方公司的意见。

The place of arbitration

仲裁地点

Bylaw 338

规章 338

1　Manual quality arbitrations can be held anywhere by agreement between the firms in dispute. If the firms cannot agree on the location for manual arbitration, such manual quality arbitrations will be held in our arbitration room.

1　争议双方协商一致，可在任何地点进行手工检测质量仲裁。如双方无法就手工检测质量仲裁的地点达成一致，则该等手工检测质量仲裁将在我们的仲裁庭进行。

2　In the event of an appeal on manual arbitration, the Directors will decide where the manual appeal will be heard.

2　如发生针对手工检测质量仲裁的申诉，则由董事会决定在何处审理手工检测质量申诉。

3　We will stamp arbitration and appeal Awards and make them effective in Liverpool, without regard to where the arbitration or appeal takes place.

3　无论仲裁或申诉在哪里进行，我们将在利物浦对仲裁和申诉裁决书盖章并使之生效。

Procedures

程序

Bylaw 339

规章 339

1　Manual quality arbitrations will be conducted on the basis of samples and decided by manual examination.

1 手工检测质量仲裁以样品为基础,并通过手工检验作出裁决。

2 Instrument testing arbitrations will be conducted on the basis of test reports. The information on the test reports will be final. The arbitrators may make an award if either of the parties fails to;

2 仪器检测质量仲裁以检测报告为基础。检测报告的结果为最终结果。如任何一方当事人未能做到以下事项,则仲裁员可以作出裁决:

agree on the allowances to be applied; or
同意适用补贴;或

agree on the interpretation of the test report as applicable to the contract; or
同意适用于合同的对检测报告的解释;或

pay an agreed allowance within 14 days (two weeks) of the test report being issued.
在检测报告发布后 14 日(两周)内支付议定的补贴。

3 Bylaws 346 and 347 do not apply for instrument testing arbitrations.

3 规章 346 和 347 不适用于仪器检测仲裁。

4 Either firm can appeal against an Award given by the arbitrator, arbitrators or umpire in line with Bylaw 352, but no further instrument tests will be conducted.

4 任何一方均可依规章 352 对独任仲裁员、仲裁员或裁定人的裁决提出申诉,但不再进行进一步的仪器检测。

Jurisdiction
司法管辖权

Bylaw 340
规章 340

Without prejudice to the provisions of the Act relating to jurisdiction, the arbitrators and umpire may rule on their own jurisdiction, that is, as to whether there is a valid arbitration agreement, whether the tribunal is properly constituted and what matters have been submitted to arbitration in accordance with the arbitration agreement.

在不违反仲裁法关于管辖规定的前提下,仲裁员和公断人可以就其司法管辖问题作出裁定,即关于是否存在有效的仲裁协议、仲裁庭的构成是否适当以及依仲裁协议将何事项提交仲裁。

Bylaw 341
规章 341

1 If one firm commences a quality arbitration and the other firm disputes

jurisdiction or the terms of the contract regarding quality, there will be a technical arbitration unless the firms agree otherwise. The technical Award will say:

1 如一方启动了一项质量仲裁，而对方对司法管辖权或涉及质量的合同条款提出争议，则须进行一项技术仲裁，双方另行协商同意除外。技术仲裁裁决书将确定：

whether we have jurisdiction,
我们是否享有司法管辖权，

what matters are subject to quality arbitration; and
质量仲裁的内容为何；以及

what contract terms apply with regard to quality.
适用什么涉及质量的合同条款。

2 A firm can challenge this Award by appeal to the Directors in the normal way.

2 一方对裁决不满的，可以通过正常途径向董事会提起申诉。

3 A quality arbitration may then take place providing the technical arbitration or appeal finds that:

3 如技术仲裁或申诉判定如下事项，则可开始进行质量仲裁：

there is a valid arbitration agreement; and
存在有效的仲裁协议；以及

our Bylaws apply.
可以适用我们的规章。

Standards
标准

Bylaw 342
规章 342

1 When we refer to any of the 'Universal Standards' for quality, we mean the Universal Standards for colour and leaf grade, adopted under the Universal Cotton Standards Agreement existing between us and the United States Department of Agriculture.

1 当我们提到适用于质量的任何"通用标准"，是指我们与美国农业部达成的"通用棉花标准协议"所采用的颜色和叶片等级全球标准。

2 The Association will hold a complete set of 'Universal Standards'. Individual Members can inspect them during our office hours. They may be used in settling arbitrations and appeals.

2 协会持有全套"全球标准"文件。个人会员可在我们上班时间查阅。标准可能被用于解决仲裁和申诉。

3 The Standards will be available for regular inspection by the Quality Appeal Panel. If they ever consider that any standard has changed, the Panel will take action.

3 质量申诉工作小组可以定期查阅标准。如他们认为有任何标准已经发生变更,则工作小组将会采取行动。

Bylaw 343

规章 343

1 'ICA Official Standards' are those that have been approved by the Directors and confirmed by the Association.

1 "ICA 官方标准"是指已经董事批准并由协会确认的标准。

2 The Association will hold the standards. Individual Members can inspect them during our office hours. They may be used in settling arbitrations and appeals.

2 协会持有该等标准文件。个人会员可在我们上班时间查阅。标准可能被用于解决仲裁和申诉。

3 The Standards will be available for regular inspection by the Quality Appeal Panel. If they ever consider that any standard has changed, the Panel will take action.

3 质量申诉工作小组可以定期查阅标准。如他们认为有任何标准已经发生变更,则工作小组将会采取行动。

4 The Directors will approve changes to the standards after considering comments of the Quality Appeal Panel. We will give each Registered Firm and Individual Member 14 days (two weeks) written notice of proposed changes. We will then confirm the changes. The new standards will come into effect the day after they have been confirmed. They will apply to contracts made on or after that date.

4 董事会将在听取质量申诉工作小组的意见后批准对标准进行变更。我们将提前 14 天(两周)以书面形式向各注册公司和个人会员通知该等拟议中的变更信息。然后我们将确认变更。新标准于确认后第二天生效。新标准将适用于当天及以后签署的合同。

5 New standards for growths or grades of cotton will be used as soon as we have confirmed them.

5 在我们确认新的棉花培植或等级标准后,随即使用该等标准。

Application of value differences to disputes
对争议采用的价差
Bylaw 344
规章 344

1 Unless Bylaw 348 or Bylaw 354 applies, or the firms in dispute agree otherwise, quality arbitration Awards will be based on the differences in value fixed by the Value Differences Committee.

1 质量仲裁裁决书将以价差委员会确定的价值差异为基础，适用规章348或354或者双方另外协商同意之情形除外。

In the case of CIF and CFR contracts, the value difference that will apply will be the difference on the date of arrival of the cotton.

对于CIF和CFR合同，所适用的价差为棉花到达日期的差异。

In the case of FOB contracts, the value difference that will apply will be the difference on the date of the bill of lading or other document of title.

对于FOB合同，所适用的价差为提单或其他产权文件日期之差异。

In all other cases, the value difference that will apply will be the difference on the day the buyer receives title to the cotton.

在所有其他情况下，所适用的价差为买方获得棉花产权之日的差异。

2 Value differences take effect from the start of the day after they are published.

2 价差自公布之日起生效。

3 If differences are not fixed, Awards will be based on the differences in value in a market appropriate to the contract. The arbitrator or arbitrators, or umpire, or Quality Appeal Committee will decide the appropriate differences.

3 如差额不固定，则裁决书将以在与合同相适应的市场价差为基础。仲裁员、多名仲裁员、裁定人或质量申诉委员会将决定相应的差额。

4 The above methods will be used to calculate an Award.

4 将使用上述方法决定裁决内容。

Bylaw 345
规章 345

1 In quality arbitrations, Awards can be shown as cash amounts or they may be shown as fractions of the appropriate currency for the weight specified in the contract.

1 在质量仲裁中，裁决书可能表现为现金金额，也可以表现为按合同规

2   In CIF and similar contracts, the Awards for grade and staple length will be shown separately. This does not apply to contracts for cotton linters or cotton waste.

2   在 CIF 和类似合同中，对等级和纤维长度的裁决将分别标示。这不适用于棉短绒或废棉合约。

'Average grade'
"平均等级"

Bylaw 346
规章 346

1   Arbitration on cotton sold as average for any particular grade will be settled by classing the different lots. Grades or fractions of grades will be sorted into those above and below the grade's standard. Whatever turns out to be average will be passed. An allowance will be made on the rest.

1   对于按某一特定等级平均级出售的棉花，将通过对不同货品的评级进行仲裁。各种等级或等级下的分等级将分类为该等级标准以上及以下。归入平均级的不作处理。其余部分计算补贴。

2   This will apply unless the buyer and the seller agree otherwise.

2   除非买卖双方另有协商同意，否则照此执行。

Classification
评级

Bylaw 347
规章 347

1   If a firm appeals against a quality arbitration Award and pays the extra set fee, the Quality Appeal Committee will issue a certificate showing the true classification breakdown for grade, colour or staple length.

1   若一方对质量仲裁裁决书提起申诉并支付额外的鉴定费，则质量申诉委员会将签发一份证书，注明详细的等级、颜色和纤维长度的评级情况。

2   American Upland Cotton

2   美国陆地棉

The colour and leaf grade of American Upland cotton will be classified under the 'Universal Standards'.

美国陆地棉的颜色和叶片等级将按"全球标准"评级。

American Pima Cotton

美国皮马棉

The grade and colour of American Pima cotton will be classified under the official cotton standards of the USA.

美国皮马棉的等级和颜色将按美国官方棉花标准评级。

In both cases, staple length will be classified under the terms of the United States Department of Agriculture standards.

在两种情况下,纤维长度均按美国农业部标准评级。

3 Non-American Cotton

3 非美国棉

In the case of a growth for which we have 'ICA Standards', grade will be classified by those standards. Staple length will be classified under the terms of the United States Department of Agriculture standards.

若对某一棉品有"ICA 标准",则按该等标准评级。纤维长度按美国农业部标准评级。

4 Anyone who wants cotton to be classified must ask at the same time as they apply for an appeal.

4 希望对棉花进行评级者必须在申请申诉的同时提出此要求。

5 Classification will only refer to the bales sampled.

5 仅对样品进行评级。

Cotton which is outside the normal quality range

超出一般质量范围的棉花

Bylaw 348

规章 348

1 In arbitrations and appeals on cotton which is outside the normal quality range of its relevant growth, the intrinsic value of the cotton will be established. That value will be taken into account in arriving at an Award. In cases where the value cannot be determined, arbitration will be based on the contract price.

1 在对超出相关品种一般质量范围的棉花进行仲裁和申诉的过程中,将确定棉花的内在价值。在作出裁决时将考虑其价值因素。若价值无法判定,则在合同价格基础上进行仲裁。

2 In arbitrations and appeals on cotton waste, linters, pickings and so on, arbitration will be based on the known value. Arbitration will be based on the contract price if the actual value cannot be established.

2 在对废棉、棉短绒、下脚等产品进行仲裁和申诉的过程中，按已知价值进行仲裁。如实际价值无法确定，则按合同价格进行仲裁。

3 The arbitrator or arbitrators, or umpire and an appointed Quality Appeal Committee can take advice or evidence from firms or individuals who are connected with the cotton trade and are experts in cotton waste, linters, pickings and so on.

3 独任仲裁员、仲裁员、公断人或任命的质量申诉委员会可以向熟悉棉花行业并对废棉、棉短绒、下脚等具备专业知识的公司或个人咨询意见或采集证据。

Anonymous arbitration

匿名仲裁

Bylaw 349

规章 349

1 Anonymous quality arbitration means that we will not disclose the names of the firms in dispute, or the arbitrators' and umpire's names.

1 匿名质量仲裁是指我们将不披露发生争议的公司的名称或仲裁员及公断人的姓名。

2 If a dispute about quality arises and both firms agree that it should go to anonymous quality arbitration, the following paragraphs are exceptions to the general arbitration procedure.

2 如品质争议双方均同意采用匿名质量仲裁，则适用以下各节，作为一般仲裁程序的例外情况。

3 Either firm can apply for anonymous arbitration by writing to the Secretary. They must explain the point at issue and give proof that the other firm is in agreement with the request.

3 任何一方均可以书面形式向秘书申请匿名仲裁。申请方必须解释争议要点并提供对方同意该请求的证明。

4 Those asking for the arbitration must give information about the status of the firms to the Secretary, to enable fees and charges to be set.

4 申请仲裁者必须向秘书提供公司身份信息，以便确定收费。

5 When the President receives the proof, he will appoint two Individual Members as arbitrators. If the arbitrators cannot agree on an Award within 21 days (three weeks) of being appointed, the President will appoint an umpire.

5 主席收到证明后，将指定两名个人会员担任仲裁员。如仲裁员于任命之后21天（三周）内不能作出裁决，则主席将任命一名裁定人。

6　The President can appoint a new arbitrator or arbitrators or umpire in either of the following situations:

6　在下列情况下，主席可以任命一名新的仲裁员、或多名新仲裁员、或一名裁定人：

if an arbitrator or umpire dies during the arbitration process, refuses or becomes unable to act; or

如有仲裁员或裁定人在仲裁期间死亡、拒绝履行职务或不能履行职务；或

if an umpire does not give his written decision on any matter referred to him by the arbitrators within seven days (one week) of him being asked to do so by either of them.

如有裁定人在收到仲裁员请求裁定的事项后七天（一周）内无法作出任何书面裁决。

7　The arbitrators and umpire will not be given the names of the firms in dispute, and the firms will not be given the arbitrators' and umpire's names.

7　仲裁员和裁定人将不会知道发生争议的公司的名称，公司也不会知道仲裁员和裁定人的姓名。

8　The Secretary will be responsible for giving any relevant selling type and samples, or the test results, and contract extracts to the arbitrators and umpire. The extracts will only be those which refer to quality. For manual arbitration he will replace the seller's type and samples identification marks with numbers before they go to the arbitrators and umpire.

8　秘书负责向仲裁员和裁定人提供一切相关的销售品种和样品、或检测结果、以及合同摘录。摘录仅包括与品质有关的部分。如系手工检测仲裁，则秘书会在提交仲裁员和裁定人之前将卖方的品种和样品识别号替换为数字编号。

9　Awards must be made on special forms.If all fees and expenses have been paid, we will send the Award to the firms in dispute.

9　裁决书必须按特定格式作出。如所有收费和费用已经付清，则我们会将裁决书送至争议各方。

Quality Arbitration Awards
质量仲裁裁决书

Bylaw 350
规章 350

1　An Award shall be made in writing on our official form, dated and signed by the arbitrator(s) or the umpire as applicable. The Chairman or Deputy Chairman and the Secretary of the appeal committee must sign an appeal Award.

1　裁决须以我们的官方格式以书面形式作出，标明日期并由仲裁员或裁定人相应签署。申诉裁决书须由申诉委员会主席或副主席及秘书签字。

2　A quality Award will not contain reasons for the Award.

2　质量裁决书中不载明裁决理由。

3　Any Award shall state that the seat of the arbitration is in England and the date by which we must receive notice of appeal.

3　裁决书须说明仲裁地点为英格兰，以及我们必须收到申诉通知的日期。

4　All Awards made under our Bylaws will be treated as having been made in England, regardless of where matters were decided, or where the Award was signed, despatched or delivered to the firms in dispute.

4　依据我们规章作出的所有裁决将视为在英格兰作出，无论系争事项在何处裁定，或者裁决书在何处签署及交寄或交付争议中的公司。

5　We will stamp every Award in our offices on the date of the Award, and apply the scale of fees laid down in Appendix C of the Rule Book.

5　我们将在裁决作出之日于我们的办公室对每一份裁决书盖章，并应用规则手册附录 C 规定的收费标准。

6　An Award will only become effective and binding when we stamp it.

6　盖章后裁决书方生效并具有约束力。

7　After we stamp an Award, we will notify all of the parties concerned.

7　裁决书盖章完毕后，我们将通知所有相关当事方。

8　The Award will only be released upon payment of the stamping fee and any outstanding fees, costs and expenses.

8　盖章费及所有应付收费、成本和费用支付完毕后，方可发出裁决书。

9　The Association will keep a copy of every Award.

9　每份裁决书协会将保留一份副本。

Interest on Awards
裁决书裁定的利息

Bylaw 351
规章 351

The arbitrator(s), umpire or Quality Appeal Committee can award simple or compound interest from such dates and at such rates as they consider just.

仲裁员、裁定人或质量申诉委员会可以裁定按照其认为公平的利率从某日起支付单利或复利。

Quality Appeals
质量申诉

Bylaw 352
规章 352

1 If either firm disagrees with an arbitrator's or arbitrators', or umpire's Award, it can appeal within the period allowed in the Award. It must send Notice of Appeal to us in writing. The reasons for appeal must be given when the appeal is made. The Chairman or Deputy Chairman of the appeal committee will then set the dates by which any further reasons or responses must be received.

1 如任何一方对独任仲裁员、仲裁员或裁定人的裁决不满，可以在裁决书载明的时限内提起申诉。申诉者必须向我们提交书面申诉通知。提起申诉时必须说明申诉理由。随后申诉委员会主席或副主席将确定接受进一步理由说明或答辩的时限。

2 We can demand an application fee set by the Directors. Details are laid down in Appendix C of the Rule book. We must receive these amounts within 14 days (two weeks) of the date of our invoice or the appeal will be dismissed.

2 我们可以要求按董事会确定的标准支付申请费。详细规定见规则手册附录C。我们必须在发票出具之日起14天（两周）内收到款项，否则申诉将被驳回。

3 This Bylaw does not apply to disputes over the costs of arbitration.

3 本规章不适用于对仲裁费的争议。

4 The appeal will be heard by a Quality Appeal Committee ('appeal committee') to be selected from the Quality Appeal Panel elected annually. The members of the Quality Appeal Panel will select a Chairman and Deputy Chairman. The Chairman and Deputy Chairman will select from the panel no less than two and no more than four of the members who are considered most qualified to judge the growth concerned to form a Quality Appeal Committee.

4 从年度选举的质量申诉工作小组中选择人员组成质量申诉委员会（下称"申诉委员会"）对申诉进行审理。质量申诉工作小组成员将选举一名主席和一名副主席。主席和副主席将从工作小组中选择至少二名至多四名被认为最具资格裁决所涉品种的成员组成质量申诉委员会。

5 The appeal committee will not hear an appeal before the end of the period allowed to appeal unless both firms agree, or both have appealed.

5 在申诉期限届满之前申诉委员会不审理申诉，除非双方同意或双方同时申诉。

6　The appeal committee can allow new evidence to be put forward covering all of the matters in dispute, unless the appeal refers to an instrument test arbitration, in which case the information contained in the last test report will be final.

6　申诉委员会可以允许提出涉及争议所有事项的新证据，除非申诉系关于仪器检测仲裁，在该等情况下最后一次检测报告所包含的信息为最终定案。

7　The appeal committee will decide the issues by a simple majority vote. Every member, including the Chairman and Deputy Chairman will have one vote. If both sides have the same number of votes, the Chairman will vote again to decide the issue.

7　申诉委员会以简单多数投票方式作出决定。所有成员，包括主席和副主席在内，均享有一票权。如双方所获票数相等，则由主席重新投票决定。

8　A Director cannot be involved in any decision about an appeal or be on an appeal committee if he has acted as an arbitrator or umpire in the dispute, or substantial injustice could result.

8　如有董事已作为仲裁员或公断人参与争议解决或可能导致重大不公正，则该董事不得涉及申诉裁决事务或加入申诉委员会。

9　An Individual Member cannot be on an appeal committee if he has acted as an arbitrator or umpire in the dispute, or substantial injustice could result.

9　如有个人会员已作为仲裁员或公断人参与争议解决，或者可能导致重大不公正，则该会员不得加入申诉委员会。

Bylaw 353
规章 353

1　Before it refers to the decision of the arbitrators, a Quality Appeal Committee must conduct an assessment of the cotton, or, in the case of instrument testing, the test report, and form an opinion. But, before making its final decision, the committee must refer to the arbitration Award.

1　在参考仲裁员的裁决之前，质量申诉委员会必须对棉花进行评估，或者，如系仪器检测，则对检测报告进行评估，并形成意见。不过，在作出最终裁定前，委员会必须参考仲裁裁决书。

2　If new arguments are offered to do with jurisdiction or the terms of the contract regarding quality, which have not been the subject of a technical arbitration or appeal, the committee will reach a decision and make an Award based on the evidence.

2　若所提出的新理由与司法管辖或关于品质的合约条款有关，且并未经过技术仲裁或申诉，则委员会在证据的基础上将作出决定并作出裁决。

3   However, in appeals against Awards under Bylaw 349:

3   不过，在依据规章349对裁决书提起的申诉中：

the names of the parties to the contract and the parties appealing will not be disclosed to the Quality Appeal Committee at any stage;

在任何阶段合约当事人和申诉人的姓名/名称都不会向质量申诉委员会披露；

if either party presents a previous appeal Award, or arbitration Award if there has been no appeal, there must also be a letter with it guaranteeing that the lot which is the subject of the appeal to us is the lot, bale for bale, which the previous Award was for; and

如有任何一方提出一份先前的申诉裁决书，或未提出申诉的仲裁裁决书，必须同时提交一封保证函，向我们保证提出申诉的系争产品就是先前裁决书中所涉及的产品；并且

the committee can refer to the arbitration or appeal decision before giving its Award, but will not be bound by them.

委员会在作出裁决之前可以参考先前的裁决书或申诉裁定，但并不受其约束。

Appeals on arbitrations conducted elsewhere
在别处对仲裁提起申诉

Bylaw 354

规章 354

1   If a manual quality arbitration was conducted under the rules of another Association, an appeal can still go to the Quality Appeal Panel. However, this must be agreed in writing by the firms in dispute.

1   如已经依据其他协会的规则进行了手工检验品质仲裁，仍可向质量申诉工作小组提起申诉。不过，该申诉必须得到争议各方的书面许可。

2   The appeal Award will be based on the value differences used for the arbitration Award, but the cotton will be judged against the appropriate 'Universal Standards' or 'ICA Standards'. If no other value differences are available, our differences will apply.

2   申诉裁决将在仲裁裁决所使用的价差基础上作出，但棉花须按相应的"全球标准"或"ICA标准"裁定。如无其他可用价差，则使用我们的价差。

3   Appeals must be lodged within the time limits laid down in the rules of the association under which the arbitration was held.

3   申诉必须在仲裁据以进行的协会规则所规定的时限内提出。

4   The samples for the appeal must be the same samples that were used in the arbitration. They must be sealed as the authentic samples and they must be signed as being so. The samples must then be sent to us. They must come with a statement saying whether the arbitration was held under natural or artificial light.

4   申诉所用的样品必须与仲裁中所用样品相同。样品必须作为真实样品封存,并须签署确认如此况。随后须将样品送达我方。样品送来时必须附一份声明,说明仲裁在自然光还是人工照明下进行。

5   If an instrument test arbitration was conducted under the rules of another association, an appeal can still go to the Quality Appeal Panel. However, this must be agreed in writing by the firms in dispute. Bylaw 352 will then apply.

5   如已经依据其他协会的规则进行了仪器检测仲裁,仍可向质量申诉工作小组提起申诉,不过该申诉必须得到争议各方的书面许可。如此将适用规章352。

Amicable settlements
和解

Bylaw 355
规章 355

1   If firms in dispute achieve a settlement prior to commencement of arbitration, but require a record in the form of an Award, they may agree jointly on appointing a sole arbitrator to make an award recording the agreed settlement.

1   如争议各方在仲裁起始前达成和解,但需要以裁决书形式作为记录,则其可以共同任命一名独任仲裁员根据议定的和解内容制作裁决书。

2   If firms settle their dispute after arbitration has commenced, they must inform us immediately. The sole arbitrator, tribunal or appeal committee will then not make any Award unless they are asked to record the settlement in the form of an Award, and they agree to do so.

2   如各方在仲裁起始后达成和解,则须立即通知我们。独任仲裁员、仲裁庭或申诉委员会将不再制作裁决书,除非当事方要求以裁决书形式记录和解内容并且独任仲裁员、仲裁庭或申诉委员会同意如此执行。

3   If the sole arbitrator, tribunal or appeal committee makes an Award, it will have the same status and effect as any other award.

3   如独任仲裁员、仲裁庭或申诉委员会依此作出了裁决书,则其效力与任何其他裁决书相同。

4   Any outstanding fees and expenses of the sole arbitrator, tribunal or appeal

committee, and any stamping charge set by us must be paid.

4 应付给独任仲裁员、仲裁庭或申诉委员会的所有收费和费用，以及所有由我们确定的盖章费都必须支付。

5 Where money has been deposited with us under Bylaw 358 (4) or Bylaw 312 (2) by way of deposit against any fees, costs or expenses in connection with or arising out of the arbitration or the appeal (as the case may be), the tribunal or appeal committee shall determine what, if any, proportion shall be refunded. Such determination shall take account of the amount of work undertaken, and/or legal fees incurred by the tribunal or appeal committee at the date they receive notice of the settlement.

5 如有依规章 358(4)或规章 312(2)之规定预留仲裁及申诉收费、成本或支出（视情况而定）押金的，仲裁庭或申诉委员会须决定退还金额的比例（如有）。该等裁定须考虑截至收到和解通知之日已进行的工作量，和/或仲裁庭或申诉委员会发生的法律费用。

Fees and Charges
收费

Application fees for arbitrations
仲裁申请费

Bylaw 356
规章 356

1 The application fees set by the Directors for arbitrations are laid down in Appendix C of the Rule Book.

1 仲裁申请费由董事会确定，并列明于规则手册附录 C。

2 A dispute may cover more than one contract, but a firm will have to pay us a separate application fee for each arbitration.

2 一项争议可能涉及数份合同，但当事方必须就每项仲裁向我们支付单独的申请费。

Application fees for appeals
申诉申请费

Bylaw 357
规章 357

1 The application fees set by the Directors for appeals are laid down in

Appendix C of the Rule Book.

1　申诉申请费由董事会确定，并列明于规则手册附录 C。

2　If they think it is appropriate, the Directors can reduce the amount of the application fee, or refund all or part of it.

2　如董事会认为适当，可以降低申请费金额或返还全部或部分申请费。

Other Fees and Charges - Technical

其他收费 – 技术

Bylaw 358

规章 358

1　Arbitrators, including technical appeal committee members, shall be entitled to charge fees which shall be fixed by reference to the total amount of time reasonably devoted by each arbitrator/technical appeal committee member to the arbitration/appeal and shall be in accordance with the following scale or such scale as shall be determined by us from time to time:

1　仲裁员，包括技术申诉委员会成员有权收取费用，收费根据下列标准或我们不时规定的标准及每位仲裁员/技术申诉委员会成员合理付出的总时间计算：

An hourly rate shall be charged up to a maximum of £150 per hour.

最高小时费率为 150 英镑每小时。

Fractions of an hour after the first hour shall be charged pro rata.

第一小时以后发生的零碎时间按比例收取。

A minimum fee of £100 shall be payable to each arbitrator.

每位仲裁员的最低收费为 100 英镑。

An additional fee of £250 per arbitration will be payable to the Chairman.

每次仲裁需额外支付 250 英镑予主席。

2　The Chairman of the tribunal and the Chairman of a technical appeal committee shall be entitled to increase the above fee scale, and charge fees at a reasonable rate within their discretion in arbitrations/appeals of extraordinary complexity and/or value.

2　仲裁庭首席仲裁员或技术申诉委员会主席有权提高上述收费标准，并可在其对特别复杂问题和/或计价问题的仲裁/申诉裁量权范围内确定合理的费率进行收费。

3　Where the tribunal or technical appeal committee find it necessary to obtain legal advice on any matter arising from an arbitration or appeal, reasonable legal fees thereby incurred will be payable as directed in the Award.

3　如仲裁庭或技术申诉委员会认为有必要就仲裁中发生的任何事项寻求法律意见的，由此发生的合理法律费用将按照裁决书指示由双方承担。

4　At any time after the receipt by us of 'the Request' and from time to time thereafter, the Chairman of the tribunal may demand that sums of money be deposited with us by any party to the dispute, by way of deposit against any fees, costs or expenses in connection with or arising out of the arbitration. Failure by any party to pay any such sums shall entitle the tribunal to suspend or discontinue the arbitration proceedings until such sums are paid.

4　我们收到"请求"之后的任何时间以及此后不定时间，仲裁庭首席仲裁员都可以要求争议的任何一方当事人向我们存入一笔款项，以作为对任何与仲裁有关或因仲裁发生的收费、成本或支出的押金。对于未支付该等款项的当事方，仲裁庭有权暂停或中止仲裁程序，直至其支付完该等款项。

5　When an Award is presented for stamping in accordance with Bylaw 309 each arbitrator or technical appeal committee member shall invoice us for all fees, clearly stating their applicable hourly rate. Arbitrators are required to submit a time sheet in a format approved by the Directors.

5　裁决书依规章 309 提交盖章时，仲裁员或技术申诉委员会成员须向我们出具发票：包含所有费用，列明适用的小时费率。仲裁员需要提交一份格式经董事批准的时间记录表。

6　The only expenses an arbitrator or technical appeal committee member shall be entitled to claim are courier fees, up to a maximum of £50.

6　独任仲裁员或者技术申诉委员会成员应有权要求的唯一费用是快递费用，最高可达 50 英镑。

7　The time sheet shall be forwarded to both parties by the Secretariat within 14 days (two weeks) of the award being released.

7　如果适用，时间记录表须在裁决书发出后 14 天（两周）内由秘书转发给双方当事人。

8　The payment of fees and expenses to arbitrators and technical appeal committee members is conditional upon the Association's receipt of the time sheet.

8　协会收到时间记录表后，方可向仲裁员和技术申诉委员会成员支付费用和开支。

9　Subject to the foregoing, arbitrators and Appeal Committee members shall be entitled to prompt payment of fees and expenses following release of the Award. If, following a review under Bylaw 359 the Directors determine that any fees or expenses are unreasonable, the arbitrators and technical appeal committee members shall act in accordance with the decision of the Directors.

9 在上述前提下,裁决书发出后仲裁员和申诉委员会成员有权要求支付费用和开支。在董事会依据规章 359 进行审核后如果发现费用或支出有不合理情形,则仲裁员和技术申诉委员会成员须按董事会的决定执行。

Bylaw 359
规章 359

1 If, once an award is released, a firm considers that the fees and expenses charged are unreasonable, it can ask the Directors to review the amounts. The Directors will decide how much is to be paid.

1 裁决书发出后,如果当事人认为收费和支出不合理,则可要求董事会重新审核金额。董事会将决定支付金额为多少。

2 We must receive notice of a request under this Bylaw within 21 days (three weeks) of the award being released.

2 本规章项下的请求通知必须于裁决发布后 21 日(三周)内提交给我们。

Bylaw 360
规章 360

1 The general principle is that costs follow the event, but subject to the overriding discretion of the tribunal and appeal committee as to which party will bear what proportion of the costs of the arbitration.

1 一般原则是根据发生情况支付费用,但仲裁庭和申诉委员会有充分权利裁定双方按何种比例承担仲裁费用,该决定具凌驾性。

2 In the exercise of that discretion the tribunal shall have regard to all the material circumstances, including such of the following as may be relevant:

2 仲裁庭在行使裁量权的过程中须考虑所有重要情况,包括以下相关方面:

Which of the issues raised in the arbitration has led to the incurring of substantial costs and which party succeeded in respect of such issues.
仲裁过程中的哪些争点招致庞大费用,哪一方在这些争点上胜诉。

Whether any claim which partially succeeded was unreasonably exaggerated.
部分胜诉的诉请是否存在不合理夸大的情形。

The conduct of the party which succeeded on any claim and any concession made by the other party.

在任何诉请上胜诉一方的行为以及对方所作的让步。

The degree of success of each party.

各方胜诉的程度。

Other Fees and Charges - Quality
其他收费-质量

Bylaw 361
规章 361

1　Quality arbitrations

1　质量仲裁

The lowest fees for quality arbitrations are laid down in Appendix C of the Rule Book however the arbitrators may charge more.

质量仲裁的最低收费标准见规则手册附录C，但仲裁员可以收取更多费用。

Both firms are liable to pay a fee. The arbitrators will apportion the fees payable by each firm.

双方均须支付费用。仲裁员将按比例分配各方应付的费用金额。

2　Quality appeals

2　质量申诉

The lowest fees for quality appeals are laid down in Appendix C of the Rule Book, however the appeal committee may charge more.

质量申诉的最低收费标准见规则手册附录C，但申诉委员会可以收取更多费用。

Each firm appealing will be liable to pay a fee. The appeal committee will apportion the fees payable by each firm.

提起申诉的各方须支付费用。申诉委员会将按比例分配各方应付的费用金额。

3　Cotton waste, linters and pickings

3　废棉、棉短绒和下脚

The quality arbitration and appeal fees on cotton waste, linters and pickings are the same as the fees for quality arbitration and appeals on cotton.

对废棉、棉短绒和下脚的质量仲裁及申诉的收费与对棉花的质量仲裁及申诉的收费相同。

4　Classifications

4　评级

The fee for classification under Bylaw 347 is laid down in Appendix C of the Rule Book. Only the firm asking for the classification will have to pay the fee.

规章 347 项下的评级费用见规则手册附录 C。评级费用仅由请求评级的一方当事人承担。

Bylaw 362

规章 362

1　If an umpire is appointed in a quality arbitration, he will receive an amount equal to 50% of the lowest fee to be paid for quality arbitration by a Principal Firm.

1　如在质量仲裁中任命了裁定人，则裁定人可以获得相当于主公司在质量仲裁中所付最低收费的 50%的费用。

2　The arbitrator whose Award/findings vary the most from that of the umpire will be liable to pay the umpire fees from his fee. If there is equal disagreement, each arbitrator will pay half. In a quality appeal, the appeal committee will decide which arbitrator has to pay the umpire.

2　仲裁员的裁决与裁定人的裁决存在最重大差异的，仲裁员将从自己的收费中向裁定人支付费用。如差异相当，则两名仲裁员各承担一半费用。在质量申诉中，由申诉委员会决定哪一名仲裁员支付裁定人费用。

Bylaw 363

规章 363

1　If, once an Award is released, a firm considers that the fees and expenses charged by the arbitrator or arbitrators, umpire or appeal committee are unreasonable then it can ask the Directors to review the amounts. The Directors will decide how much is to be paid.

1　裁决书发出后，如果一方公司认为独任仲裁员及仲裁员、裁定人或申诉委员会收取的费用和支出不合理，则可要求董事会重新审核金额。董事会将决定支付金额为多少。

2　We must receive notice of a request under this Bylaw within 14 days (two weeks) of notice of fees and expenses being given or the Award being released, whichever is the earlier.

2　本规章项下之请求通知，须于费用通知发出或裁决书发布后 14 日（两周）内（以先到者为准）发送给我们。

Stamping charges
盖章费
Bylaw 364
规章 364

1 The stamping charges are laid down in Appendix C of the Rule Book. The rate to be paid will be in line with the firm's registration status on the date of the contract giving rise to the dispute. If a firm has been suspended or expelled from registration, or has been refused re-registration since arbitration was commenced, it must pay the non-registered rate.

1 盖章费标准见规则手册附录 C。盖章费费率将按当事公司在引起争议的合同签署之日的注册身份计算。如当事方已经遭到暂停或开除会员资格处理，或者在仲裁开始以后被拒绝再注册，则其必须按非注册公司费率付费。

2 Quality arbitrations and appeals
2 质量仲裁和申诉

In a quality arbitration both firms will be liable to pay a stamping charge but the arbitrators will apportion the charge payable by each firm.

在质量仲裁中，双方公司都有义务支付盖章费，但仲裁员将确定各方的付费比例。

In a quality appeal under Bylaw 354 each firm appealing will be liable to pay any stamping charge but the appeal committee will apportion the charge payable by each firm.

在规章 354 项下的质量申诉中，各申诉公司有义务支付盖章费，但申诉委员会将确定各方的付费比例。

Liability for payment of fees
支付费用的义务
Bylaw 365
规章 365

If a Principal Firm appoints an arbitrator or umpire for one of its subsidiary firms that is not a registered firm, and the non-registered firm fails to pay, the Principal Firm will be liable for any arbitration, umpire and stamp fees due.

如有主公司为其非注册附属公司任命了一名仲裁员或裁定人，而该非注册公司未能支付费用，则主公司有义务支付到期的仲裁、裁定和盖章费用。

Unfulfilled awards and defaulting parties
未执行的裁决书和违约方

Reporting
报告
Bylaw 366
规章 366

1 If the Association receives written advice from a party to an Award, ("the Reporting party") or from their representative that an Award has not been complied with by the other party to the Award ("the alleged defaulter"), the Directors are to be informed.

1 如协会收到裁决书一方当事人（下称"报告方"）或其代表的书面意见，称对方（"违规嫌疑人"）未遵守裁决书，则应通知董事会。

2 Before acting on such advice, the Secretary shall write to the alleged defaulter notifying them of the Directors' intention to list their name unless, within a period of 14 days (two weeks), the alleged defaulter provides them with compelling reasons not to do so. The Directors shall consider any reasons submitted by the alleged defaulter before deciding whether or not the information received from the Reporting Party should be circulated.

2 在对该等意见采取行动之前，秘书须书面通知违规嫌疑人董事会有意将其列入名单，除非违规嫌疑人在 14 天（两周）内提供有力的理由说明为何不应如此执行。董事会在决定是否传阅从报告方收到的信息前，须考虑违规嫌疑人提出的任何理由。

3 The Directors may pass on the name of the defaulting party to Individual Members, Member Firms, Member Associations of the Committee for International Co-operation between Cotton Associations (CICCA) or any other organisation or person by any method it chooses, including the listing of the name of the defaulter and appropriate details in the publicly accessible area of the Association's website.

3 董事会可以采用任何方式将违规方的名称通告给个人会员、成员公司、棉花协会国际合作委员会成员协会或任何其他组织或个人，包括将违规方名称和适当的细节列于协会网站公共开放区域。

4 If the Directors so decide, this information and any other appropriate information will be circulated on a list of unfulfilled Awards, to be known as the 'ICA List of Unfulfilled Awards'.

4 如董事会作出决定,则该信息和任何其他适当信息将列入一项裁决未执行名单,称为"ICA 未执行裁决书名单"。

Advisory Notices
忠告性通知

5　The Directors may also at any time circulate to Individual Members, Member Firms, and Member Associations of the Committee for International Co-operation between Cotton Associations (CICCA), an Advisory Notice advising them of any entity which appears to be related to, or utilised by a defaulter. Such Advisory Notice shall also be displayed in that area of the Association's website restricted to Individual Members and Member Firms.

5　董事会也可以在任何时候向个人会员、成员公司以及棉花协会国际合作委员会(CICCA)成员协会发出建议通知,告知其某一实体可能与违规者有关或被其利用。该等建议通知同时在协会网站上仅限个人会员和成员公司浏览的区域内展示。

6

a　Where the party requesting the issue of an Advisory Notice is not the Reporting party who has provided the advice referred to in paragraph (1) above ("the Advising party") the Secretary will write to the Reporting party notifying them of the request and seeking comments within seven days (one week).

a　如请求发布建议通知的一方(下称"建议方")并非上文第(1)节所述提出意见的报告方,则秘书将在七天(一周)内将相关情况以书面方式通知给报告方并寻求其意见。

b　After receipt of comments, if any, from the Reporting party, the Secretary may write to the defaulter and other parties that it proposes to name in the Advisory Notice, informing them of the proposed contents of the Advisory Notice and asking them to provide evidence to rebut the contents of the same within 14 days (two weeks).

b　收到报告方意见后(如有),秘书可以致信违规方和建议通知中提及的该方,告知其拟议中建议通知的内容,并要求其在 14 天(两周)内提供为自己辩护的证据。

c　The Directors will consider any comments or evidence received under paragraph (6a) and paragraph (6b) above and will decide whether or not an Advisory Notice ought to be issued.

c　董事会将考虑依上文第(6a)和(6b)节规定收到的意见及证据,并将决定是否需要发布建议通知。

7   The Reporting party has responsibility for the accuracy of the information supplied directly to the ICA under this Bylaw and shall indemnify and hold harmless the Association and its Directors from and against all liabilities, damages, costs and expenses incurred by them or either of them by reason of any inaccuracy in such information. The reporting party shall inform the Association immediately should the Award be settled to enable the party to be removed from the List of Unfulfilled Awards.

7   报告方应对其依本规章直接向ICA提供的信息的准确性负责，如因该等信息不准确导致协会及其董事发生任何责任、损害、成本和支出的，报告方须承担赔偿责任并确保不发生损害。如就裁决书达成和解，则报告方须立即通知协会，将相关方从裁决未执行名单中撤除。

8   The Advising party has responsibility for the accuracy of the information supplied directly to the ICA under this Bylaw with regard to paragraph (5) and paragraph (6a) above and shall indemnify and hold harmless the Association and its Directors from and against all liabilities, damages, costs and expenses incurred by them or either of them by reason of any inaccuracy in such information.

8   建议方应对其依本规章及上文第(5)和(6a)节直接向ICA提供的信息的准确性负责，如因该等信息不准确导致协会及其董事发生任何责任、损害、成本和支出的，建议方须承担赔偿责任并确保不发生损害。

9   The parties to any arbitration shall be deemed to have consented to the Directors taking the action set out in this Bylaw.

9   参与任何仲裁的当事方须视为同意董事会采取本规章所述的行动。

Appendix C
附录 C

A summary of our fees and charges
酬金及费用总结

These fees and charges are effective from the day this Rule Book comes into force until we say otherwise.

以下收费项目自本规则手册生效之日起适用，直至我们另行通知。

Fees and charges for Technical Arbitrations and Appeals
技术仲裁及申诉费用

Please note that the amount to be paid in each case will be in line with the

firm's registration status on the date of the contract giving rise to the dispute
请注意,须支付的金额将依据争议涉及合约的签署当日公司的注册状况。

TECHNICAL ARBITRATIONS
技术仲裁

Application fees	
Principal Firms and Related Companies	No fee
Principal Firms and Related Companies but not registered on the date of the contract	£500.00
Association Member Firms	£2,500.00
Non-registered firms that apply for membership at the time of applying for arbitration	Annual registration fee + £500.00
Non-registered firms (including those firms whose application forregistration has been refused)	£10,000.00
Other arbitration fees	
An hourly rate shall be charged by the arbitrators, up to a maximum of £150.00.	
Fractions of an hour after the first hour shall be charged pro rata.	
A minimum fee of £100.00 shall be payable to each arbitrator.	
An additional fee of £250.00 per arbitration will be payable to the Chairman.	
The only expenses an arbitrator shall be entitled to claim are courier fees, up to a maximum of £50.00.	

申请费	
主公司及其关联公司	无须付费
在合约签署之日还未注册的主公司及其关联公司	£500.00
协会成员公司	£2,500.00
非注册公司,但在申请仲裁时同时申请注册	年度注册费 + £500.00
非注册公司(包括注册申请被拒绝的公司)	£10,000.00
其他仲裁费用	
仲裁员按每小时收费,最高£150.00。	
第1小时之后,小时的一部分则按以上收费率计算。	
每名仲裁员最低收费为£100.00。	
每次仲裁需额外支付£250.00予主席。	
仲裁员有权申领的唯一费用为快递费,最多不超过£50.00。	

TECHNICAL APPEALS
技术申诉

Application fees	
Principal Firms and Related Companies	No fee
Non-registered firms	£2,000.00
Association Member Firms	£500.00

申请费	
主公司及其关联公司	无须付费
非注册公司	£2,000.00
协会成员公司	£500.00

Other appeal fees

The chairman of the appeal committee shall decide the hourly rate to be charged by the appeal committee members, up to a maximum of £150.00.

Fractions of an hour after the first hour shall be charged pro rata.

A minimum fee of £100.00 shall be payable.

An additional fee of £250.00 per arbitration will be payable to the Chairman.

The Association will charge as its fees 25% of the technical appeal committee's total fees.

其他申诉费用

付给申诉委员会会员的每小时费率由申诉委员会主席决定,最高£150.00。

第1小时之后,小时的一部分则按以上收费率计算。

最低收费为£100.00。

每仲裁250.00英镑额外的费用将支付予主席。

协会将收取技术申诉委员会总费用的25%。

STAMPING AND NOTARISATION OF TECHNICAL AWARDS
技术仲裁裁决书盖章及公证

Stamping charges	
Principal Firms and Related Companies	£400.00
Association Member Firms	£600.00
Non-registered firms	£800.00
Technical appeal awards	No charge
Notarisation and legalisation of Awards	
All firms	£300.00

盖章费	
主公司及其关联公司	£400.00
协会成员公司	£600.00
非注册公司	£800.00
技术申诉裁决书	无须付费
裁决书公证	
所有公司	£300.00

Fees and charges for Small Claims Technical Arbitrations and Appeals
小额索赔技术仲裁及申诉费用

SMALL CLAIMS TECHNICAL ARBITRATIONS
小额索赔技术仲裁

Application fees	
Principal Firms and Related Companies	No fee
Principal Firms and Related Companies but not registered on the date of the contract	£250.00
Association Member Firms	£1,250.00
Non-registered firms cannot apply for Small Claims arbitration unless they apply for membership at the time of applying for arbitration. Non-registered firms that apply for membership at the time of applying for arbitration. If your application for registration with us is refused your application for Small Claims arbitration will also be refused.	Annual registration fee + £250.00
Other arbitration fees	
An hourly rate shall be charged by the Sole Arbitrator, up to a maximum of £150.00.	
Fractions of an hour after the first hour shall be charged pro rata.	
A minimum fee of £100.00 shall be payable.	
The only expenses an arbitrator will be entitled to claim are courier fees, up to a maximum of £50.00.	

申请费	
主公司及其关联公司	无须付费
在合约签署之日还未注册的主公司及其关联公司	£250.00
协会成员公司	£1,250.00
非注册公司不得申请小额赔技术仲裁，除非其在申请仲裁时同时申请注册。非注册公司，但在申请仲裁时同时申请注册。如果您的注册申请被拒绝，您的小额索赔仲裁也将被拒绝。	年度注册费 + £250.00
其他仲裁费用	
独任仲裁员按每小时收费，最高£150.00。	
第1小时之后，小时的一部分则按以上收费率计算。	
最低收费为£100.00。	
仲裁员有权要求的唯一费用是快递费用，最多50.00英镑。	

SMALL CLAIMS TECHNICAL APPEALS
小额索赔技术申诉

Application fees	
Principal Firms and Related Companies	No fee
Non-registered firms	£1,000.00
Association Member Firms	£250.00
Other appeal fees	
The chairman of the appeal committee shall decide the hourly rate to be charged by the appeal committee members, up to a maximum of £150.00.	
Fractions of an hour after the first hour shall be charged pro rata.	
A minimum fee of £100.00 shall be payable.	
The Association will charge as its fees 25% of the Small Claims appeal committee's total fees.	

申请费	
主公司及其关联公司	无须付费
非注册公司	£1,000.00
协会成员公司	£250.00
其他申诉费用	
付给申诉委员会会员的每小时费率由申诉委员会主席决定，最高£150.00。	
第1小时之后，小时的一部分则按以上收费率计算。	
最低收费为£100.00。	
协会将收取小额索赔申诉委员会总费用的25%。	

STAMPING AND NOTARISATION OF SMALL CLAIMS TECHNICAL AWARDS
小额索赔技术仲裁裁决书盖章及公证

Stamping charges	
Principal Firms and Related Companies	£400.00
Association Member Firms	£600.00
Non-registered firms	£800.00
Small Claims appeal awards	No charge
Notarisation and legalisation of Awards	
All firms	£300.00

盖章费	
主公司及其关联公司	£400.00
协会成员公司	£600.00
非注册公司	£800.00
小额索赔申诉裁决书	无须付费
裁决书公证	
所有公司	£300.00

Fees and charges for Quality Arbitrations and Appeals
质量仲裁及申诉费用
QUALITY ARBITRATION
质量仲裁

Application fees	
Registered Firms	No fee
Non-registered Firms	No fee
Quality arbitration, appeal and classification	
The lowest amount the arbitrators or appeal committee will charge for very bale represented by the samples provided is given below. They may charge more. If the samples provided represent less than 50 bales, they will charge for 50 bales.	

续表

Quality Arbitration	
Registered Firms	£0.35
Non-registered Firms	£1.00
Quality Appeal	
Registered Firms	£0.65
Non-registered Firms	£1.95
Classification	
For grade, colour and staple	£1.00
For grade and colour only	£0.65
For staple only	£0.65

申请费	
注册公司	无须付费
非注册公司	无须付费
质量仲裁、申诉及评级	
仲裁员或申诉委员会对于所提供样品中的每个棉包收取的最低费用如下。可能收取更多费用。如果提供样品代表少于 50 个棉包,按 50 个棉包收费。	
质量仲裁	
注册公司	£0.35
非注册公司	£1.00
质量申诉	
注册公司	£0.65
非注册公司	£1.95
评级	
等级、颜色及纤维长度	£1.00
仅对等级及颜色	£0.65
仅对纤维长度	£0.65

STAMPING AND NOTARISATION OF QUALITY AWARDS AND APPEAL AWARDS
质量仲裁裁决书和申诉裁决书盖章及公证

Stamping charges	
The amount we will charge both firms for every bale represented by the samples provided is given below. If the samples provided represent less than 50 bales, we will charge for 50 bales.	
Principal Firms and Related Companies	£0.03
Association Member Firms	£0.12
Non-registered firms	£0.24
Notarisation and legalisation of Awards	
All firms	£300.00

盖章费	
对于所提供样品中的每个棉包向双方公司收取的费用如下。如果提供样品代表少于 50 个棉包,按 50 个棉包收费。	
主公司及其关联公司	£0.03
协会成员公司	£0.12
非注册公司	£0.24
裁决书公证	
所有公司	£300.00

Section 4   Administration Bylaws
第 4 部分   行政管理规章
Membership and registration
会员资格与注册

Bylaw 400
规章 400

Applications for membership must be made on forms approved by the Directors. The forms are available from the Secretary.

会员资格的申请必须以董事批准的表格形式提出。表格可向秘书索取。

Bylaw 401
规章 401

Individual Members and Registered Firms must write to the Secretary at once if any of the information presented to the Association in their application changes. If the Secretary asks an Individual Member or Registered Firm to confirm that the information they gave in their application is still correct, they must reply immediately.

如果个人会员和注册公司通过申请表提交给协会的信息发生变更,则应立即书面告知秘书。如果秘书向个人会员或注册公司确认其在申请表中提供的信息是否仍然正确,个人会员或注册公司应立即回答。

Bylaw 402
规章 402

If the Directors suspend a Registered Firm, we will treat it as a non-registered firm during the time it is suspended.

如果董事会暂停某一注册公司的资格,则被暂停资格者在暂停期间按非注册待遇处理。

Bylaw 403
规章 403

The conditions for registration are laid down in the Articles of Association.

注册的条件规定于协会章程中。

Bylaw 404
规章 404

1   Each year Member Firms will pay the registration fee set by the Directors.

1　成员公司每年均须支付董事会确定的注册费。

2　All Member Firms are entitled to receive a current copy of our Bylaws and Rules and all later amendments.

2　所有成员公司均有权获得协会当前的规章和规则及所有嗣后修正文件的副本。

3　The Directors may cancel the registration of a Member Firm but will refund the registration fee paid, proportionate to the unexpired period in the year in which cancellation is effected.

3　董事会可以注销某一成员公司的注册,但将按资格撤销当年未到期期间的比例返还已支付的注册费。

Bylaw 405
规章 405

1　A Principal Firm is either a Merchant or a Producer or Mill.

1　主公司应为经营商、生产商或工厂。

Applications for registration must be proposed and seconded by Individual Members of the Association.

注册申请必须由协会个人会员提出并获得附议。

Each firm will have at least one Individual Member.

每家公司会有至少一名个人会员。

Principal Firms may apply to register any of their related companies as a Dependent Related Company. There is no limit on the number of Related Companies a Principal Firm may register, but no more than five will pay the fee set by the Directors. The relationship between Principal Firms and Related Companies will be kept confidential.

主公司可以申请将自己的关联企业注册为独立关联公司或非独立关联公司。主公司可注册的关联公司数量不限,但经董事会确定支付会费的不超过五家。主公司与关联公司的关系将保密。

2　An Affiliate Industry Firm is a firm or organisation that provides a service to the cotton trade.

2　关联行业公司是指向棉花业提供服务的公司或组织。

Applications for registration must be proposed and seconded by Individual Members of the Association.

注册申请必须由协会个人会员提出并获得附议。

Each firm will have at least one Individual Member.
每家公司会有至少一名个人会员。

Affiliate Industry Firms may apply to register any of their related companies as a Dependent Related Company. There is no limit on the number of Related Companies an Affiliate Industry Firm may register, but no more than five will pay the fee set by the Directors. The relationship between Affiliate Industry Firms and Related Companies will be kept confidential.

关联行业公司可以申请将其任何关联公司注册为非独立关联公司。关联行业公司可注册的关联公司数量不限，但经董事会确定支付会费的不超过五家。关联行业公司与关联公司的关系将得到保密。

3 An Agent Firm is any firm that provides an agency service so as to bring a Principal Firm into contractual relationships with other parties.

3 代理公司是指提供代理服务、在主公司与他方之间建立合同关系的公司。

Applications for registration must be proposed and seconded by Individual Members of the Association.

注册申请必须由协会个人会员提出并获得附议。

Agent Firms will not be entitled to have an Individual Member.

代理公司无权享有个人会员资格。

4 An Affiliated Association is any recognised association related to the cotton industry that declares its support of the principles of the ICA and its Bylaws and Rules.

4 附属协会是指任何与棉花业有关的、得到行业认可，并且宣称支持 ICA 的主要原则及其规章和规则的协会。

Applications for registration must be made in writing to the Directors.

注册申请须以书面形式向董事会提出。

5 An Association Member Firm is any producer or mill that is also a member of an Affiliated Association.

5 协会成员公司是指同时参加某一附属协会的生产商或工厂。

Applications for registration must be proposed and seconded by Individual Members of the Association.

注册申请必须由协会个人会员提出并获得附议。

Association Member Firms will not be entitled to have an Individual Member.

协会成员公司无权拥有个人会员资格。

Bylaw 406
规章 406

1 An Individual Member, Principal Firm, Related Company or Association Member Firm cannot resign if:

1 如有下列情形，则个人会员、主公司、关联公司或协会成员公司不得退会：

he or it is involved in arbitration arising out of a contract governed by International Cotton Association Bylaws or Rules or ICA arbitration; or

其正涉及一项由受国际棉花协会规章或规则或 ICA 仲裁规则管辖的合同引起的仲裁；或

there is an unfulfilled quality or technical arbitration or appeal award against them, made under our Bylaws.

其有一项根据本会规章对其提出的质量未达标或尚未履行完毕技术仲裁或上诉裁决。

2 Paragraph (1) does not take away the Directors' right to suspend or expel an Individual Member or Member Firm found guilty of an offence at any time under the Articles.

2 (1)节并未剥夺董事会依据章程在任何时候于个人会员或成员公司依据章程被认定为违规的情况下暂停或剥夺其会员资格的权利。

3 The Directors may cancel the registration of an Individual Member and may refund the registration fee paid, proportionate to the unexpired period in the year in which cancellation is effected.

3 董事会可以注销注册的个人会员，但可按资格撤销当年未到期期间的比例返还已支付的注册费。

4 If any Individual Member or Registered Firm resigns, but the Directors do not accept the resignation, the Individual Member or Registered Firm will lose all rights and privileges that they get from membership or registration. They will not be able to withdraw from or avoid arbitration arising from contracts they have entered into.

4 如有任何个人会员或注册公司退会，但董事会不予接受，则该个人会员或注册公司将丧失其因会员资格或注册所获得的一切权利和特权。他们也不能撤回或逃避因已经签署的合同而发生的仲裁。

5 The loss of rights and privileges will not prevent another firm seeking

arbitration on claims arising out of existing contracts.

5　丧失权利和特权并不妨碍其他公司依据已有的合同提起索赔的仲裁。

Elections
选举

General
一般规定
Bylaw 407
规章 407

Each year there will be an election for President, First Vice-President, Second Vice-President, and Ordinary Directors. The procedure is as follows:

会长、第一副会长、第二副会长、普通董事和委员会委员每年选举一次。选举程序如下：

1　A notice of the election will be sent to each Individual Member who is entitled to vote at least 35 days (five weeks) before the Annual General Meeting. Nominations must be sent to the President within 14 days (two weeks) of thenotice going out.

1　在年度大会召开前至少 35 天（五周）向每一名个人会员发出选举通知。提名必须在选举通知发出后 14 日（两周）内送至会长处。

2　Individual Members who are entitled to vote can put names forward to be elected as President, First Vice-President, Second Vice-President or as an Ordinary Director. The names must be put forward in writing by a proposer and seconder. Before any candidates are put forward, they must give their permission and be willing to serve.

2　享有投票权的个人会员可以提名会长、第一副会长、第二副会长以及普通董事。提名必须由提名人和附议人以书面形式提出。任何候选人被提名之前，必须同意提名并愿意承担服务工作。

3　If there are as many candidates as vacancies, those candidates will be taken as being elected.

3　如候选人人数与空缺职位相等，则候选人作为选上处理。

4　Voting lists will be sent out at least 21 days (three weeks) before the Annual General Meeting. They will give the candidates', proposer's and seconder's names. They will go to each Individual Member entitled to vote. Voting is done by putting the voter's initials against chosen names. The lists must be sent to the President. This must be done within 14 days (two weeks) of the lists going out.

4 投票名单应在年度大会召开前至少21天（三周）前发出。列明候选人、提名人和附议人的姓名。投票名单会发给每一位有投票权的个人会员。投票人应将自己的姓名缩写标注于所选择的人名之上，以此完成投票。投票名单须发给会长。该程序必须在投票名单发出后14天（两周）内完成。

5 Individual Members must vote for at least two thirds of the vacancies.

5 个人会员必须对至少三分之二的空缺职位投票。

6 Any vote not made according to these instructions will not count

6 不遵守以上说明的选票将视为废票。

7 The President and Secretary will determine the result of the voting. The President's decision will be final.

7 会长和秘书将决定投票的结果。会长的决定是最终决定。

8 If two or more candidates get the same numbers of votes, the President will make a deciding vote.

8 如果两名或两名以上候选人获得的票数相当，则由会长作出决定票。

9 The President has the final say on:

9 会长对以下事项享有最终决定权：

the validity of nominations;

提名的有效性；

the number of votes; and

票数；以及

all questions or disputes relating to the election.

一切有关选举的问题或争议。

10 If more candidates apply than there are vacancies, those with the highest number of votes will be elected.

10 如候选人人数多余空缺职位，则获得选票数最多者当选。

11 If there are not enough candidates, the Directors can appoint qualified Individual Members to fill the vacancies. Those appointed by the Directors will hold office for the same time and as if they had been elected.

11 如果候选人不足，则董事会可以指定合格的个人会员填补空缺职位。经董事会指定任职者的任命与选举产生者一致。

12 The Secretary will post the results on the Association's website.

12 秘书应在协会网站公布选举结果。

13 Newly elected Officers and Ordinary Directors will take office from the time the results are announced at the Annual General Meeting. Until then, the retiring Officers and Ordinary Directors will stay in office.

13 新选出的职员、普通董事和委员会委员自选举结果于年度大会宣布之时开始任职。在此之前，即将卸任的职员、普通董事和委员将保持其职位。

14 All Officers and Directors in office when these Bylaws are adopted will be recognised as elected and constituted under these Bylaws. They will stay in office until they retire under the election Bylaws.

14 本规章生效时在任的所有职员、董事和委员会委员视为按规章选举任职。他们将继续任职，直至根据选举规章规定卸任。

Casual vacancies on the Board of Directors
董事会和委员会的临时职务空缺
Bylaw 408
规章 408

If, between Annual General Meetings, we are short of a Director we will hold an election as described in Bylaw 407. The Directors will say when the notice of election is to be given and when the voting list is to go out and be returned.

如在两次年度大会之间，有董事职位空缺，则应按规章 407 之规定进行选举。董事会确定选举通知的发出时间以及选票的发出及收回时间。

Bylaw 409
规章 409

The replacement Individual Member elected to fill a vacancy on the Board of Directors will only stay in office for as long as the original would have done.

改选个人会员填补董事会空缺者，任职期间为原任者剩余任期。

Committees
委员会

General
一般规定
Bylaw 410
规章 410

Individual Members who are entitled to do so can put their own names forward to serve on Members' Committees. They do not need to be proposed or seconded.

Committees and their Chairmen will be appointed annually by the Directors.

有权任职的个人会员可以提名自己进入会员委员会服务。他们不需要经提名或附议。各委员会和其主席由董事每年委任。

Bylaw 411

规章 411

Committees must act efficiently but can run in any way they choose, including:

委员会须有效运作，但可自行选择运行方式，包括：

meetings;

会议；

telephone discussions;

电话讨论；

teleconferences; and

电话会议；以及

videoconferences.

视频会议。

Bylaw 412

规章 412

1 The below committees will comprise the number of persons as stipulated in the table. A quorum is the lowest number of members of the committee needed to be present before any valid business can be performed.

1 下列委员会构成人数应符合表中规定。法定人数是指构成委员会有效运作的最少委员人数。

	Appointed members	Persons needed to form a quorum
Arbitration Strategy Committee	As determined by the Directors	5
Rules Committee	12	5
Preliminary Investigation Committee	See Bylaw 413	4
Value Differences Committee	See Bylaw 414	5

	指定会员	构成法定人数的人数
仲裁委员会	董事会主席确定	5
规则委员会	12	5
初步调查委员会	见规章 413	4
价值差异委员会	见规章 414	5

2   Representatives of CICCA Member-Associations may be appointed to serve on the Rules Committee whenever common regulations are under consideration under Article 105.3. But, they cannot be Chairman or Deputy Chairman of the Committee unless they are an Individual Member of the ICA.

2   根据章程105.3，如涉及一般管制事项，则可以任命CICCA会员协会代表参加规则委员会工作。但是，他们不能成为主席或委员会副主席，除非他们是在ICA的个人会员。

3   The President, First Vice-President and Second Vice-President will automatically be members of Members' Committees. This does not apply to the Preliminary Investigation Committee or to a Quality Appeal Committee.

3   会长，第一副会长和第二副会长将自动成为议员委员会的成员。这并不适用于初步调查委员会，或质量申诉委员会。

4   Membership of committees will only last for one year. When members retire, they can be appointed again.

4   委员会成员的任职仅为一年。任职期满，委员可连任。

Preliminary Investigation Committee
初步调查委员会
Bylaw 413
规章413

The Preliminary Investigation Committee will be constituted and its proceedings regulated according to the following provisions:
初步调查委员会的组成和程序依如下规定：

(a)   The Committee will be appointed by the Directors, from an approved panel. The approved panel will comprise:

(a)   委员会由董事会从经批准小组中选任。经批准小组人员包括：

nine Individual Members of the Association. The Individual Members shall have held office as a President, First Vice-President, Second Vice-President, Treasurer or Ordinary Director of the Association, but shall have ceased to hold such office, and any member of the said panel who shall be elected or re-elected to any such office, shall, ipso facto, cease to be a member of the said panel.

九名协会个人会员。该等个人会员应当曾经担任过协会会长、第一副会长、第二副会长、财务主管或普通董事，但现已卸任，并且该小组的任何人员若当选或重新当选上述职位，则自然停止担任上述小组的职位。

up to eight Associate Directors of the Association,
其中联席董事不超过八人，

up to two nominees of other Member-Associations of the Committee for International Cooperation between Cotton Associations (CICCA) who have held or hold office as a director of their Association,
不超过两名棉花协会国际合作委员会（CICCA）的其他会员协会的被提名人，他们为曾任或现任其协会董事，

up to three independent individuals from outside the cotton and allied textile trades, who shall be appointed by the Directors.
棉花和纺织业以外独立人士不超过三人，由董事会指定。

(b) The Directors will appoint a Committee comprising:

(b) 董事会指定建立一个委员会，人员组成包括：

a Chairman, who shall be an Individual Member of the Association and shall have held office as President of the Association, up to six individuals from the approved panel, including an independent individual. A majority of the members of the Committee must be Individual Members of the Association.

主席一人，应为协会个人会员，且曾经担任过协会会长，从经批准小组选任不超过六人，包括一名独立人士。协会个人会员须占委员会委员的多数。

(c) The Directors will have power at any time and from time to time to appoint any qualified person as a member of the panel to fill any casual vacancy among the elected Individual Members, but any member of the said panel so appointed shall hold office only until the next following Annual General Meeting of the Association, and shall then be eligible for election.

(c) 董事会有权在任何时候及不时指定任何适格人员担任工作小组成员，以填补选举产生的个人会员出现的临时空缺，但以如此方式任命的小组成员任职期仅至下一届协会年度大会召开时，届时另行选举。

Value Differences Committee
价差委员会
Bylaw 414
规章 414

1 The Value Differences Committee will comprise up to 4 members appointed by us, up to 4 members appointed by Bremer Baumwollboerse and up to 8 further Individual Members appointed by the Directors from those expressing interest.

1 价差委员会将由我们委任的最多 4 名成员、由 Bremer Baumwollboerse 委任的最多 4 名成员以及由对此有意的机构的董事会委任的最多 8 名成员组成。

2 The Value Differences Committee can agree to add Individual Members or non-Members to the committee. The people they nominate will have the same voting rights as appointed members.

2 价差委员会可以同意向委员会增加个人会员或非会员。接受任命的人员拥有与选举产生者同样的投票权。

3 The Value Differences Committee will consult at least once in each four-week period. The Chairman can call meetings more often.

3 价差委员会至少每四周开会商讨一次。主席可以更频繁地召开会议。

4 As long as the Chairman approves, members of the Value Differences Committee can ask an alternate to attend. The alternate:

4 只要主席批准，价值差异委员会成员可以要求由代理人出席会议。代理人：

must be from the same firm as the member;
必须来自同一家公司的成员；

may be an Individual Member or a person other than an Individual Member; and
可能是个人会员或除个人会员除外；和

can vote at committee meetings.
可在委员会会议上投票。

Quality Appeal Panel
质量申诉工作小组

Bylaw 415
规章 415

1 A Quality Appeal Committee can agree to add any Individual Member to the committee to advise them on cotton submitted to them. The person drafted on will be seen as a committee member when judging that case.

1 质量申诉委员会可以同意向委员会增加任何个人会员，以就提交到委员会的棉花产品提供咨询意见。在裁判案件过程中加入委员会者视为委员会成员。

2 Each firm cannot have more than one vote at any of the Quality Appeal Committee meetings. A representative of the American Cotton Shippers Association may be appointed to serve on Quality Appeal Committees whenever 'American

Cotton', American/Pima varieties, or other cotton which has been traded by a member of the American Cotton Shippers Association is concerned. But, he cannot be Chairman or Deputy Chairman of a committee.

2　在质量申诉委员会会议中，各公司投票数不得超过一票。如有涉及美国棉商协会会员的"美国棉"、美国/皮马各品种或其他棉花交易，则美国棉花承运人协会可以指定一名代表加入质量申诉委员会。但其不得担任某委员会的主席或副主席。

3　This Bylaw does not apply to contracts for the shipment of American cotton from any place in the United States of America.

3　本规章不适用于从美利坚合众国任何地区装运美国棉的承运合同。

Bylaw 416
规章 416

No more than two members of the same firm may be appointed from the Quality Appeal Panel to any one Quality Appeal Committee.

从质量申诉工作小组受命进入任何一个质量申诉委员会者，同一家公司的不得超过两人。

Bylaw 417
规章 417

Candidates for membership of the Quality Appeal Panel must work in the cotton trade.

质量申诉工作小组成员候选人必须为棉花行业从业者。

Disciplinary Procedures
纪律程序

Bylaw 418
规章 418

1　A Member Firm that enters into a contract for the purchase or sale of raw cotton or for the provision of services with or on behalf of an individual, firm or company listed on the ICA List of Unfulfilled Awards (that contract being concluded on or after the day following notification of the listing of the company) or entering into a contract for the purchase or sale of raw cotton or for the provision of services with the intention of circumventing the ICA List of Unfulfilled Awards, shall be liable to a penalty of:

1　如有成员公司代表列入 ICA 未执行裁决书名单的个人或公司签署原棉

购销合同或服务合同（合同在该公司列入名单的通知下达当日或之后签订）或与其签署此合同，或以规避 ICA 裁决未执行名单为目的签署原棉购销合同或服务合同的，须受下列处罚：

  a denial of arbitration services
  a 拒绝提供仲裁服务
  b caution
  b 警告
  c censure
  c 谴责
  d payment of a fine, not exceeding £25,000
  d 罚款，金额不超过 25,000 美元
  e suspension
  e 暂停资格
  f expulsion
  f 开除

or any combination thereof, as the Preliminary Investigation Committee or Directors shall decide.

或根据初步调查委员会或董事会的决定同时处以几种处罚。

  2 Individual Members and Member Firms will be subject to the provisions and procedures laid down in the Articles.

  2 个人会员和成员公司应遵守本章程规定和程序。

  3 If a Member Firm wishes to trade with a party against whom it has an outstanding award listed on the ICA List of Unfulfilled Awards with the sole purpose of settling that award then that Member Firm will be required to advise the Directors in writing of that intention. Within seven days (one week) of entering into a contract or contracts for that purpose, the Member Firm shall provide the Directors with the date, reference number and estimated date of fulfilment of that contract. Subject to compliance with the above, the provisions of paragraph (1) above shall not apply to that contract or contracts.

  3 如有成员公司希望与有未履行裁决列入 ICA 未执行裁决书名单的当事方进行交易，其唯一目的是履行裁决，则该成员公司将须向董事会书面说明其意图。以此为目的的一项或多项合同签署后七日（一周）内，该成员公司须向董事会提供该合同的日期、参考编号和预计履行完毕日。如符合上述规定，则

第(1)节之规定不适用于该等合同。

4 If a Member Firm has an outstanding contract with a party whose name subsequently appears on the ICA List of Unfulfilled Awards, within seven days (one week) of the listing, the Member Firm shall provide the Directors with the date, reference number and estimated date of fulfilment of that contract. Subject to compliance with the above, the provisions of paragraph (1) above shall not apply to that contract or contracts.

4 如果成员公司仍有未完全履行的合同，而该合同的另一方随后列入在ICA未执行裁决书名单内，在七日（一周）内，该成员公司须向董事会提供该合同的日期、参考编号和预计履行完毕日。如符合上述规定，则上文第(1)节之规定不适用于该等合同。

5 Any Member Firm whose conduct is the subject of the investigation by the Preliminary Investigation Committee shall be entitled at his, its or their own expense to

5 任何成员公司，其行为正接受初步调查委员会审查者，有权自担费用以：

a give evidence personally;

a 亲自提交证据；

b obtain any professional or expert assistance and for that purpose to have any legal representative, accountant or expert present at the hearing but without any right of audience;

b 聘请专家协助或为此目的聘请任何法律代表、会计师或专家出席听证会，但不享有发言权；

c call any accountants or experts as witnesses;

c 传召任何会计师或专家担任证人；

d call any witness or witnesses and to produce any books or documents which he or it may consider material to the case;

d 传召任何证人以及提供任何其认为对案件审理重要的书籍或文件；

e appoint any Individual Member of the Association, who shall be willing so to act, to assist him or it in his or its case, to examine witnesses and to address the Directors on his or its behalf.

e 在本人愿意的情况下任命任何协会个人会员，协助案件审理，盘问证人并代表自己向董事会发表意见。

6 If a Member Firm disagrees with the decision of the Preliminary

Investigation Committee it can appeal to the Directors but it must do so within 14 days (2 weeks) of the decision having been notified. There is no further right of appeal for Affiliate Industry Firms or Related Companies should they disagree with the decision of the Directors. Individual Members and Principal Firms may appeal any decision of the Directors before Individual Members and shall be entitled at his, its or their own expense to the rights set out in paragraph (5) above.

6 如成员公司不同意初步调查委员会做出的裁决，可以向董事会提起上诉，但须在裁决下达后14天（两周）内提出。关联行业公司或关联公司对董事会裁决不满的不享有继续上诉的权利。个人会员和主公司可以对董事会的裁决提起上诉，并有权自担费用行使上文第(5)节规定的权利。

7 The said Committee and the Directors hearing an appeal shall be at liberty to have their Solicitor present at the investigation for the purpose of advising them on legal or technical matters and to assist them in drawing up their decision in writing.

7 所述委员会和董事会在审理上诉案件的过程中，有权聘请律师进行调查，以便在起草书面裁决时就法律或技术问题得到协助。

8 No Director who has participated in a Committee investigating a case shall take any part in an investigation by the Directors relating to that case or in any appeal hearing pertaining to that case.

8 曾参与委员会案件调查的董事不得参加与其有关联的案件调查工作或该案的上诉审理工作。

9 The Preliminary Investigation Committee will determine by whom the costs of the investigation are to be borne.

9 初步调查委员会将决定调查费用由哪一方承担。

❺ 进口棉花检验监督管理办法
5 Administrative Measures on Inspection and Supervision of Import Cotton

第一章 总 则
Chapter Ⅰ General Provisions

第一条 为了加强进口棉花检验监督管理，提高进口棉花质量，维护正常贸易秩序，根据《中华人民共和国进出口商品检验法》(以下简称商检法)及其实施条例的规定，制定本办法。

Article 1　For enhancing the administration on inspection and supervision of import cotton, improving import cotton quality, and maintaining normal trade order, these Measures are formulated in accordance with provisions of the *Law of the People's Republic of China on Import and Export Commodity Inspection* (hereinafter referred to as the *Commodity Inspection Law*) and its implementation regulations.

第二条 本办法适用于进口棉花的检验监督管理。

Article 2　These Measures shall apply to administration on inspection and supervision of import cotton.

第三条 国家质量监督检验检疫总局（以下简称国家质检总局）主管全国进口棉花的检验监督管理工作。

Article 3　The General Administration of Quality Supervision, Inspection & Quarantine (hereinafter referred to as AQSIQ) takes principle charge of the administration on inspection and supervision of import cotton nationwide.

国家质检总局设在各地的出入境检验检疫机构（以下简称检验检疫机构）负责所辖地区进口棉花的检验监督管理工作。

The entry-exit inspection and quarantine authorities established by AQSIQ in various localities (hereinafter referred to as inspection and quarantine authority) take the charge of administration on inspection and supervision of import cotton within the areas under their jurisdictions.

第四条 国家对进口棉花的境外供货企业（以下简称境外供货企业）实施质量信用管理，对境外供货企业可以实施登记管理。

Article 4　The State shall apply a system of quality credit administration on overseas suppliers of import cotton (hereinafter referred to as the overseas supplier), and may take registration of them.

第五条　检验检疫机构依法对进口棉花实施到货检验。

Article 5　The inspection and quarantine authority shall conduct arrival inspection for import cotton in accordance with relevant laws.

第二章　境外供货企业登记管理
Chapter Ⅱ　Administration on Registration of Overseas Suppliers

第六条　为了便利通关，境外供货企业按照自愿原则向国家质检总局申请登记。

Article 6　To facilitate customs clearance, overseas suppliers may apply for registration to AQSIQ voluntarily.

第七条　申请登记的境外供货企业（以下简称申请人）应当具备以下条件：

Article 7　An overseas supplier applying for registration (hereinafter referred to as the applicant) shall meet the following requirements:

（一）具有所在国家或者地区合法经营资质；

(1) With legal qualification of business operation in the country or region where it locates;

（二）具有固定经营场所；

(2) With fixed business operation premises;

（三）具有稳定供货来源，并有相应质量控制体系；

(3) With stable sources of supply, and with corresponding quality control systems;

（四）熟悉中国进口棉花检验相关规定。

(4) Familiar with relevant provisions of China on import cotton inspection.

第八条　申请人申请登记时应当向国家质检总局提交下列书面材料：

Article 8　When making application to AQSIQ for registration, the applicant shall submit the following written documents:

（一）进口棉花境外供货企业登记申请表（以下简称登记申请表）；

(1) Registration Application Form for Overseas Supplier of Import Cotton (hereinafter referred to as the Registration Application Form);

（二）合法商业经营资质证明文件复印件；

(2) Copies of certificates for legal business operation;

（三）组织机构图及经营场所平面图；
(3) Organizational chart and business premises plan;
（四）质量控制体系的相关材料；
(4) Related documents on quality control systems;
（五）质量承诺书。
(5) Letter of Quality Guarantee.
以上材料应当提供中文或者中外文对照文本。
The above mentioned documents shall be in Chinese or in Chinese-foreign language bilingually.
第九条 境外供货企业可以委托代理人申请登记。代理人申请登记时，应当提交境外供货企业的委托书。
Article 9 An overseas supplier may entrust an agent to make application for registration. When making such application, the letter of entrustment issued by the overseas supplier shall be provided.
第十条 国家质检总局对申请人提交的申请，应当根据下列情形分别作出处理：
Article 10 AQSIQ shall deal with applications submitted by applicants respectively in consideration of the following circumstances:
（一）申请材料不齐全或者不符合法定形式的，应当当场或者自收到申请材料之日起 5 个工作日内一次告知申请人需要补正的全部内容；逾期不告知的，自收到申请材料之日起即为受理；
(1) Where the application documents are incomplete or do not meet the statutory forms, the applicant shall be informed in a way of one-for-all all contents to be supplemented and/or corrected on site or within 5 working days from the date of receiving the documents; if no such information is given within the time limit, the application shall be regarded as being accepted from the date of receiving them;
（二）申请材料齐全、符合规定形式，或者申请人按照国家质检总局的要求提交全部补正材料的，应当受理。
(2) Where the application documents are complete and meet the statutory forms, or all the supplemented and/or corrected documents have been submitted by the applicant in accordance with the requirements of AQSIQ, the application shall be accepted;
（三）申请人自被告知之日起 20 个工作日内未补正申请材料，视为撤销申请；申请人提供的补正材料仍不符合要求的，不予受理，并书面告知申请人。

(3) Where the applicant failed to submit the supplemented and/or corrected documents within 20 working days from the date of being informed, the application shall be regarded as being revoked; where the supplemented and/or corrected documents submitted by the applicant still failed to meet the requirements, the application shall not be accepted, and the applicant shall be informed in written form.

第十一条 受理当事人提交的申请后，国家质检总局应当组成评审组，开展书面评审，必要时开展现场评审。上述评审应当自受理之日起3个月内完成。

Article 11 Following acceptance of the application, AQSIQ shall set up an assessment group to carry out assessment of the documents, and when necessary, site assessment shall be carried out. The above mentioned assessment shall be completed within 3 months from the date of acceptance.

第十二条 经审核合格的，国家质检总局应当对境外供货企业予以登记，颁发《进口棉花境外供货企业登记证书》（以下简称登记证书）并对外公布。

Article 12 For conformed ones following examination, AQSIQ shall make registration and issue the Certificate of Registration for Overseas Supplier of Import Cotton (hereinafter referred to as the Registration Certificate) and open it to the public.

第十三条 经审核不合格的，国家质检总局对境外供货企业不予登记，并书面告知境外供货企业。

Article 13 For non-conformed ones following examination, AQSIQ shall not make registration and inform them in written form.

第十四条 登记证书有效期为3年。

Article 14 The validation of the Registration Certificate is 3 years.

第十五条 不予登记的境外供货企业自不予登记之日起2个月后方可向国家质检总局重新申请登记。

Article 15 Overseas suppliers not accepted for registration can only make re-application to AQSIQ 2 months later from the date of being informed.

第十六条 已登记境外供货企业的名称、经营场所或者法定代表人等登记信息发生变化的，应当及时向国家质检总局申请变更登记，提交本办法第八条规定的登记申请表及变更事项的证明材料，国家质检总局应当自收到变更登记材料之日起30个工作日内作出是否予以变更登记的决定。

Article 16 Where any registered information, such as name of the enterprise, business operation premises or legal person, etc., changes, the overseas supplier shall apply timely for alteration to AQSIQ. When doing so, such relevant documents as the Registration Application Form specified in Article 8 of these Measures and the verification materials for alteration shall be submitted. AQSIQ shall make a decision on whether to accept the alteration or not within 30 working days from the date of receiving the documents for alteration.

第十七条 需要延续有效期的，已登记境外供货企业应当在登记证书有效期届满3个月前向国家质检总局申请复查换证，复查换证时提交本办法第八条规定的材料，国家质检总局应当在登记证书有效期届满前作出是否准予换证的决定。

Article 17 Where extension of validity is demanded, the registered overseas supplier shall apply to AQSIQ 3 months prior to the expiration for review and renewal of the certificate, when making such application, documents specified in Article 8 of these Measures shall be submitted. AQSIQ shall make the decision on whether to issue a new certificate or not before the date of expiration.

到期未申请复查换证的，国家质检总局予以注销。

For those fail to make application for review and renewal of the certificate at the expiration, AQSIQ shall revoke their certificates.

第三章 质量信用管理
Chapter Ⅲ Quality Credit Administration

第十八条 国家质检总局对境外供货企业实行质量信用管理。直属检验检疫局根据进口棉花的实际到货质量和境外供货企业的履约情况，对境外供货企业的质量信用进行评估，并上报国家质检总局。

Article 18 AQSIQ shall apply a system of quality credit administration on overseas suppliers. The inspection and quarantine bureau directly under AQSIQ shall make quality credit assessment on the overseas suppliers based on their actual arrival quality of import cotton as well as their behaviors of honoring contracts, and report the result to AQSIQ.

第十九条 按照质量信用，境外供货企业分为A、B、C三个层级：

Article 19 Overseas suppliers are divided into three levels i. e. A, B and C according to their quality credit:

知之日起 15 个工作日内，向作出评估结果的直属检验检疫局提出书面申辩，并提交相关证明材料。经复核，原评估结果有误的，予以更正。

Article 23 Where the overseas supplier disagrees to the determination of preliminary assessment, a written argument and relevant supporting documents shall be submitted to the inspection and quarantine bureau directly under AQSIQ making the determination, within 15 working days from the date of receiving the written notice. The inappropriate determination shall be corrected following re-examination and verification.

无异议或者期限届满未申辩的，直属检验检疫局确定最终评估结果，书面告知境外供货企业，同时上报国家质检总局。

Where no disagreement or no argument is ever submitted until the date of expiration, the inspection and quarantine bureau directly under AQSIQ shall make final determination of assessment, notify the overseas supplier in written form, and report to AQSIQ.

第二十四条 国家质检总局根据评估结果及时调整境外供货企业质量信用层级，并通知检验检疫机构及相关单位。

Article 24 AQSIQ shall timely adjust the quality credit level of overseas supplier based on the result of the assessment, and notify the inspection and quarantine authority as well as relevant agencies.

第二十五条 实施质量信用评估过程中发生复验、行政复议或者行政诉讼的，应当暂停评估。待复验、行政复议或者行政诉讼结束后，继续组织评估。

Article 25 The quality credit assessment shall be suspended where there is re-inspection, administrative reconsideration or administrative proceedings. It shall not be continued until the completion of the re-inspection, administrative reconsideration or administrative proceedings.

第二十六条 国家质检总局对获得登记的境外供货企业质量信用层级按下列方式进行动态调整：

Article 26 AQSIQ shall dynamically adjust the quality credit levels of registered overseas suppliers in the following way:

（一）A 级境外供货企业进口的棉花发生本办法第二十条所列情形的，境外供货企业的质量信用层级由 A 级降为 B 级；

(1) For Level A overseas supplier, the quality credit level shall be downgraded from Level A to Level B, where any circumstance listed in Article 20 of these

Measures occurred;

（二）自直属检验检疫局书面通知境外供货企业质量信用层级之日起 5 个月内，从 B 级境外供货企业进口的棉花发生本办法第二十条所列情形的，境外供货企业的质量信用层级由 B 级降为 C 级；如未发生本办法第二十条所列情形的，质量信用层级由 B 级升为 A 级；

(2) For Level B overseas supplier, within 5 months from the date when the inspection and quarantine bureau directly under AQSIQ notified its quality credit level in written form, where any circumstance listed in Article 20 of these Measures occurred, its quality credit level shall be downgraded from Level B to Level C; Where no such circumstance occurred, it shall be upgraded from Level B to Level A;

（三）自直属检验检疫局书面通知境外供货企业质量信用层级之日起 5 个月内，从 C 级境外供货企业进口的棉花未发生本办法第二十条所列情形的，境外供货企业（不含未在国家质检总局登记的企业）的质量信用层级由 C 级升为 B 级。

(3) For Level C overseas supplier (excluding those not yet registered at AQSIQ), within 5 months from the date when the inspection and quarantine bureau directly under AQSIQ notified the overseas supplier of its quality credit level in written form, where no circumstance listed in Article 20 of these Measures occurred, its quality credit level shall be upgraded from Level C to Level B.

第四章　进口检验
Chapter Ⅳ　Import Inspection

第二十七条　进口棉花的收货人或者其代理人应当向入境口岸检验检疫机构报检。报检时，除提供规定的报检单证外，已登记境外供货企业应当提供《进口棉花境外供货企业登记证书》（复印件）。

Article 27　The consignee of import cotton or its agent shall apply for inspection to inspection and quarantine authority at entry port. When applying, in addition to the required inspection documents and sheets, the Certificate of Registration for Overseas Supplier of Import Cotton (copy) shall also be provided if the overseas supplier is a registered one.

第二十八条　检验检疫机构根据境外供货企业的质量信用层级，按照下列方式对进口棉花实施检验：

Article 28   The inspection and quarantine authority shall carry out inspection on the import cotton based on the quality credit level of the overseas supplier in the following way:

（一）对A级境外供货企业的棉花，应当在收货人报检时申报的目的地检验，由目的地检验检疫机构按照国家质检总局制定的检验检疫行业标准实施抽样检验；

(1) To the cotton from level A overseas supplier, the inspection shall be conducted at the destination declared by the consignee when applying, sampling inspection shall be carried out by the local inspection and quarantine authority in accordance with the inspection and quarantine sectoral standards developed by AQSIQ;

（二）对B级境外供货企业的棉花，应当在收货人报检时申报的目的地检验，由目的地检验检疫机构实施两倍抽样量的加严检验；

(2) To the cotton from level B overseas supplier, the inspection shall be conducted at the destination declared by the consignee when applying, tightened inspection with double sampling shall be carried out by the local inspection and quarantine authority;

（三）对C级境外供货企业的棉花，检验检疫机构在入境口岸实施两倍抽样量的加严检验。

(3) To the cotton from level C overseas supplier, tightened inspection with double sampling shall be carried out by the inspection and quarantine authority at entry port.

第二十九条 实施进口棉花现场检验工作的场所应当具备以下条件：

Article 29   Checking site for on-spot inspection of import cotton shall meet the following conditions:

（一）具有适合棉花存储的现场检验场地；

(1) With an on-spot inspection site suitable for cotton storage;

（二）配备开箱、开包、称重、取样等所需的设备和辅助人员；

(2) With equipment and supporting staff for container-opening, unpacking, weighing and sampling, etc.;

（三）其他检验工作所需的通用现场设施。

(3) Other general facilities for on-spot inspection.

第三十条 检验检疫机构对进口棉花实施现场查验。查验时应当核对进口棉花批次、规格、标记等，确认货证相符；查验包装是否符合合同等相关要求，有无包装破损；查验货物是否存在残损、异性纤维、以次充好、掺杂掺假等情况。对集装箱装载的，检查集装箱铅封是否完好。

Article 30 The inspection and quarantine authority shall carry out on-spot check to the import cotton, checking lot number, specification and mark, etc. so as to confirm that the goods and certificates matched; checking whether the package meet related requirements of the contract, etc. or is damaged; checking whether the goods is damaged, with foreign fibers, shoddy and/or adulterated, etc.. For those loaded in containers, check the seals.

第三十一条 检验检疫机构按照国家质检总局的相关规定对进口棉花实施数重量检验、品质检验和残损鉴定，并出具证书。

Article 31 The inspection and quarantine authority shall carry out inspection on quantity/weight and quality as well as damage survey of the import cotton in accordance with relevant provisions of AQSIQ, and issue certificates.

第三十二条 进口棉花的收货人或者发货人对检验检疫机构出具的检验结果有异议的，可以按照《进出口商品复验办法》的规定申请复验。

Article 32 Where the consignee or consignor of import cotton disagrees to the inspection result issued by the inspection and quarantine authority, he/she may apply for re-inspection in accordance with provisions of the *Measures for the Re-inspection of Import and Export Commodities*.

第五章 监督管理
Chapter V Supervision and Administration

第三十三条 境外供货企业质量控制体系应当持续有效。

Article 33 The quality control system of overseas supplier shall be consistent and effective.

国家质检总局可以依法对境外供货企业实施现场核查。

AQSIQ may carry out on-spot examination and verification to overseas suppliers in accordance with relevant laws.

第三十四条 收货人应当建立进口棉花销售、使用记录以及索赔记录，检验检疫机构可以对其记录进行检查，发现未建立记录或者记录不完整的，书面通知收货人限期整改。

Article 34 The consignee shall establish records on sales, use and claims of import cotton, on which the inspection and quarantine authority may carry out check. For those without or with incomplete records, a written notice for rectification and correction within a limited period of time shall be delivered.

第三十五条 检验检疫机构应当建立质量信用评估和检验监管工作档案。国家质检总局对质量信用评估和检验监管工作进行监督检查。

Article 35 The inspection and quarantine authority shall establish filing system on quality credit assessment as well as the inspection and supervision. AQSIQ shall carry out supervision and inspection on the work of quality credit assessment as well as the inspection and supervision.

第三十六条 已登记境外供货企业发生下列情形之一的，国家质检总局撤销其登记。境外供货企业自撤销之日起6个月后方可向国家质检总局重新申请登记：

Article 36 Where any of the following circumstances occurs, AQSIQ shall revoke the registration of registered overseas supplier. Re-registration can only be applied to AQSIQ 6 months beyond the date of revocation:

（一）提供虚假材料获取登记证书的；

(1) Submitted false documents for obtaining the Registration Certificate;

（二）在国家质检总局组织的现场检查中被发现其质量控制体系无法保证棉花质量的；

(2) The quality control system is found unable to guarantee the quality of cotton upon on-spot inspection organized by AQSIQ;

（三）C级已登记境外供货企业发生本办法第二十条所列情形的；

(3) The level C registered overseas supplier involved in any of the circumstances listed in Article 20 of these Measures;

（四）不接受监督管理的。

(4) Refuse to receive supervision and administration.

第六章 法律责任
Chapter Ⅵ Legal Liabilities

第三十七条 收货人发生下列情形之一的，有违法所得的，由检验检疫机构处违法所得3倍以下罚款，最高不超过3万元；没有违法所得的，处1万元以下罚款：

Article 37 Where the consignee engages in any of the following misbehaviors and gets illegal gains, the inspection and quarantine authority shall punish him/her with a penalty up to 3 times of the sum of illegal gains and the maximal penalty shall not exceed RMB 30,000 Yuan. The consignee that gets no illegal gains but involves in any of the following misbehaviors shall be punished with a penalty of no more than RMB 10,000 Yuan;

（一）书面通知限期整改仍未建立进口棉花销售或者使用记录以及索赔记录的；

(1) Not establish records on sales, use and claims of import cotton though the rectification and correction within a limited period of time is notified in written from;

（二）不如实提供进口棉花的真实情况造成严重后果的；

(2) Be dishonest to provide true information of import cotton thus causing serious consequences;

（三）不接受监督管理的。

(3) Refuse to receive supervision and administration.

第三十八条 有其他违反相关法律、行政法规行为的，检验检疫机构依照相关法律、行政法规追究其法律责任。

Article 38 For those behaviors violating relevant laws and administrative regulations, the inspection and quarantine authority shall investigate their legal liabilities accordingly.

第三十九条 检验检疫机构的工作人员滥用职权，故意刁难当事人，徇私舞弊，伪造检验检疫结果的，或者玩忽职守，延误出证的，按照《中华人民共和国进出口商品检验法实施条例》第五十九条规定依法给予行政处分；构成犯罪的，依法追究刑事责任。

Article 39 Where a staff member of the inspection and quarantine authority abuses his/her power to intentionally create difficulties for the parties, commits illegalities for personal interests or by fraudulent means to falsify inspection and quarantine results; or neglects his/her duty to delay the issue of certificates, he/she shall be given an administrative sanction in accordance with provisions of Article 59 of the *Regulations on Implementation of the Law of the People's Republic of China on Import and Export Commodity Inspection*; if a crime is constituted, the criminal liability shall be investigated for in accordance with laws.

第七章 附 则
Chapter Ⅶ Supplementary Provisions

第四十条 进口棉花的动植物检疫、卫生检疫按照法律法规及相关规定执行。

Article 40 The animal and plant quarantine as well as the health quarantine of import cotton shall be conducted in accordance with laws, regulations and relevant provisions.

第四十一条 香港、澳门和台湾地区的棉花供货企业的登记管理和质量信用评估管理按照本办法执行。

Article 41 Registration and quality credit assessment administration for cotton suppliers from Hong Kong, Macao and Taiwan region shall be conducted in accordance with these Measures.

第四十二条 从境外进入保税区、出口加工区等海关特殊监管区域的进口棉花，按照相关规定执行。

Article 42 Import cotton entering into areas under special customs control such as bonded zones, export processing areas, etc. from abroad shall be conducted in accordance with relevant provisions.

第四十三条 本办法由国家质检总局负责解释。

Article 43 These Measures shall be interpreted by AQSIQ.

第四十四条 本办法自 2013 年 2 月 1 日起施行。

Article 44 These Measures shall be implemented as of February 1, 2013.

❻ 棉花质量监督管理条例
6 Regulations on Supervision & Administration of Cotton Quality

第一章 总则
Chapter I General Provisions

第一条 为了加强对棉花质量的监督管理,维护棉花市场秩序,保护棉花交易各方的合法权益,制定本条例。

Article 1 These Regulations are formulated for the purposes of strengthening supervision and administration of cotton quality, ensuring the order of cotton market and protecting the legitimate rights and interests of all parties in the cotton trade.

第二条 棉花经营者(含棉花收购者、加工者、销售者、承储者,下同)从事棉花经营活动,棉花质量监督机构对棉花质量实施监督管理,必须遵守本条例。

Article 2 Cotton business operators (including cotton purchasers, processors, sellers and warehouse operators, the same below) who engage in cotton business activities and cotton quality supervision organizations which exercise the cotton quality supervision and administration must comply with these Regulations.

第三条 棉花经营者从事棉花加工经营活动,应当按照国家有关规定取得资格认定。

Article 3 Cotton business operators who engage in such cotton business activities as purchase, processing, sale or storage shall obtain approval for their respective business qualifications according to the relevant provisions of the State.

棉花经营者应当建立、健全棉花质量内部管理制度,严格实施岗位质量规范、质量责任及相应的考核办法。

Cotton business operators shall establish and perfect their internal cotton quality management system and strictly implement the post-based quality specifications, quality responsibilities and relevant assessment methods.

第四条 国务院质量监督检验检疫部门主管全国棉花质量监督工作,由其所属的中国纤维检验机构负责组织实施。

Article 4 The quality supervision, inspection and quarantine department of the State Council shall take full charge of the cotton quality supervision work throughout the country, and its subordinate unit, the China fiber inspection institution, shall be responsible to organize the implementation.

省、自治区、直辖市人民政府质量监督部门负责本行政区域内棉花质量监督工作。设有专业纤维检验机构的地方,由专业纤维检验机构在其管辖范围内对棉花质量实施监督;没有设立专业纤维检验机构的地方,由质量监督部门在其管辖范围内对棉花质量实施监督(专业纤维检验机构和地方质量监督部门并列使用时,统称棉花质量监督机构)。

Quality supervision departments of people's governments of provinces, autonomous regions and municipalities directly under the Central Government shall be responsible for the cotton quality supervision work within their respective administrative areas. Where specialized fiber inspection institutions have been established, they shall exercise the cotton quality supervision within the areas under their jurisdiction. Where specialized fiber inspection institutions have not been established, the quality supervision departments shall exercise the supervision of cotton quality within the areas under their jurisdiction (when the specialized fiber inspection institution and the local quality supervision department are used concurrently, they shall be collectively referred to as the cotton quality supervision organization).

第五条 地方各级人民政府及其工作人员不得包庇、纵容本地区的棉花质量违法行为,或者阻挠、干预棉花质量监督机构依法对棉花收购、加工、销售、承储中违反本条例规定的行为进行查处。

Article 5 Local people's governments at various levels and their staff members may not shelter or connive at illegal cotton quality acts within their respective areas or obstruct or interfere with cotton quality supervision organizations which, according to law, investigate into and deal with the acts of infringing the provisions of these Regulations in cotton purchase, processing, sale and storage.

第六条 任何单位和个人对棉花质量违法行为,均有权检举。

Article 6 All units and individuals shall be entitled to report illegal cotton quality acts.

第二章 棉花质量义务
Chapter II Cotton Quality Obligations

第七条 棉花经营者收购棉花,应当建立、健全棉花收购质量检查验收制度,具备品级实物标准和棉花质量检验所必备的设备、工具。

Article 7 Cotton business operators who engage in cotton purchase shall establish and perfect their respective cotton quality inspection and acceptance system and have the quality grading standards in kind and necessary equipment and tools for the cotton quality inspection.

棉花经营者收购棉花时,应当按照国家标准和技术规范,排除异性纤维和其他有害物质后确定所收购棉花的类别、等级、数量;所收购的棉花超出国家规定水分标准的,应当进行晾晒、烘干等技术处理,保证棉花质量。

Cotton business operators who purchase cotton shall identify the class, grade and quantity of the cotton they purchased according to national standards and technical specifications after they remove heterogeneous fibers and other harmful substances therefrom. If the moisture of the cotton they purchased exceeds the level specified by the State, they shall make such technical treatments as sunning and stoving in order to assure the cotton quality.

棉花经营者应当分类别、分等级置放所收购的棉花。

Cotton business operators shall stockpile the purchased cotton by classes and grades.

第八条 棉花经营者加工棉花,必须符合下列要求:

Article 8 Cotton business operators who process cotton must meet the following requirements:

(一)按照国家标准,对所加工棉花中的异性纤维和其他有害物质进行分拣,并予以排除;

(1) to clean up and remove heterogeneous fibers and other harmful substances from the cotton to be processed according to national standards;

(二)按照国家标准,对棉花分等级加工,并对加工后的棉花进行包装并标注标识,标识应当与棉花质量相符;

(2) to process cotton by grades according to national standards and to bale and mark the cotton they processed, and to guarantee that the marking shall be consistent with the cotton quality; and

（三）按照国家标准，将加工后的棉花成包组批放置。

(3) to stockpile the processed cotton by bales, lots and batches according to national standards.

棉花经营者不得使用国家明令禁止的皮辊机、轧花机、打包机以及其他棉花加工设备加工棉花。

Cotton business operators who process cotton may not use lap rollers, cotton gins, baling machinery and other cotton processing equipment that are expressly prohibited by the State.

第九条 棉花经营者销售棉花，必须符合下列要求：

Article 9 Cotton business operators who sell cotton must meet the following requirements:

（一）每批棉花附有质量凭证；

(1) to attach the quality certificate to every batch of the cotton to be sold;

（二）棉花包装、标识符合国家标准；

(2) to guarantee that the packing and marking for the cotton to be sold shall conform to national standards;

（三）棉花类别、等级、重量与质量凭证、标识相符；

(3) to guarantee that the class, grade and weight of the cotton to be sold shall be consistent with the quality certificate and mark; and

（四）经公证检验的棉花，附有公证检验证书，其中国家储备棉还应当粘贴公证检验标志。

(4) to attach the testing certificate to the cotton to be sold which has been tested and certified, and to stick the testing sign on the national reserve cotton.

第十条 棉花经营者承储国家储备棉，应当建立、健全棉花入库、出库质量检查验收制度，保证入库、出库的国家储备棉的类别、等级、数量与公证检验证书、公证检验标志相符。

Article 10 Cotton business operators who store the national reserve cotton shall establish and perfect the incoming and outgoing cotton quality inspection and acceptance system, and guarantee that the class, grade and quantity of the incoming and outgoing national reserve cotton shall be consistent with the testing certificate and sign.

棉花经营者承储国家储备棉，应当按照国家规定维护、保养承储设施，保证国家储备棉质量免受人为因素造成的质量变异。

Cotton business operators who store the national reserve cotton shall maintain and keep the storage facilities according to the provisions of the State, and guarantee that the national reserve cotton quality may not be varied due to artificial factors.

棉花经营者不得将未经棉花质量公证检验的棉花作为国家储备棉入库、出库。

Cotton business operators may not take the cotton whose quality has not been tested and certified into or out of warehouses as the national reserve cotton.

政府机关及其工作人员,不得强令棉花经营者将未经棉花质量公证检验的棉花作为国家储备棉入库、出库。

Governmental organs and their staff members may not force cotton business operators to take the cotton whose quality has not been tested and certified into or out of warehouses as the national reserve cotton.

第十一条　棉花经营者收购、加工、销售、承储棉花,不得伪造、变造、冒用棉花质量凭证、标识、公证检验证书、公证检验标志。

Article 11　Cotton business operators who purchase, process, sell or store cotton may not forge, alter or counterfeit the cotton quality certificate or mark or the testing certificate or sign.

第十二条　严禁棉花经营者在收购、加工、销售、承储等棉花经营活动中掺杂掺假、以次充好、以假充真。

Article 12　Cotton business operators shall, in their cotton business activities such as purchase, processing, sale and storage, be strictly prohibited to make any adulteration, to take the poor as the good or to mix the spurious with the genuine.

第三章　棉花质量监督
Chapter III　Cotton Quality Supervision

第十三条　国家实行棉花质量公证检验制度。

Article 13　The State shall implement the cotton quality testing and certification system.

前款所称棉花质量公证检验,是指专业纤维检验机构按照国家标准和技术规范,对棉花的质量、数量进行检验并出具公证检验证书的活动。

Cotton quality testing and certification mentioned in the proceeding paragraph means the activities of specialized fiber inspection institutions to inspect the cotton quality and quantity and to issue the testing certificate according to national standards and technical specifications.

第十四条 棉花经营者向用棉企业销售棉花，交易任何一方在棉花交易结算前，可以委托专业纤维检验机构对所交易的棉花进行公证检验；经公证检验后，由专业纤维检验机构出具棉花质量公证检验证书，作为棉花质量、数量的依据。

Article 14 After the cotton business operator sells cotton to the cotton-using enterprise, either of them may, before the closing of their cotton transaction, request the specialized fiber inspection institution to test and certify the cotton they transacted. After such testing and certification, the specialized fiber inspection institution shall issue the testing certificate of cotton quality that shall serve as the cotton quality and quantity criteria.

第十五条 国家储备棉的入库、出库，必须经棉花质量公证检验；经公证检验后，由专业纤维检验机构出具棉花质量公证检验证书，作为国家财政支付存储国家储备棉所需费用的依据。

Article 15 National reserve cotton must, when entering into or leaving from warehouses, be subject to the cotton quality testing and certification. After testing and certification, the specialized fiber inspection institution shall issue the testing certificate of cotton quality that shall serve as the criteria for allocating the storage cost for the national reserve cotton by the fiscal department of the State.

经公证检验的国家储备棉，由专业纤维检验机构粘贴中国纤维检验机构统一规定的公证检验标志。

After the national reserve cotton is tested and certified, the specialized fiber inspection institution shall stick the testing sign specified by the fiber inspection institution thereon.

第十六条 专业纤维检验机构进行棉花质量公证检验，必须执行国家标准及其检验方法、技术规范和时间要求，保证客观、公正、及时。专业纤维检验机构出具的棉花质量公证检验证书应当真实、客观地反映棉花的质量、数量。

Article 16 Specialized fiber inspection institutions must, in testing and certifying the cotton quality, implement national standards as well as their testing methods, technical specifications and time requirements, and guarantee objectiveness, fairness and timeliness. The testing certificate of cotton quality issued by the specialized fiber inspection institution shall, truly and objectively, indicate the cotton quality and quantity.

棉花质量公证检验证书的内容应当包括：产品名称、送检（委托）单位、批号、包数、检验依据、检验结果、检验单位、检验人员等内容。

The testing certificate of cotton quality shall contain the followings: product name, applicant (requestor) unit, lot and batch numbers, quantity of bales, testing criteria, testing result, testing unit, tester and others.

棉花质量公证检验证书的格式由国务院质量监督检验检疫部门规定。

The quality supervision, inspection and quarantine department of the State Council shall provide the pattern of the testing certificate of cotton quality.

第十七条 专业纤维检验机构实施棉花质量公证检验不得收取费用，所需检验费用按照国家有关规定列支。

Article 17 Specialized fiber inspection institutions may not collect any fees for testing and certifying cotton quality. The necessary expenses therefore shall be itemized and paid according to the relevant provisions of the State.

第十八条 国务院质量监督检验检疫部门在全国范围内对经棉花质量公证检验的棉花组织实施监督抽验，省、自治区、直辖市人民政府质量监督部门在本行政区域内对经棉花质量公证检验的棉花组织实施监督抽验。

Article 18 The quality supervision, inspection and quarantine department of the State Council shall organize the supervision and sampling inspection over the cotton whose quality is tested and certified throughout the country. Quality supervision departments of people's governments of provinces, autonomous regions and municipalities directly under the Central Government shall organize the supervision and sampling inspection over the cotton whose quality is tested and certified within their respective administrative areas.

监督抽验的内容是：棉花质量公证检验证书和公证检验标志是否与实物相符；专业纤维检验机构实施的棉花质量公证检验是否客观、公正、及时。

Such supervision and sampling inspection shall include the followings: whether or not the testing certificate and sign of cotton quality are consistent with the real object; and whether or not the cotton quality testing and certification by the specialized fiber inspection institution are objective, fair and timely.

监督抽验所需样品从公证检验的留样中随机抽取，并应当自抽取样品之日起 10 日内作出检验结论。

Samples needed in such supervision and sampling inspection shall be taken randomly from the samples left after the cotton quality testing and certification, and the inspection conclusion shall be made within 10 days from the date of taking samples.

第十九条 棉花质量监督机构对棉花质量公证检验以外的棉花,可以在棉花收购、加工、销售、承储的现场实施监督检查。

Article 19 Cotton quality supervision organizations may, at the cotton purchasing, processing, selling or storing site, exercise the supervision and inspection over the cotton other than those whose quality is tested and certified.

监督检查的内容是:棉花质量、数量和包装是否符合国家标准;棉花标识以及质量凭证是否与实物相符。

Such supervision and inspection shall include the followings: whether or not the cotton quality, quantity and packing conform to national standards; whether or not the cotton marking and quality certificate are consistent with the real object.

第二十条 棉花质量监督机构在实施棉花质量监督检查过程中,根据违法嫌疑证据或者举报,对涉嫌违反本条例规定的行为进行查处时,可以行使下列职权:

Article 20 Cotton quality supervision organizations may perform the following functions and powers in the course of exercising the supervision and inspection over the cotton quality when they, upon evidences and reports, investigate into and deal with the acts suspected of infringing the provisions of these Regulations:

(一)对涉嫌从事违反本条例的经营活动的场所实施现场检查;

(1) to perform the on-spot inspection over the site of the cotton business activity suspected of infringing the provisions of these Regulations;

(二)向棉花经营单位的有关人员调查、了解与涉嫌从事违反本条例的经营活动有关的情况;

(2) to investigate and inquire the persons concerned of the cotton business operation units for information in connection with the cotton business activity suspected of infringing the provisions of these Regulations;

(三)查阅、复制与棉花经营有关的合同、单据、账簿以及其他资料;

(3) to consult and copy the contracts, documents, books and other materials in connection with the cotton business operation; and

(四)对涉嫌掺杂掺假、以次充好、以假充真或者其他有严重质量问题的棉花以及专门用于生产掺杂掺假、以次充好、以假充真的棉花的设备、工具予以查封或者扣押。

(4) to seal up or seize the cotton suspected of making adulteration, taking the poor as the good, mixing the spurious with the genuine or having other serious quality problems, and the equipment and tools used for producing the above-mentioned cotton.

第二十一条 棉花质量监督机构根据监督检查的需要,可以对棉花质量进行检验;检验所需样品按照国家有关标准,从收购、加工、销售、储备的棉花中随机抽取,并应当自抽取检验样品之日起3日内作出检验结论。

Article 21 According to the need of supervision and inspection, cotton quality supervision organizations may inspect the cotton quality. The samples needed in such inspection shall be taken randomly from the cotton purchased, processed, sold or reserved according to relevant national standards, and the inspection conclusion shall be made within three days from the date of taking samples.

依照前款规定进行的检验不得收取费用,所需检验费用按照国家有关规定列支。

No fees may be collected in the inspection performed according to the provisions of the preceding paragraph. The inspection expenses shall be itemized and paid according to the relevant provisions of the State.

第二十二条 棉花经营者、用棉企业对依照本条例进行的棉花质量公证检验和棉花质量监督检查中实施检验的结果有异议的,可以自收到检验结果之日起5日内向省、自治区、直辖市的棉花质量监督机构或者中国纤维检验机构申请复检;省、自治区、直辖市的棉花质量监督机构或者中国纤维检验机构应当自收到申请之日起7日内作出复检结论,并告知申请人。棉花经营者、用棉企业对复检结论仍有异议的,可以依法向人民法院提起诉讼。

Article 22 The cotton business operator or cotton-using enterprise may, if disagreeing with the inspection conclusion in the cotton quality testing and certification or the cotton quality supervision and inspection performed according to these Regulations, apply for a re-inspection to the cotton quality supervision organization of the province, autonomous region or municipality directly under the Central Government or the China fiber inspection institution within five days from the date of receiving the inspection conclusion. The cotton quality supervision organization of the province, autonomous region or municipality directly under the Central Government or the China fiber inspection institution shall make the re-inspection conclusion and notify the applicant within seven days from the date of receiving the application. The cotton business operator or cotton-using enterprise may, if disagreeing with the re-inspection conclusion, initiate the action before the people's court according to law.

第二十三条 经国务院质量监督检验检疫部门认可的其他纤维检验机构，可以受委托从事棉花质量检验业务。具体办法由国务院质量监督检验检疫部门会同国务院有关部门规定。

Article 23 Other fiber inspection institutions acknowledged by the quality supervision, inspection and quarantine department of the State Council may engage in cotton quality inspection services upon authorization. Concrete measures therefor are to be provided by the quality supervision, inspection and quarantine department of the State Council jointly with relevant departments of the State Council.

第四章 罚则
Chapter IV Penalty Provisions

第二十四条 棉花经营者收购棉花，违反本条例第七条第二款、第三款的规定，不按照国家标准和技术规范排除异性纤维和其他有害物质后确定所收购棉花的类别、等级、数量，或者对所收购的超出国家规定水分标准的棉花不进行技术处理，或者对所收购的棉花不分类别、等级置放的，由棉花质量监督机构责令改正，可以处3万元以下的罚款。

Article 24 Where the cotton business operator who purchases cotton, in violation of the provisions of Paragraph 2 or 3 of Article 7 of these Regulations, fails to identify the class, grade or quantity of the purchased cotton after removing heterogeneous fibers and other harmful substances according to national standards and technical specifications, or fails to make technical treatments for the purchased cotton whose moisture exceeds the level specified by the State, or fails to stockpile the purchased cotton by classes and grades, the cotton quality supervision organization shall order him to make corrections and may impose a fine of less than 30,000 yuan thereon.

第二十五条 棉花经营者加工棉花，违反本条例第八条第一款的规定，不按照国家标准分拣、排除异性纤维和其他有害物质，不按照国家标准对棉花分等级加工、进行包装并标注标识，或者不按照国家标准成包组批放置的，由棉花质量监督机构责令改正，并可以根据情节轻重，处10万元以下的罚款；情节严重的,由原资格认定机关取消其棉花加工资格。

Article 25 Where the cotton business operator who processes cotton, in violation of the provisions of Paragraph 1 of Article 8 of these Regulations, fails to clean up and remove heterogeneous fibers and other harmful substances according to national standards, or fails to process, bale or mark the cotton by grades according to national standards, or fails to stockpile cotton by bales, batches and lots according to national standards, the cotton quality supervision organization shall order him to make corrections and may, depending on the seriousness of the circumstances, impose a fine of less than 100,000 yuan thereon; if the circumstances are serious, the original qualification-approving authority shall revoke his business qualification for cotton processing.

棉花经营者加工棉花,违反本条例第八条第二款的规定,使用国家明令禁止的棉花加工设备的,由棉花质量监督机构没收并监督销毁禁止的棉花加工设备,并处非法设备实际价值2倍以上10倍以下的罚款;情节严重的,由原资格认定机关取消其棉花加工资格。

Where the cotton business operator who processes cotton, in violation of the provisions of Paragraph 2 of Article 8 of these Regulations, uses the cotton processing equipment that is prohibited by the State, the cotton quality supervision organization shall confiscate and supervise to destroy the cotton processing equipment prohibited and concurrently, impose a fine of more than two times but less than ten times the actual value of that equipment; if the circumstances are serious, the original qualification-approving authority shall revoke his business qualification for cotton processing.

第二十六条 棉花经营者销售棉花,违反本条例第九条的规定,销售的棉花没有质量凭证,或者其包装、标识不符合国家标准,或者质量凭证、标识与实物不符,或者经公证检验的棉花没有公证检验证书、国家储备棉没有粘贴公证检验标志的,由棉花质量监督机构责令改正,并可以根据情节轻重,处10万元以下的罚款。

Article 26 Where the cotton business operator who sells cotton, in violation of the provisions of Article 9 of these Regulations, fails to attach the quality certificate to the cotton to be sold, or fails to bale or mark the cotton according to the national standard, or fails to guarantee that the quality certificate or mark is consistent with the real object, or fails to attach the testing certificate to the cotton tested and certified, or fails to stick the testing sign on the national reserve cotton, the cotton quality supervision organization shall order him to make corrections and may, depending on the seriousness of the circumstances, impose a fine of less than 100,000 yuan thereon.

第二十七条　棉花经营者承储国家储备棉，违反本条例第十条第一款、第二款、第三款的规定，未建立棉花入库、出库质量检查验收制度，或者入库、出库的国家储备棉实物与公证检验证书、标志不符，或者不按照国家规定维护、保养承储设施致使国家储备棉质量变异，或者将未经公证检验的棉花作为国家储备棉入库、出库的，由棉花质量监督机构责令改正，可以处10万元以下的罚款；造成重大损失的，对负责的主管人员和其他直接责任人员给予降级以上的纪律处分；构成犯罪的，依法追究刑事责任。

Article 27　Where the cotton business operator who stores the national reserve cotton, in violation of the provisions of Paragraph 1, 2 or 3 of Article 10 of these Regulations, fails to establish the incoming and outgoing cotton quality inspection and acceptance system, or fails to guarantee that the testing certificate or sign is consistent with the real object of the incoming or outgoing national reserve cotton, or fails to maintain and keep the storage facilities according to the provisions of the State, thus causing a quality variance of the national reserve cotton, or takes the cotton not being tested and certified into or out of warehouse as the national reserve cotton, the cotton quality supervision organization shall order him to make corrections and may impose a fine of less than 100,000 yuan thereon; if a serious loss is caused, disciplinary sanctions above demotion shall be given to the principal person responsible and other persons directly responsible therefore; if a crime is constituted, criminal responsibility shall be investigated according to law.

第二十八条　棉花经营者隐匿、转移、损毁被棉花质量监督机构查封、扣押的物品的，由棉花质量监督机构处被隐匿、转移、损毁物品货值金额2倍以上5倍以下的罚款；构成犯罪的，依法追究刑事责任。

Article 28　Where the cotton business operator hides, shifts or destroys the goods or articles sealed up or seized by the cotton quality supervision organization, the cotton quality supervision organization shall impose a fine of more than two times but less than five times the value of the goods or articles so hidden, shifted or destroyed; if a crime is constituted, criminal responsibility shall be investigated according to law.

第二十九条 棉花经营者违反本条例第十一条的规定,伪造、变造、冒用棉花质量凭证、标识、公证检验证书、公证检验标志的,由棉花质量监督机构处 5 万元以上 10 万元以下的罚款;情节严重的,移送工商行政管理机关吊销营业执照;构成犯罪的,依法追究刑事责任。

Article 29 Where the cotton business operator, in violation of the provisions of Article 11 of these Regulations, forges, alters or counterfeits the cotton quality certificate or mark, or the testing certificate or sign, the cotton quality supervision organization shall impose a fine of more than 50,000 but less than 100,000 yuan; if the circumstances are serious, the case shall be transferred to the administrative department for industry and commerce for the revocation of his business licence; if a crime is constituted, criminal responsibility shall be investigated according to law.

第三十条 棉花经营者违反本条例第十二条的规定,在棉花经营活动中掺杂掺假、以次充好、以假充真,构成犯罪的,依法追究刑事责任;尚不构成犯罪的,由棉花质量监督机构没收掺杂掺假、以次充好、以假充真的棉花和违法所得,处违法货值金额 2 倍以上 5 倍以下的罚款,并移送工商行政管理机关依法吊销营业执照。

Article 30 The cotton business operator who, in violation of the provisions of Article 12 of these Regulations, makes an adulteration, takes the poor as the good or mixes the spurious as the genuine, thus constituting a crime, shall be investigated for criminal responsibility according to law; if no crime is constituted, the cotton quality supervision organization shall confiscate the above-mentioned cotton and ill-gotten revenues therefrom, impose a fine of more than two time but less than five times the value of the illegal goods; and transfer the case to the administrative department for industry and commerce for the revocation of his business license according to law.

第三十一条 专业纤维检验机构违反本条例第十六条的规定,不执行国家标准及其检验方法、技术规范或者时间要求,或者出具的棉花质量公证检验证书不真实、不客观的,由国务院质量监督检验检疫部门或者地方质量监督部门责令改正;对负责的主管人员和其他直接责任人员依法给予降级或者撤职的行政处分。

Article 31 Where the specialized fiber inspection institution, in violation of the provisions of Article 16 of these Regulations, fails to implement the national standard, testing method, technical specification or time requirement, or fails to issue the authentic and objective testing certificate of cotton quality, the quality supervision, inspection and quarantine department of the State Council or the local quality supervision department shall order it to make corrections; and give administrative sanctions of demotion or dismissal from posts to the principal person responsible and other persons directly responsible according to law.

第三十二条 专业纤维检验机构违反本条例第十七条的规定收取公证检验费用的,由国务院质量监督检验检疫部门或者地方质量监督部门责令退回所收取的公证检验费用;对负责的主管人员和其他直接责任人员依法给予记大过或者降级的行政处分。

Article 32 Where the specialized fiber inspection institution, in violation of the provisions of Article 17 of these Regulations, collects a fee for testing and certification, the quality supervision, inspection and quarantine department of the State Council or the local quality supervision department shall order it to return back the amount collected; and give administrative sanctions of major demerit or demotion to the principal person responsible and other persons directly responsible according to the law.

第三十三条 专业纤维检验机构未实施公证检验而编造、出具公证检验证书或者粘贴公证检验标志,弄虚作假的,由国务院质量监督检验检疫部门或者地方质量监督部门对负责的主管人员和其他直接责任人员依法给予降级或者撤职的行政处分;构成犯罪的,依法追究刑事责任。

Article 33 Where the specialized fiber inspection institution fails to perform the testing and certification but falsifies and issues the testing certificate or sticks the testing sign, thus practicing fraud, the quality supervision, inspection and quarantine department of the State Council or the local quality supervision department shall give administrative sanctions of demotion or dismissal from posts to the principal person responsible and other persons directly responsible according to law; if a crime is constituted, criminal responsibility shall be investigated according to law.

第三十四条 政府机关及其工作人员违反本条例第十条第四款的规定,强令将未经公证检验的棉花作为国家储备棉入库、出库的,对负责的主管人员和其他直接责任人员依法给予降级或者撤职的行政处分。

Article 34 Where the governmental organ and its staff members, in violation of the provisions of Paragraph 4 of Article 10 of these Regulations, force to take the cotton not tested and certified into or out of warehouse as the national reserve cotton, administrative sanctions of demotion and dismissal from posts shall be given to its principal person responsible and other persons directly responsible according to law.

第三十五条 政府机关及其工作人员包庇、纵容本地区的棉花质量违法行为，或者阻挠、干预棉花质量监督机构依法对违反本条例的行为进行查处的，依法给予降级或者撤职的行政处分；构成犯罪的，依法追究刑事责任。

Article 35 Where the governmental organ and its staff members shelter or connive at the illegal cotton quality act within its area, or obstruct or interfere with the cotton quality supervision organization which, according to law, investigates and deals with the act of infringing the provisions of these Regulations, administrative sanctions of demotion or dismissal from posts shall be given thereto according to law; if a crime is constituted, criminal responsibility shall be investigated according to law.

第三十六条 本条例第二十八条、第三十条规定的棉花货值金额按照违法收购、加工、销售的棉花的牌价或者结算票据计算；没有牌价或者结算票据的，按照同类棉花市场价格计算。

Article 36 The value of the cotton specified in Article 28 or 30 of these Regulations shall be calculated according to the quotation or document of settlement of the cotton illegally purchased, processed or sold; if no quotation or document of settlement is available, it shall be calculated according to the market price of the same kind of cotton.

第三十七条 依照本条例的规定实施罚款的行政处罚，应当依照有关法律、行政法规的规定，实行罚款决定与罚款收缴分离，收缴的罚款必须全部上缴国库。

Article 37 In respect of administrative penalties of fines imposed according to the provisions of these Regulations, the decision-making and collection of fines shall be separated in accordance with the provisions of the relevant laws and administrative regulations, and all fines collected must be turned over to the State Treasury.

第五章 附则
Chapter V Supplementary Provisions

第三十八条 毛、绒、茧丝、麻类纤维的质量监督管理，比照本条例执行。

Article 38 Quality supervision and administration of wool, velvet, and natural silk and hemp-type fibers shall be governed by applying mutatis mutandis these Regulations.

第三十九条 本条例自公布之日起施行。

Article 39 These Regulations shall be effective as of the date of promulgation.

❼ 世界主要产棉国家和地区
7 Major Cotton Producing Country and Region in the World

序号 No.	英文 English	中文 Chinese	序号 No.	英文 English	中文 Chinese
1	Argentina	阿根廷	23	Mali	马里
2	Australia	澳大利亚	24	Mexico	墨西哥
3	Azerbaijan	阿塞拜疆	25	Mozambique	莫桑比克
4	Burkina Faso	布基纳法索	26	Nigeria	尼日利亚
5	Benin	贝宁	27	Peru	秘鲁
6	Brazil	巴西	28	Pakistan	巴基斯坦
7	Ivory Coast	象牙海岸	29	Paraguay	巴拉圭
8	Cameroon	喀麦隆	30	Romania	罗马尼亚
9	China	中国	31	Sudan	苏旦
10	Cuba	古巴	32	Senegal	塞内加尔
11	Algeria	阿尔及利亚	33	Syrian Arab Republic	叙利亚
12	Ecuador	厄瓜多尔	34	Chad	乍得
13	Egypt	埃及	35	Togo	多哥
14	Spain	西班牙	36	Tajikistan	塔吉克斯坦
15	Ethiopia	埃塞俄比亚	37	Turkmenistan	土库曼斯坦
16	Greece	希腊	38	Turkey	土耳其
17	Guatemala	危地马拉	39	Tanzania	坦桑尼亚
18	India	印度	40	United States	美国
19	Kenya	肯尼亚	41	Uruguay	乌拉圭
20	Kyrgyzstan	吉尔吉斯斯坦	42	South Africa	南非
21	Kazakhstan	哈萨克斯坦	43	Zambia	赞比亚
22	Morocco	摩洛哥	44	Zimbabwe	津巴布韦

8 棉花生长常用农药名称
8 Name of Cotton Pesticides

中文 Chinese	英文 English	中文 Chinese	英文 English
2,4-滴丁酯	2,4-D	单嘧磺隆	monosulfuron
2 甲 4 氯钠	MCPA-Na	甲基二磺隆	mesosulfuron-methyl
五氯酚钠	PCP-Na	苄嘧磺隆	bensulfuron-methyl
禾草灵	diclofop	吡嘧磺隆	pyrazosulfuron-ethyl
燕麦灵	barban	醚磺隆	cinosulfuron
恶唑禾草灵	fenoxaprop-ethyl	乙氧嘧磺隆	ethoxysulfuron
喹禾灵	Quizalofop-ethyl	四唑嘧磺隆	azimsulfuron
精喹禾灵	fluazifop-p-butyl	环丙嘧磺隆	cyclosulfamuron
吡氟禾草灵	fluazifop	砜嘧磺隆	rimsulfuron
精吡氟禾草灵	fluazifop-p-butyl	氯嘧磺隆	chlorimuron-ethyl
氟吡乙禾灵	haloxyfop-ethoxy ethyl	胺苯磺隆	ethametsulfuron
高效氟吡甲禾灵	haloxyfop-R-methyl	甲嘧磺隆	sulfometuron-methyl
喹禾糖醋(喹唑糠酯)	quizlofop-p-tefuryl	啶嘧磺隆	flazasulfuron
恶草酸	propaquizafop	伏草隆	fluometuron
氰氟草酯	cyhalofop-butyl	绿麦隆	chlorothluron
稀禾啶	sethoxydim	敌草隆	diuron
烯草酮	clethodim	咪唑乙烟酸	imazethapyr
吡喃草酮	tepraloxydim	甲氧咪唑烟酸	imazamox
噻吩磺隆	thifensulfuron-methyl	咪唑喹啉酸	imazaquin
苯磺隆	tribenuron-methyl	咪唑烟酸	imazapyr
氯磺隆	chlorsulfuron	双草醚	bispyribac-sodium
甲磺隆	metsulfuron-methyl	双嘧双苯醚	pyribenzoxim
酰嘧磺隆	amidosulfuron	唑嘧磺草胺	flumetsulam
唑嘧氟磺酯	florasulam	甲草胺	alacholr
乙氧氟草醚	oxyflurofen	异丙甲草胺	metolachlor
氟磺胺草醚	fomesafen	精异丙甲草胺	S-metolachlor
三氟羧草醚	acifluorfen-sodiuma	乙草胺	acetochlor
乙羧氟草醚	fluoroglycofen-ethyl	丙草胺	pretilachlor
乳氟禾灵	lactofen	异丙草胺	propisochlor
氟烯草酸	flumiclorac-pentyl	敌草胺	napropamide
丙炔氟草胺	flumioxazin	苯噻酰草胺	mefenacet
吡草醚	pyraflufen-ethyl	四唑草胺	fetrazamide
恶草酮	oxadiazon	吡氟酰草胺	dlflufenican

续表

中文 Chinese	英文 English	中文 Chinese	英文 English
氟唑草酮	carfentrazone-ethyl	地乐胺	dibutralin
2甲4氯乙硫酯	MCPA-thioethyl	禾草丹	thiobencarb
麦草畏	dicamba	禾草敌	molinate
氯氟吡氧乙酸	fluroxypyr	哌草丹	dimepiperate
三氯吡氧乙酸	trichlopyr	甜菜宁	phenmedipham
草除灵	benazolin	野麦畏	triallate
二氯喹啉酸	quinclorac	灭草敌	vernolate
杀草隆	dimuron	威百亩	metam-sodium
莠去津	atrazine	氟乐灵	trifluralin
氰草津	cyanazine	仲丁灵	butralin
莠灭净	ametryn	二甲戊灵	pendimethalin
嗪草酮	metribuzin	草甘膦	glyphosate
环嗪酮	hexazinone	莎稗磷	anilofos
溴苯腈	bromoxynil	双丙氨磷	bialaphos
辛酰溴苯腈	bromoxynil octanoate	草铵膦	gulfosinate-ammonium
丁草特	butylate	丁草胺	butachlor
百草枯	paraquat	西玛津	simazine
敌草快	diquat	西草净	simetryn
灭草松	bentazone	阔叶散	Harmony
野燕枯	difenzoquat	哒草特	pyridate
异恶草松	clomazone	环庚草醚	cinmethylin
异恶唑草铜	isoxaflutole	环草特	cyeloate
恶嗪草铜	oxaziclomefone	甲羧除草醚	bifenox
双苯酰草胺	diphenamid	除草醚	nitrofen

⑨ 世界主要产棉国和地区分级标准
9 Grading Standards of the World's Major Cotton-Producing Countries and Region

国别 Country	品级标准 Grading Standards			
	陆地棉品级标准 Upland Cotton Grading Standards			
	类别 Classify	中文全称 Full Name in Chinese	英文全称 Full Name in English	英文简称 English Abbreviations
美国 America	白棉 White Cotton	上级	Good Middling	G.M
			Strict Middling Plus	S.M.P
		次上级	Strict Middling	S.M
			Middling Plus	M.P
		中级	Middling	M
			Strict Low Middling Plus	S.L.M.P
		次中级	Strict Low Middling	S.L.M
			Low Middling Plus	L.M.P
		下级	Low Middling	L.M
			Strict Good Ordinary Plus	S.G.O.P
		次下级	Strict Good Ordinary	S.G.O
			Good Ordinary Plus	G.O.P
		平级	Good Ordinary	G.O
		级外	Below Good Ordinary	B.G.O
	色棉 Colored Cotton	淡点污	Light Spotted	Lt.Sp
		点污	Spotted	Sp
		淡黄染	Tinged	Tg
		黄染	Yellow Stained	Y.S
		灰	Gray	G
	皮马棉品级标准 PIMA Grading Standards			
	品级 Grade		符号 Symbol	代码 Code
	1		AP1	10
	2		AP2	20
	3		AP3	30
	4		AP4	40
	5		AP5	50
	6		AP6	60

续表

国别 Country	品级标准 Grading Standards		
乌兹别克斯坦 Uzbekistan	等级 Grade	名称 Name	
^	I	Birinchi	
^	II	Ikkinchi	
^	III	Uchinchi	
^	IV	Turtinchi	
^	V	Beshinchi	
^	根据疵点和杂质的多少按实物标准把每等棉花又细分为： According to the number of defects and impurities, material standard subdivide each grade into: 1(Oliy, Hihgest)、2(Yakshi, Good)、3(Urta, Middle)、4(Oddiy, Ordinary)、5(Iflos, Trashy)		
巴基斯坦 Pakistan	陆地棉品级标准 Upland Cotton Grading Standards		
^	全称 Full Name	简称 Abbreviation	
^	Super Fine	S.F	
^	Fine to Super Fine	F/S.F	
^	Fine	F	
^	Fully Good to Fine	F.G/F	
^	Fully Good	F.G	
^	Good to Fully Good	G/F.G	
^	Good	G	
^	注：如高于 S.F 等级为 Choice，低于 G 等级为 Fair。 Notes: Grade Choice if exceed S.F, Grade Fair if below G.		
^	德西棉品级标准 Desi Cotton Grading Standards		
^	信德德西棉 Sind Desi	巴哈瓦尔 Bahauzlpur Desi	旁遮普德西棉 Punjab Desi
^	Choice	—	—
^	*Super Fine	Super Fine	Super Fine
^	Fine to S. F.	* Fine	* Fine
^	—	Fully Good to F	Fully Good to F
^	注：*为标准级 Notes: * means standard level		

续表

国别 Country	品级标准 Grading Standards		
	品级名称 Grade	1/2 级品级名称 1/2 Grade	简称 Abbreviations
埃及 Egypt	Extra		EX
		Fully Good to Extra	F.G/EX
	Fully Good		F.G
		Good to Fully Good	G/F.G
	Good		G
		Fully Good Fair to Good	F.G.F/G
	Fully Good Fair		F.G.F
		Good Fair to Fully Good Fair	G.F/F.G.F
	Good Fair		G.F
		Fully Fair to Good Fair	F.F/G.F
	Fully Fair		F.F
印度 India		通常采用小样成交 Usually by sample	
澳大利亚 Australia		通常采用美棉标准或者小样成交 Usually by American standrds or sample	
希腊 Greece		通常采用美棉标准或者小样成交 Usually by American standrds or sample	
西班牙 Spain		通常采用美棉标准 Usually by American standrds	
墨西哥、巴西等美洲国家 Mexico, Brazil and other American Countries		通常采用美棉标准或者小样成交 Usually by American standrds or sample	
布基纳法索、贝宁、津巴布韦、坦桑尼亚等非洲国家 Burkina Faso, Benin, Zimbabwe, Tanzania and other African Countries		通常采用小样成交 Usually by sample	

参考文献

[1] United States Department of Agriculture. Agriculture Marketing Service. Agriculture Handbook 566, April.
[2] 英汉纺织工业词汇. 上海市纺织工业局《英汉纺织工业词汇》编写组, 北京：中国纺织出版社. 2004
[3] A NEW ENGLISH-CHINESE DICTIONARY.《新英汉词典》编写组. 上海：上海译文出版社. 2009
[4] USTER HVI Instruction Manual. Uster Technologies.
[5] 进出口纺织服装词汇. 李新实 卢艳光 主编. 北京：化学工业出版社. 2007
[6] 世界棉花产销与检验. 钱林森 主编. 济南：山东科学技术出版社. 1996
[7] COTTON PURCHASE CONTRACT. China National Textiles Import and Export Corporation.
[8] COTTON PURCHASE CONTRACT. Cotton Purchase Contract of China Cotton Association.
[9] 中国国际经济贸易仲裁委员会仲裁规则. 中国国际贸易促进委员会/中国国际商会.
[10] Bylaws and Rules of The International Cotton Association Limited. International Cotton Association Ltd.
[11] 进口棉花检验监督管理办法. 国家质量监督检验检疫总局令2013年第151号. 北京：中国质检出版社, 中国标准出版社. 2014
[12] 棉花质量监督管理条例. 中华人民共和国国务院令2006年第470号. 北京：中国法制出版社.
[13] Regulations on Supervision & Administration of Cotton Quality. Maoting Editor. 检验检疫服务网. http://en.ciqcid.com/Laws/Administrative/zjxgxzfg/45955.htm.
[14] 国内外棉花综述与检验检疫. 王新 主编. 北京：国家质检总局检验监管司. 2008
[15] 农药品种手册精编. 张敏恒 主编. 北京：全国农药信息总站/化学工业出版社. 2013